INTRODUCTION TO HIGH-ENERGY ASTROPHYSICS

High-energy astrophysics covers cosmic phenomena that occur under the most extreme physical conditions. It explores the most violent events in the Universe: the explosion of stars, matter falling into black holes, and gamma-ray bursts – the most luminous explosions since the Big Bang. Driven by a wealth of new observations, the last decade has seen a large leap forward in our understanding of these phenomena.

Exploring modern topics of high-energy astrophysics, such as supernovae, neutron stars, compact binary systems, gamma-ray bursts, and active galactic nuclei, this textbook is ideal for undergraduate students of high-energy astrophysics. It is a self-contained, up-to-date overview of this exciting field of research. Assuming a familiarity with basic physics, it introduces relevant concepts, such as gas dynamics and radiation processes, in an instructive way. An extended appendix gives an overview of some of the most important high-energy astrophysics instruments, and each chapter ends with exercises.

STEPHAN ROSSWOG is Professor of Astrophysics at the School of Engineering and Science, Jacobs University Bremen (formerly International University Bremen), Germany. He is a member of the German Physical Society (DPG) and the International Astronomical Union and is the recipient of a Particle Physics and Astronomy Research Council (PPARC) Advanced Fellowship.

MARCUS BRÜGGEN is Professor of Astrophysics at the School of Engineering and Science, Jacobs University Bremen, Germany. He is a member of the International Astronomical Union and Deutsche Astronomische Gesellschaft and a recipient of the Blackwell Prize from the Royal Astronomical Society.

INTRODUCTION TO HIGH-ENERGY ASTROPHYSICS

STEPHAN ROSSWOG AND MARCUS BRÜGGEN

Jacobs University Bremen, Germany

CAMBRIDGE UNIVERSITY PRESS
Cambridge, New York, Melbourne, Madrid, Cape Town, Singapore, São Paulo

Cambridge University Press
The Edinburgh Building, Cambridge CB2 8RU, UK

Published in the United States of America by Cambridge University Press, New York

www.cambridge.org
Information on this title: www.cambridge.org/9780521857697

First published 2007

Printed in the United Kingdom at the University Press, Cambridge

A catalog record for this publication is available from the British Library

Library of Congress Cataloging in Publication Data

Rosswog, Stephan, 1968–
Introduction to high-energy astrophysics / Stephan Rosswog and Marcus Brüggen.
p.   cm.
Includes bibliographical references and index.
ISBN-13: 978-0-521-85769-7 (hardback)
1. Astrophysics.   2. Nuclear astrophysics.   3. Cosmochemistry.
I. Brüggen, Marcus,1972–   II. Title.
QB461.R67   2007
523.01′9–dc22
2007008118

ISBN-13:   978-0-521-85769-7 hardback

Contents

Preface

What is high-energy astrophysics? The term is customarily used for a large set of different astrophysical phenomena that in some sense involve "high" energies. In some cases, the detected particles, for example, cosmic or gamma rays, are of particularly high energy. In other cases, for example, in radio astronomy, the detected photons may be of low energy but are produced by very energetic electrons. Admittedly, the definition of high-energy astrophysics is vague and imprecise. In fact, almost every astrophysical object features aspects that involve high energies. Typically, high-energy astrophysics revolves around phenomena that involve physics under the most extreme conditions. The matter in the center of a neutron star, for example, is much denser than that in an atomic nucleus. Active galactic nuclei harbor black holes at their centers, with masses a billion times greater than the mass of the Sun. Such supermassive black holes accelerate jets to velocities greater than 99% of the speed of light and display a variety of special relativistic effects. These are just two examples of the kinds of objects that we describe in this book.

The past decade has been a very exciting time for high-energy astrophysics. New instruments, such as X-ray and gamma-ray observatories, have revolutionized our view of the high-energy universe. Some mysteries have been solved; many new ones have arisen. The novel views of many high-energy phenomena that have come about in the past few years should quickly enter the classroom, which is one of the prime motivations behind the writing of this book. Although in recent years words such as *supernovae* and *black holes* have become familiar and ubiquitous, even in the nonscientific realm, these intriguing and glamorous constituents of our universe are frequently under-represented in university physics education. In this book, we wish to convey the fascination that arises from studying the physics of these extraordinary phenomena.

The field of high-energy astrophysics has become so vast that it is impossible to give a proper introduction to all its subdisciplines. In this spirit, we also have to remark that inevitably some interesting fields had to be left out. For this introductory text, we have chosen five areas of focus: supernovae, neutron stars, compact binaries, gamma-ray bursts, and active galactic nuclei.

We suggest two ways of using this book in a university course. First, one can go through this book front to back, starting with the physics chapters on special relativity and gas and radiation processes. Then, armed with the necessary physical gadgetry, the reader can dive into the colorful physical phenomena of high-energy astrophysics introduced in Chapters 4–8.

An alternative route would be to start right away with the astrophysics and cover the physical groundwork as one goes along. Here, we would recommend sticking to the order of Chapters 4–8, although other permutations are certainly possible. For this second route, in each chapter we have given plenty of cross-references to the relevant foundation chapters where the student can read up on the necessary physics. The latter approach may be more suitable for a graduate course, where students may be expected to have covered much of Chapters 1–3 anyway, whereas the former approach may be more suited for an undergraduate course. In any case, we are always happy to receive feedback on how this book is being employed in teaching.

For this undergraduate text, we generously borrowed from existing books and published articles. Little is in a strict sense original work. Only rarely do we credit the originators of the various theories and models that we describe in this book, and only sometimes do we depict the often tortuous ways in which certain astrophysical paradigms have come into being. We felt that a proper, historical referencing of the many topics covered in this book would disrupt the flow of the text and be of only limited use to the student. Instead, we give suggestions for further reading at the end of every chapter.

Finally, one word about the sore subject of units. Although most undergraduate courses in physics will use SI units, in astrophysics the most common set of units remains the cgs system. Therefore, we stick to cgs units in this book. Moreover, in various places we use additional units that are particularly helpful such as, for example, the parsec as a unit of distance, the electron volt as a unit of energy, and the solar mass as a unit of mass. These units are defined where used.

Many people have helped us. Thanks go to Matthias Hoeft, Matthias Lieben-dörfer, Elke Roediger, and Aurora Simionescu for proofreading the often half-finished bits of this book. Particular thanks go to Eva Stueeken and Daniel Price. Finally, we thank the Max-Planck Institut für Astrophysik in Garching for its hospitality.

1

Special relativity

1.1 Introduction

In this chapter, we briefly review the basics of special relativity and provide a
short summary of tensor calculus. We assume that the reader is familiar with the
fundamental ideas and concepts of the special theory of relativity (SR). More
complete introductions to special relativity and tensor calculus can be found in,
for example, *A First Course in General Relativity* by Schutz or *Gravitation* by
Misner, Thorne, and Wheeler. A good summary is provided in *Radiative Processes
in Astrophysics* by Rybicki and Lightman.

Practically every mechanical process that we encounter in our daily lives can be
described in terms of Newtonian theory. In astrophysics, however, many systems
are relativistic so that applying Newtonian physics can lead to completely wrong
answers. The Lorentz factor, given by

$$\gamma = \sqrt{\frac{1}{1 - (v/c)^2}}, \tag{1.1}$$

where v is the velocity and c the speed of light, quantifies the importance of special
relativistic effects. In a sense, γ measures how close the velocity is to the speed
of light: $\gamma = 1$ for $v = 0$ and $\gamma \to \infty$ for $v \to c$. As an example, jets that are
emitted from supermassive black holes in the centers of galaxies (see Chapter 8)
have Lorentz factors of up to $\sim$30, corresponding to 99.94% of the speed of
light. The most violent explosions in the Universe since the Big Bang, gamma-ray
bursts (see Chapter 7) accelerate material to Lorentz factors of several hundreds.
Electrons spiraling around the magnetic field lines of pulsars possess Lorentz factors of $\sim 10^7$. The cosmic rays that continuously bombard the Earth's atmosphere
contain protons with energies of up to $\sim 10^{20}$ eV, which correspond to Lorentz factors of $\gamma = E/m_p c^2 = 10^{20}$ eV/938 MeV $\approx 10^{11}$. Therefore, neglecting special

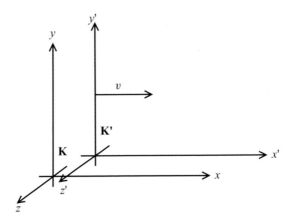

Figure 1.1   Coordinate systems used for the Lorentz transformations: the x-axes are aligned, and the relative velocity between the two frames is v.

relativity in astrophysics can lead to completely wrong interpretations of observations.

1.2   Lorentz transformations

The special theory of relativity is closely related to the notion of *inertial frames*. An inertial frame is a reference frame in which every body is either at rest or moves with a constant velocity along a straight line. In particular, a body viewed from an inertial frame accelerates only when a physical force is applied. In the absence of a net force, a body at rest remains at rest and a body in motion continues to move uniformly. In SR, the set of time and space coordinates, (ct, x, y, z),[1] labels a *space-time event* or simply an event. The coordinates of an event measured in two reference frames, **K** and **K'**, that have a constant relative velocity, v, are related via a *Lorentz transformation*. Unless otherwise stated, we assume that the relative velocity is along the x-axis and that the x-axes of both frames point in the same direction (see Fig. 1.1). In this case, the Lorentz transformation between the coordinates of an event in **K** labeled with (ct, x, y, z) and coordinates of the same event in **K'** labeled with (ct', x', y', z') reads as follows:

$$t' = \gamma \left(t - \frac{vx}{c^2} \right) \qquad\qquad (1.2)$$

$$x' = \gamma(x - vt) \qquad\qquad (1.3)$$

---

[1]   We use arrows for three-vectors only; lengths of vectors are denoted just by a letter. Components of four-vectors are labeled by either super- or subscripts. Four-vectors as geometrical objects are framed by brackets.

$$y' = y \tag{1.4}$$

$$z' = z. \tag{1.5}$$

Often, the inverse transformation from $\mathbf{K}'$ to $\mathbf{K}$ is needed. It can be easily be obtained by exchanging primed and unprimed quantities and v with $-v$:

$$t = \gamma \left(t' + \frac{vx'}{c^2} \right) \tag{1.6}$$

$$x = \gamma(x' + vt') \tag{1.7}$$

$$y = y' \tag{1.8}$$

$$z = z'. \tag{1.9}$$

1.3   Special relativistic effects

In this section, we discuss important effects that are direct consequences of the Lorentz transformations.

1.3.1   Length contraction

Consider a person sitting in frame $\mathbf{K}'$ carrying a rod oriented along the x-axis with length $L_0 \equiv x_2' - x_1'$. What length would an observer sitting in $\mathbf{K}$ measure for the same rod? Equation (1.3) yields

$$L_0 = x_2' - x_1' = \gamma(x_2 - x_1) = \gamma L, \tag{1.10}$$

where $L = x_2 - x_1$ and the ends of the rod have been measured simultaneously in each frame ($t_2' = t_1'$ and $t_2 = t_1$). This means that in its rest frame, the rod appears to be longer by a factor of γ than in a moving frame, or, seen from the frame that moves relative to the rod, its *length is contracted*.

1.3.2   Time dilation

Assume that you have a clock located at the origin of the system $\mathbf{K}'$ and that the time interval between two ticks of the clock is $T_0 = t_2' - t_1'$. Then, an observer in system $\mathbf{K}$ measures (see Eq. [1.6])

$$T = t_2 - t_1 = \gamma \left(t_2' + \frac{vx_2'}{c^2} \right) - \gamma \left(t_1' + \frac{vx_1'}{c^2} \right) = \gamma \left(t_2' - t_1' \right) = \gamma T_0, \tag{1.11}$$

as $x_2' = x_1' = 0$. This means that *the time interval appears to be stretched* by a factor of γ with respect to the object's rest frame.

This effect is observable for unstable elementary particles, for example, in accelerators or cosmic rays. To illustrate this muons produced by cosmic rays in the Earth's atmosphere can only be detected on the ground because their lifetimes are increased by this special relativistic effect. (See Exercise 1 at the end of this chapter.)

1.3.3   Proper time as a Lorentz invariant

Quantities that do not change under a Lorentz transformation are called *Lorentz invariants*. An important such quantity is the *proper time* τ defined via

$$d\tau^2 = dt^2 - \frac{1}{c^2}(dx^2 + dy^2 + dz^2), \tag{1.12}$$

where (dt, dx, dy, dz) is measured in some arbitrary coordinate system. A clock carried by an observer at a fixed location $(dx = dy = dz = 0)$ shows $d\tau^2 = dt^2$, therefore the name *proper time*. By transforming $d\tau$ from Eq. (1.12) to a frame $\mathbf{K}'$ using Eqs. (1.2)–(1.5), we find that $d\tau' = d\tau$, that is, $d\tau$ is indeed a Lorentz invariant. So how is $d\tau$ related to the general time coordinate dt? The relation can be found by just realising that $d\tau$ measures the time increment of a resting clock. Therefore, time dilation according to Eq. (1.11) yields $\gamma d\tau = dt$. The same result can be obtained more formally by

$$d\tau = \sqrt{dt^2 - \frac{(dx^2 + dy^2 + dz^2)}{c^2}} = dt\sqrt{1 - \left(\frac{v}{c}\right)^2} = \gamma^{-1}dt. \tag{1.13}$$

1.3.4   Transformation of velocities

With the Lorentz transformation given by Eqs. (1.2)–(1.5), we can calculate how velocities transform. If the velocity between our two Lorentz frames is again v and an object has a velocity of $\vec{u} = \frac{d\vec{x}}{dt}$, where $\vec{x} = (x, y, z)$, or $\vec{u}' = \frac{d\vec{x}'}{dt'}$ in the respective frames, the relation, between $\vec{u}$ and $\vec{u}'$ can be easily found by means of Eqs. (1.6)–(1.9):

$$u_x = \frac{dx}{dt} = \frac{\gamma(dx' + vdt')}{\gamma(dt' + vdx'/c^2)} = \frac{dx'/dt' + v}{1 + (v/c^2)(dx'/dt')} = \frac{u'_x + v}{1 + vu'_x/c^2}. \tag{1.14}$$

Completely analogously one finds for the other components:

$$u_y = \frac{u'_y}{\gamma(1 + vu'_x/c^2)} \quad \text{and} \quad u_z = \frac{u'_z}{\gamma(1 + vu'_x/c^2)}. \tag{1.15}$$

In comparison with the x-component, in the last two equations, there is no factor γ in the nominator to cancel the one in the denominator, which is a result of the x- and y-components being unaffected by the Lorentz transformation. Of course,

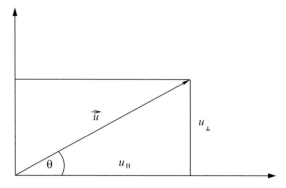

Figure 1.2   For the special relativistic transformation of velocities: splitting the velocity $\vec{u}$ into a component parallel ($u_{\parallel}$) and perpendicular ($u_{\perp}$) to the velocity between the two frames, v.

for low velocities where $vu'_x \ll c^2$ and $\gamma \approx 1$, the equations reduce to the usual Galilean transformation of velocities.

If one splits up the velocity into components parallel and antiparallel to v, $u_{\parallel}$, and $u_{\perp}$, the velocity transformation can be written compactly as

$$u_{\parallel} = \frac{u'_{\parallel} + v}{1 + \frac{vu'_{\parallel}}{c^2}} \quad \text{and} \quad u_{\perp} = \frac{u'_{\perp}}{\gamma \left(1 + \frac{vu'_{\parallel}}{c^2}\right)}. \tag{1.16}$$

Because these components transform differently, the angles appear to be different in different frames. Consider a frame **K** in which an object is moving with velocity $\vec{u}$ (see Fig. 1.2). The angle between $\vec{u}$ and the velocity between our two frames, $\vec{v}$, is given by

$$\tan \theta = \frac{u_{\perp}}{u_{\parallel}} = \frac{u'_{\perp}/\gamma(1 + vu'_{\parallel}/c^2)}{(u'_{\parallel} + v)/(1 + vu'_{\parallel}/c^2)} = \frac{u'_{\perp}}{\gamma(u'_{\parallel} + v)}, \tag{1.17}$$

where we have inserted Eq. (1.16). If the primed velocity components are now expressed via the angle θ' with respect to the $\vec{v}$, $u'_{\parallel} = u' \cos \theta'$, and $u'_{\perp} = u' \sin \theta'$, we have

$$\tan \theta = \frac{u' \sin \theta'}{\gamma(u' \cos \theta' + v)}. \tag{1.18}$$

This equation describes the *relativistic aberration of light*.

Aberration is an effect that also occurs at nonrelativistic speeds. Imagine standing under an umbrella, and rain is falling straight down from the sky. When you start to walk, you tilt your umbrella slightly forward to protect yourself from the rain,

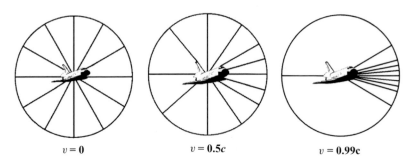

$v = 0$ $v = 0.5c$ $v = 0.99c$

Figure 1.3   Shown are light rays from 12 fixed, distant stars as they appear to a spaceship. In the leftmost panel, the spaceship is at rest; in the middle it travels with 0.5c and in the rightmost panel with 0.99c. In the rightmost panel, almost all photons arrive from the front, even if they stem from stars that are actually behind the spacecraft (that is why some people refer to this effect as the *paranoid effect*).

the more so the faster you walk. It seems like the rain is falling from a position in the sky in front of you, rather than from directly above.

A very similar effect occurs for motion at relativistic speeds. Assume that a spaceship is at rest with respect to a set of very distant stars and that the stars are distributed isotropically, so you see the same number of stars per solid angle in each direction. As the spaceship starts to increase its speed, more and more stars seem to lie ahead of it; the stars seem to pile up in forward direction. This effect is illustrated in Fig. 1.3.

1.3.5   Relativistic beaming

This velocity transformation law leads to an interesting effect called *relativistic beaming*. Beaming plays an important role in the interpretation of observations of, say, active galactic nuclei (Chapter 8) or gamma-ray bursts (Chapter 7) (see Exercise 2) and also in radiation processes such as synchrotron radiation.

It is instructive to consider a photon that moves upward in a frame $\mathbf{K}'$, that is, it has $\theta' = \pi/2$. If we insert this into Eq. (1.18), we see that in frame $\mathbf{K}$ the angle is given by ($u' = c$):

$$\tan \theta = \frac{u' \sin \theta'}{\gamma (u' \cos \theta' + v)} = \frac{c}{\gamma v}. \tag{1.19}$$

Therefore, for very large velocities the angle becomes very small, and for ultrarelativistic motion, $v \approx c$, we have

$$\theta \approx \tan \theta \approx \frac{1}{\gamma}, \tag{1.20}$$

nonrelativistic velocity   relativistic velocity

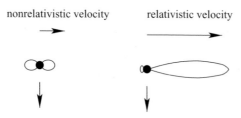

Figure 1.4   Change of a dipole pattern (acceleration perpendicular to the indicated velocity) due to relativistic beaming (left: nonrelativistic, right: relativistic velocity).

which implies that the photon is beamed in forward direction. For a source that radiates isotropically in its rest frame, half of the photons (those with angles $|\theta'| \leq \pi/2$) will be beamed into a cone with an half-opening angle given by the Lorentz factor, $\theta \approx 1/\gamma$. The effect of relativistic beaming on an electron emitting a typical dipole pattern is shown in Fig. 1.4.

1.3.6   Doppler effect

The Doppler effect describes the change of frequency as measured by an observer who moves relative to the source. As an example, think of a car that is passing you: as the car is approaching, you hear a higher frequency. Once it has passed you, the frequency is lower. In the nonrelativistic case, the frequency at the source and the observer are related by $\omega_{\text{obs}} = \omega_{\text{source}} (1 - v/c)^{-1}$, where v is the relative velocity.

For a rapidly moving object that emits a periodic signal, such as an electromagnetic wave, we have to apply the special relativistic version of the Doppler effect. We must account for both the previously discussed time dilation, a purely special relativistic effect, and the geometric effect that the source has moved between two pulses.

Assume a source moves at velocity v and emits in its rest frame $\mathbf{K}'$ pulses at a period T' and frequency $\omega_{\text{source}} = \omega' = 2\pi/T'$. What is the frequency observed by an observer in frame $\mathbf{K}$? First, the observer sees the time interval stretched because of the relativistic time dilation, $\Delta t = \gamma T' = \gamma(2\pi/\omega')$. In addition, the source has moved the distance l from 1 to 2 between two pulses (see Fig. 1.5). The light emitted at point 2 has to travel a shorter distance than the light coming from point 1. Therefore, the observed period, the time between the two arriving light pulses as measured by an observer in $\mathbf{K}$, will be Δt minus the time it took to travel the

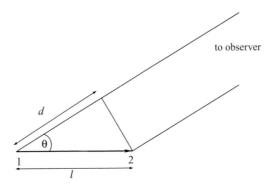

Figure 1.5   A rapidly moving, periodically emitting source travels during one period from point 1 to 2.

extra distance $d = l \cos \theta = v \Delta t \cos \theta$. The observer will measure

$$\Delta t_{\text{obs}} = \Delta t - \frac{d}{c} = \Delta t \left[1 - \left(\frac{v}{c} \right) \cos \theta \right] \tag{1.21}$$

and interpret this as a frequency

$$\omega_{\text{obs}} = \frac{2\pi}{\Delta t_{\text{obs}}} = \frac{2\pi}{\Delta t \left[1 - (v/c) \cos \theta \right]}$$

$$= \frac{\omega_{\text{source}}}{\gamma \left[1 - (v/c) \cos \theta \right]} = \mathcal{D} \cdot \omega_{\text{source}}, \tag{1.22}$$

where $\mathcal{D}$ is called the *Doppler factor*. This is the *relativistic Doppler formula*. The γ in the denominator accounts for the relativistic time dilation; the second term in the bracket corrects for the light-travel effect that occurs also in the nonrelativistic case. In nonrelativistic physics, motions perpendicular to the line of sight do not cause a frequency shift. This is different in the relativistic case. Even for $\theta = \pi/2$, a frequency shift occurs: $\omega_{\text{obs}} = \omega_{\text{source}}/\gamma$. As we had seen before, the Lorentz factor occurs because of relativistic time dilation. This purely relativistic effect is called the *transverse Doppler effect*.

1.4   Basics of tensor calculus

A tensor is the generalization of the concept of a vector and can be thought of as a set of numbers, for example, a matrix, with a well-defined behavior under a change of the coordinate basis.

1.4.1 The metric tensor

An important example of a tensor is the metric tensor. It can be used to measure distances via a *scalar product*, which associates a number with two vectors. The metric tensor essentially determines how to assign the number to the vectors.

Let us start with the well-known scalar product of real-valued, three-dimensional vectors. Consider two vectors $\vec{x} = (x^1, x^2, x^3)$ and $\vec{y} = (y^1, y^2, y^3) \in \mathbb{R}^3$, where, as usual, the components are just the projections onto the basis vectors, for example, $\vec{x} = (x^1, x^2, x^3) = x^1 \hat{e}_1 + x^2 \hat{e}_2 + x^3 \hat{e}_3$. We use the *Einstein summation convention* according to which we sum over a repeated upper and lower index, that is, the expression $a_j b^j$ stands for $\sum_{j=1}^{3} a_j b^j$. Of course, j is just a *dummy index*: $a_j b^j$ is exactly the same as $a_i b^i$. To avoid conflicts, it is sometimes necessary to rename dummy indices. For example, $\left(\sum a_i x^i \right) \left(\sum b_i y^i \right)$ should be written as $a_i x^i b_j y^j$. With this rule, the vector $\vec{x}$ can be written as

$$\vec{x} = x^i \hat{e}_i. \tag{1.23}$$

The scalar product of the vectors $\vec{x}$ and $\vec{y}$ is then given as

$$\vec{x} \cdot \vec{y} = (x^i \hat{e}_i) \cdot (y^j \hat{e}_j) = (\hat{e}_i \cdot \hat{e}_j) x^i y^j. \tag{1.24}$$

Note that it is important here to use two different summation indices to distinguish the two sums. If we know which numbers are assigned to the products of the basis vectors, $\hat{e}_i \cdot \hat{e}_j$, we have, via Eq. (1.24), a rule for the scalar products of general vectors. This is the information contained in the metric tensor. Therefore, one defines the components of the metric tensor as

$$g_{ij} \equiv \hat{e}_i \cdot \hat{e}_j, \tag{1.25}$$

and we can now write

$$\vec{x} \cdot \vec{y} = g_{ij} x^i y^j \equiv G(\vec{x}, \vec{y}). \tag{1.26}$$

Of course, our Cartesian basis vectors in $\mathbb{R}^3$ are unit vectors, that is, of length unity, and they are mutually perpendicular to each other. Thus, the metric tensor is in this case just the unit matrix

$$g = (g_{ij}) = \begin{pmatrix} 1 & 0 & 0 \\ 0 & 1 & 0 \\ 0 & 0 & 1 \end{pmatrix}. \tag{1.27}$$

At this stage, it may seem somewhat cumbersome to write the simple scalar product in this way, but this allows a very smooth transition to the special relativistic case.

In relativity, four-vectors with one time and three space components play a prominent role. We have already encountered an important four-vector at the beginning of this chapter, the space-time point (ct, x, y, z), which in tensor calculus is simply written as (x^0, x^1, x^2, x^3).

We follow here the convention that Latin indices refer to the space components and run from 1 to 3, whereas Greek indices refer to space-time components and run from 0 to 3. Therefore, $a_j b^j = \sum_{j=1}^3 a_j b^j$ is different from $a_\mu b^\mu = \sum_{\mu=0}^3 a_\mu b^\mu$. Like in $\mathbb{R}^3$, the components of a four-vector are just its projections onto the basis vectors:

$$X^\mu = (x^0, x^1, x^2, x^3) = x^0 \hat{e}_0 + x^1 \hat{e}_1 + x^2 \hat{e}_2 + x^3 \hat{e}_3 = x^\mu \hat{e}_\mu. \tag{1.28}$$

As in the three-dimensional case, the *metric tensor* can be thought of as a machine with two input slots that produces a number out of two vectors via a scalar product. It is symmetric

$$G(u, v) = G(v, u) \tag{1.29}$$

and linear

$$G(u, \xi v + \psi w) = \xi G(u, v) + \psi G(u, w), \tag{1.30}$$

where u, v, w can be either three- or four-vector vectors and $\xi, \psi \in \mathbb{R}$. The *components of the metric tensor* are again defined as the scalar products of the basis vectors:

$$g_{\mu\nu} \equiv G(\hat{e}_\mu, \hat{e}_\nu) = g_{\nu\mu}, \tag{1.31}$$

where at the last equal sign we have used the symmetry property. With these conventions, the scalar product of the vectors u and v is given as

$$G(u, v) = G(u^\alpha \hat{e}_\alpha, v^\beta \hat{e}_\beta) = u^\alpha v^\beta G(\hat{e}_\alpha, \hat{e}_\beta) = u^\alpha v^\beta g_{\alpha\beta}, \tag{1.32}$$

where we have made use of Eqs. (1.30) and (1.31). The inverse matrix of $g_{\alpha\beta}$ is denoted $g^{\alpha\beta}$ and fulfills

$$g^{\alpha\lambda} g_{\lambda\beta} = \delta^\alpha{}_\beta, \tag{1.33}$$

where $\delta^\alpha{}_\beta$ is the *Kronecker delta*,[2] which has the value of 1 for $\alpha = \beta$ and 0 otherwise. Generally, a tensor $T^{\mu\nu}$ is said to be *symmetric* if $T^{\mu\nu} = T^{\nu\mu}$ and *antisymmetric* if $T^{\mu\nu} = -T^{\nu\mu}$. It can easily be shown that if a tensor is (anti-) symmetric in one coordinate system, this is also true in any other coordinate system.

The special relativistic scalar product is similar to the scalar product of Eq. (1.26). We have already seen an example of such a special relativistic scalar product (see

---

[2] After the German mathematician Leopold Kronecker (1823–1891).

Eq. [1.12]): the square of the proper time increment, $d\tau$, is the scalar product of an event vector increment, dx^μ, with itself:

$$c^2 d\tau^2 = -\eta_{\mu\nu} dx^\mu dx^\nu \equiv -ds^2, \tag{1.34}$$

where we have denoted the special relativistic version of the metric tensor $g_{\mu\nu}$ as $\eta_{\mu\nu}$:

$$(\eta_{\mu\nu}) = \begin{pmatrix} -1 & 0 & 0 & 0 \\ 0 & 1 & 0 & 0 \\ 0 & 0 & 1 & 0 \\ 0 & 0 & 0 & 1 \end{pmatrix}. \tag{1.35}$$

We have also introduced the four-dimensional generalization of the line element, $ds^2 = \eta_{\mu\nu} dx^\mu dx^\nu = -c^2 dt^2 + dx^2 + dy^2 + dz^2$.

1.4.2 Co- and contravariant vectors

In tensor calculus, one distinguishes between *contra-* and *covariant* vectors. Everything that is known in elementary vector calculus as a "vector" becomes a contravariant vector in tensor calculus, and its components are denoted with upper indices like x^μ. A covariant vector, also known as *one-form* or *differential form*,[3] maps a (contravariant) vector into the real numbers $f : \mathbb{R}^3 \to \mathbb{R}$. Like contravariant vectors, covariant vectors also have components; they are denoted with lower indices, for example, x_μ.

A contravariant vector can be visualized as an arrow; a covariant vector or one-form can be imagined as a sequence of planes (see Fig. 1.6). A covariant vector that operates on a contravariant vector (arrow) produces a scalar. In this picture, the action of the covariant vector on the contravariant vector can now be thought of as counting the number of times the arrow pierces the sequence of surfaces. For example, if the arrow corresponding to some vector, u, pierces the surfaces corresponding to the one-form f 3.3 times, we have $f(u) = 3.3$.

Co- and contravariant vectors can be transformed into each other with the help of the metric tensor. This process is also known as *raising* and *lowering* indices. To get an idea of what this means, think of a vector $\vec{x} \in \mathbb{R}^3$. How can we use the metric tensor, g, to make a one-form out of $\vec{x}$? Think of the metric tensor as a machine with two open slots, where you can insert vectors to output a number, $G : \mathbb{R}^3 \times \mathbb{R}^3 \to \mathbb{R}$. If we leave one slot open, $G(\vec{x}, \ )$, this object does exactly what we want a one-form to do: it takes a vector and produces, via the scalar product/metric tensor, a number. In the language of tensor calculus, this is called

---

[3] This may be familiar as *bra* vectors from quantum mechanics, *ket* vectors correspond to contravariant vectors.

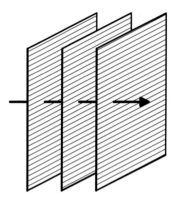

Figure 1.6   A (contravariant) vector can be visualized as an arrow; a one-form
can be imagined as a sequence of planes. The one-form applied to the vector
corresponds in this picture to the number of times the arrow pierces the planes.

"lowering an index," and it is written as

$$v_\alpha = g_{\alpha\beta} v^\beta.$$ (1.36)

The "raising of an index" works completely analogously, but now we have to use
the inverse of the metric tensor,

$$v^\alpha = g^{\alpha\beta} v_\beta.$$ (1.37)

Let us now look at the change of the components of a contravariant vector under a
change of the base vectors. The contravariant vector transforms according to

$$x'^\mu = \Lambda^\mu_\nu x^\nu,$$ (1.38)

where $\Lambda^\mu_\nu = \partial x'^\mu / \partial x^\nu$ is the matrix that describes this change of basis. An example
could be the Lorentz transformation Eqs. (1.2)–(1.5), now written in matrix form
with $\beta = v/c$ as

$$(x'^\mu) = \begin{pmatrix} ct' \\ x' \\ y' \\ z' \end{pmatrix} = \begin{pmatrix} \gamma & -\gamma\beta & 0 & 0 \\ -\gamma\beta & \gamma & 0 & 0 \\ 0 & 0 & 1 & 0 \\ 0 & 0 & 0 & 1 \end{pmatrix} \begin{pmatrix} ct \\ x \\ y \\ z \end{pmatrix}.$$ (1.39)

The matrix Λ^μ_ν depends only on the velocities, not on the coordinates. Therefore,
$dx'^\mu = \Lambda^\mu_\nu dx^\nu$, or

$$\Lambda^\mu_\nu = \frac{\partial x'^\mu}{\partial x^\nu}.$$ (1.40)

We know how a contravariant vector transforms (see Eq. [1.38]), so how does the
corresponding covariant vector transform? This requires some index gymnastics.

Start from two known things, the raising of an index,

$$x^\nu = \eta^{\nu\lambda} x_\lambda, \tag{1.41}$$

where we have used the special relativistic metric tensor $\eta^{\nu\lambda}$, and the transformation law, Eq. (1.38). On the one hand, we have

$$x'^\mu = \Lambda^\mu_\nu x^\nu = \Lambda^\mu_\nu \eta^{\nu\lambda} x_\lambda, \tag{1.42}$$

and on the other

$$x'^\mu = \eta^{\mu\rho} x'_\rho. \tag{1.43}$$

Therefore,

$$\eta^{\mu\rho} x'_\rho = \Lambda^\mu_\nu \eta^{\nu\lambda} x_\lambda. \tag{1.44}$$

If we multiply both sides with $\eta_{\alpha\mu}$ and use $\eta_{\alpha\mu}\eta^{\mu\rho} = \delta^\rho_\alpha$, we have

$$\eta_{\alpha\mu}\eta^{\mu\rho} x'_\rho = \delta^\rho_\alpha x'_\rho = x'_\alpha = \eta_{\alpha\mu}\Lambda^\mu_\nu \eta^{\nu\lambda} x_\lambda. \tag{1.45}$$

We now introduce the abbreviation

$$\tilde\Lambda^\lambda_\alpha \equiv \eta_{\alpha\mu}\Lambda^\mu_\nu \eta^{\nu\lambda}, \tag{1.46}$$

so that Eq. (1.45) can be written as

$$x'_\alpha = \tilde\Lambda^\lambda_\alpha x_\lambda. \tag{1.47}$$

This is the transformation law for the components of a covariant vector.

The matrix $(\tilde\Lambda^\lambda_\alpha)$ is the inverse of the matrix (Λ^α_λ). This can be seen as follows. Start from the invariance of the line element:

$$\eta_{\mu\nu} dx^\mu dx^\nu = \eta_{\alpha\beta} dx'^\alpha dx'^\beta = \eta_{\alpha\beta}\Lambda^\alpha_\mu \Lambda^\beta_\nu dx^\mu dx^\nu. \tag{1.48}$$

By comparing the coefficients of $dx^\mu dx^\nu$ in the first and last expression, one finds

$$\eta_{\mu\nu} = \eta_{\alpha\beta}\Lambda^\alpha_\mu \Lambda^\beta_\nu. \tag{1.49}$$

We can use this together with Eq. (1.46) to find

$$\tilde\Lambda^\lambda_\alpha \Lambda^\alpha_\rho = (\eta_{\alpha\mu}\Lambda^\mu_\nu \eta^{\nu\lambda})\Lambda^\alpha_\rho = \eta_{\alpha\mu}\Lambda^\mu_\nu \Lambda^\alpha_\rho \eta^{\nu\lambda} = \eta_{\nu\rho}\eta^{\nu\lambda} = \delta^\lambda_\rho. \tag{1.50}$$

With these results at hand, the motivation for *co-* and *contravariant* becomes obvious. Assume we change from one set of basis vectors $\{\hat e_\alpha\}$ to another $\{\hat e'_\beta\}$ that is related by

$$\hat e'_\mu = \tilde\Lambda^\lambda_\mu \hat e_\lambda. \tag{1.51}$$

Then a covariant vector transforms like the basis vectors,

$$x'_\mu = \tilde{\Lambda}^\lambda_\mu x_\lambda, \tag{1.52}$$

and is therefore *covariant* ("like the basis"), while the contravariant components transform with the inverse matrix,

$$x'^\mu = \Lambda^\mu_\lambda x^\lambda \tag{1.53}$$

("contrary to the basis").

Let us have a look at a general expression of the form $A_\mu B^\mu$. How does it transform if we change the basis?

$$A'_\mu B'^\mu = \left(\tilde{\Lambda}^\rho_\mu A_\rho\right)\left(\Lambda^\mu_\lambda B^\lambda\right). \tag{1.54}$$

Generally, matrix multiplications are not commutative, that is, the order in which matrices are multiplied matters. Because tensor calculus does the bookkeeping for us, the quantities in this equation can just be considered as the matrix components and therefore just numbers, so we can freely rearrange the equation into

$$A'_\mu B'^\mu = \tilde{\Lambda}^\rho_\mu \Lambda^\mu_\lambda A_\rho B^\lambda = \delta^\rho_\lambda A_\rho B^\lambda = A_\lambda B^\lambda = A_\mu B^\mu. \tag{1.55}$$

This means that we know, without the need for any calculation, that an expression of type $A_\mu B^\mu$ is a Lorentz invariant if the A and B are tensors of the type indicated by the position of their indices. Therefore, from just writing down the proper time in the form of Eq. (1.34), we could have known that $d\tau$ is Lorentz invariant.

1.4.3   General tensors

So far we have encountered several types of tensors, for example, *zeroth-rank tensors* also known as *scalars* or Lorentz invariants of type $A_\mu B^\mu$ such as the proper time (see Eq. [1.34]). We also distinguished between co- and contravariant vectors, which are *first-rank tensors*.[4] We have also encountered the metric tensor $g_{\mu\nu}$, which is said to be a twice-covariant tensor (two lower indices) and its inverse, $g^{\mu\nu}$, which is twice contravariant (two upper indices). The position of the index tells us how the tensor transforms: for each lower index, there is a matrix $\tilde{\Lambda}^\rho_\mu$, and for each upper index, there is a matrix Λ^μ_λ. That is, $g_{\mu\nu}$ transforms like

$$g'_{\mu\nu} = \tilde{\Lambda}^\alpha_\mu \tilde{\Lambda}^\beta_\nu g_{\alpha\beta}, \tag{1.56}$$

and $g^{\mu\nu}$ like

$$g'^{\mu\nu} = \Lambda^\mu_\alpha \Lambda^\nu_\beta g^{\alpha\beta}. \tag{1.57}$$

---

[4] The number of indices of a tensor gives its rank; the positions of the indices determine the behavior under basis transformation.

There are also second-rank tensors of mixed type T^μ_ν that transform like

$$T'^\mu_\nu = \Lambda^\mu_\alpha \tilde{\Lambda}^\beta_\nu T^\alpha_\beta. \tag{1.58}$$

An example of such a second-rank tensor of mixed type is a mapping between two vectors in a vector space, such as a rotation. Such a rotation could be given as $R : \mathbb{R}^2 \times \mathbb{R}^2 \to \mathbb{R}^2$, in components $X^\mu = R^\mu_\nu X^\nu$, with a rotation matrix

$$(R^\mu_\nu) = \begin{pmatrix} \cos\varphi & -\sin\varphi \\ \sin\varphi & \cos\varphi \end{pmatrix}. \tag{1.59}$$

A tensor T^μ_ν could also be used to produce a number out of a covariant vector X_μ and a contravariant vector Y^ν: $T^\mu_\nu X_\mu Y^\nu \in \mathbb{R}$. Operations such as raising and lowering of indices of *general* tensors work completely analogously to the case of first-rank tensors. For example, we can produce a mixed tensor of rank two out of a twice-covariant tensor: $T^\mu_\nu = g^{\mu\lambda} T_{\lambda\nu}$.

1.4.4 Gradients

A prototype of a covariant vector is the gradient $\partial_\alpha \equiv \frac{\partial}{\partial x^\alpha}$. According to the chain rule, it transforms as

$$\frac{\partial}{\partial x'^\alpha} = \frac{\partial x^\beta}{\partial x'^\alpha} \frac{\partial}{\partial x^\beta}. \tag{1.60}$$

We know that a contravariant tensor transforms like $x'^\alpha = \Lambda^\alpha_\rho x^\rho$ (see Eq. [1.38]). If we multiply both sides of this equation with $\tilde{\Lambda}^\beta_\alpha$ we have

$$x'^\alpha \tilde{\Lambda}^\beta_\alpha = \tilde{\Lambda}^\beta_\alpha \Lambda^\alpha_\rho x^\rho = \delta^\beta_\rho x^\rho = x^\beta, \tag{1.61}$$

or

$$\frac{\partial x^\beta}{\partial x'^\alpha} = \tilde{\Lambda}^\beta_\alpha. \tag{1.62}$$

Therefore Eq. (1.60) becomes

$$\partial'_\alpha = \tilde{\Lambda}^\beta_\alpha \partial_\beta; \tag{1.63}$$

that is, ∂_α transforms exactly like a covariant vector. Similarly, we can take the gradient of a contravariant vector (or general tensor)

$$\partial_\mu A^\nu \equiv A^\nu_{,\mu}, \tag{1.64}$$

where we have introduced the comma notation for the derivative. It is now straightforward to show (see Exercise 12) that $A^\nu_{,\mu}$ transforms like a second-rank tensor of mixed type.

1.4.5   Important four-vectors

The postulates that the speed of light is constant and that the laws of physics must be the same in any inertial frame, regardless of position or velocity, are at the heart of special relativity. We had earlier already introduced a first four-vector $(x^\mu) = (ct, x, y, z)$ that labels an event in four-dimensional space-time. It turns out to be very convenient to define several further four-vectors.

In classical mechanics, the three-vectors velocity and momentum are given as $\vec{v} = d\vec{x}/dt$ and $\vec{p} = m\vec{v}$. Completely analogous to this, one defines the *four-velocity* in special relativity as

$$(U^\mu) \equiv \left(\frac{dx^\mu}{d\tau}\right)$$

(1.65)

and the *four-momentum* as

$$(p^\mu) \equiv m(U^\mu),$$

(1.66)

where m is the particle's rest mass (we treat the case of particles with zero rest mass later). The four-velocity can then be written as

$$(U^\mu) = \left(\frac{dx^\mu}{d\tau}\right) = \gamma\left(\frac{dx^\mu}{dt}\right) = \gamma\left[\frac{d(ct)}{dt}, \frac{dx}{dt}, \frac{dy}{dt}, \frac{dz}{dt}\right] = \gamma\,(c, \vec{v}),$$

(1.67)

where we have used $dt = \gamma d\tau$ (see Eq. [1.13]). By applying the definition, one finds the relation $U_\mu U^\mu = -c^2$:

$$U_\mu U^\mu = \frac{dx_\mu}{d\tau}\frac{dx^\mu}{d\tau} = \frac{\eta_{\mu\nu}dx^\nu dx^\mu}{d\tau d\tau} = \frac{ds^2}{d\tau^2} = -c^2,$$

(1.68)

where we have used Eq. (1.34). For the four-momentum, we find

$$(p^\mu) \equiv m(U^\mu) = (\gamma mc, \gamma m\vec{v}).$$

(1.69)

If we expand the Lorentz factor for moderate velocities into a binomial series,

$$\gamma = \left[1 - \left(\frac{v}{c}\right)^2\right]^{-1/2} \approx 1 + \frac{1}{2}\left(\frac{v}{c}\right)^2 + O\left[\left(\frac{v}{c}\right)^3\right],$$

(1.70)

we find that the zero component of the four-momentum is given by

$$p^0 = \gamma mc \approx \frac{1}{c}\left(mc^2 + \frac{1}{2}mv^2\right).$$

(1.71)

In words, the zero component of the four-momentum is just the rest mass energy plus the classic kinetic energy divided by the speed of light. The spatial part of the four-momentum is $\gamma m\vec{v}$, so just the relativistically increased mass times the three-velocity, or the special relativistic three-momentum. Therefore, we can write

$$(p^\mu) = \left(\frac{E}{c}, \vec{p}\right). \tag{1.72}$$

We can now calculate the square of the four-momentum

$$p_\mu p^\mu = \left(\eta_{\mu\nu} p^\nu\right) p^\mu = \frac{-E^2}{c^2} + \vec{p}^2 = -m^2 c^2, \tag{1.73}$$

where the minus sign comes from the (0,0)-component of the metric tensor (Eq. [1.35]) and we have used $E^2 = p^2 c^2 + m^2 c^4$. Therefore, the rest mass is, as expected, a Lorentz invariant. So what about photons? Moving at the speed of light, their line element is

$$ds^2 = -c^2 dt^2 + d\vec{x}^2 = 0 = -d\tau^2 c^2; \tag{1.74}$$

that is, we cannot define the four-momentum for photons via the four-velocity as $dx^\mu/d\tau$. But we can still define the four-momentum as in Eq. (1.72) with $E = h\nu$ and $\vec{p} = (E/c)\hat{e}_p$, where ν is the photon frequency and $\hat{e}_p$ the unit vector pointing in the direction of the photon propagation. As a quick check, we can calculate $p_\mu p^\mu = -E^2/c^2 + p^2 = 0$, so the rest mass of the photon is zero, as it should be.

It is often useful to express physical quantities as a function of the components of four-vectors. For example, from $E = \gamma mc^2$ and $\vec{p} = \gamma m\vec{v}$, we immediately find

$$\vec{v} = \frac{\vec{p}c^2}{E}. \tag{1.75}$$

Analogous to the three-dimensional quantities, one defines the four-acceleration,

$$a^\mu = \frac{dU^\mu}{d\tau}, \tag{1.76}$$

and the four-force,

$$F^\mu = ma^\mu = \frac{dp^\mu}{d\tau}. \tag{1.77}$$

The four-acceleration is always perpendicular to the four-velocity in the sense that their scalar product is zero:

$$U_\mu a^\mu = U_\mu \frac{d}{d\tau} U^\mu = \frac{1}{2} \frac{d}{d\tau}(U_\mu U^\mu) = \frac{1}{2} \frac{d}{d\tau}(-c^2) = 0, \tag{1.78}$$

where we have made use of Eq. (1.68).

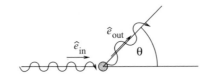

Figure 1.7   Geometry of Compton scattering.

1.4.6   Compton scattering

As an example, we can apply four-vectors to the scattering of photons off electrons, a process known as Compton scattering. Take a photon that scatters off an electron, which causes its direction to change by an angle θ with its original direction, as illustrated in Fig. 1.7. The special case, where the scattering can be regarded as fully elastic (i.e., the photon energy does not change) is called Thomson scattering and is discussed in Chapter 3. In reality, the photon transfers some of its energy to the electron and therefore loses energy. As a result, the photon wavelength is going to change. Let us work out what the change of its wavelength is.

The individual laws of conservation of energy and momentum can be combined into the single law of conservation of the four-momentum. To solve the Compton scattering problem, we apply the following strategy: (i) write down four-momentum conservation, (ii) solve the equation for the uninteresting quantity, here the momentum of the electron after the collision (we are interested in the momentum of the photon after the collision), (iii) square this equation and use $p_\mu p^\mu = -m^2 c^2$, (iv) write down the four-momentum in a specific frame, (v) insert the components into the equation obtained in step (iii), and (vi) solve for the quantity of interest.

If we denote the four-momentum of the electron with p_e and that of the photon with p_γ, four-momentum conservation is expressed as

$$(p_e)^\mu + (p_\gamma)^\mu = (p'_e)^\mu + (p'_\gamma)^\mu, \tag{1.79}$$

where the primes label the quantities after the collision. The "uninteresting quantity" is the electron momentum after the collision:

$$(p'_e)^\mu = (p_e)^\mu + (p'_\gamma)^\mu - (p'_\gamma)^\mu. \tag{1.80}$$

Squaring and applying $p_\mu p^\mu = -m^2 c^2$ causes the masses of the electrons and photons to drop out, and we are left with

$$(p_e)_\mu (p_\gamma)^\mu - (p_e)_\mu (p'_\gamma)^\mu - (p_\gamma)_\mu (p'_\gamma)^\mu = 0, \tag{1.81}$$

where the products are between four-momenta. If we now go to the lab frame where the electron is initially at rest, the momenta are

$$(p_e)^\mu = (m_e c, 0), \qquad (p_\gamma)^\mu = \left(\frac{h}{\lambda}, \frac{h}{\lambda}\hat{e}_{in}\right), \qquad (p'_\gamma)^\mu = \left(\frac{h}{\lambda'}, \frac{h}{\lambda'}\hat{e}_{out}\right) \quad (1.82)$$

and the products are

$$(p_e)_\mu (p_\gamma)^\mu = -\frac{m_e c h}{\lambda}, \qquad (p_e)_\mu (p'_\gamma)^\mu = -\frac{m_e c h}{\lambda'},$$

$$(p_\gamma)_\mu (p'_\gamma)^\mu = -\frac{h^2}{\lambda \lambda'} + \frac{h^2}{\lambda \lambda'}\hat{e}_{in} \cdot \hat{e}_{out}. \qquad\qquad (1.83)$$

If we use $\hat{e}_{in} \cdot \hat{e}_{out} = \cos\theta$ and solve for $\lambda' - \lambda$, we finally have

$$\Delta\lambda = \lambda' - \lambda = \frac{h}{m_e c}(1 - \cos\theta). \qquad (1.84)$$

Interestingly, the shift in wavelength is independent of the wavelength itself; it only depends on the scattering angle. The quantity in front of the bracket in Eq. (1.84), $\lambda_C = h/m_e c$, is called the Compton wavelength of the electron. By plugging in numbers, we find $\lambda_C = 2.4 \cdot 10^{-10}$ cm. For visible light, $\lambda \sim 500$ nm and the change in wavelength after scattering amounts to only ~ 0.005 nm, which is tiny. Therefore, the elastic Thomson scattering is a good approximation. However, in the X-ray part of the spectrum, at wavelengths of ~ 0.05 nm, the effect is already 10% in a single scattering. We return to this process in Chapter 3.

1.4.7 Absorption of a photon by an isolated electron

Can an isolated electron absorb a photon? Assuming that this is possible, the four-momentum conservation reads

$$(p_\gamma)^\mu + (p_e)^\mu = (p'_e)^\mu. \qquad (1.85)$$

Again, square the result, and use $p_\mu p^\mu = -m_e^2 c^2$ to find

$$(p_\gamma)_\mu (p_e)^\mu = 0. \qquad (1.86)$$

If we go into the frame where the electron is initially at rest, $((p_e)^\mu) = (m_e c, 0)$, and $((p_\gamma)^\mu) = (E/c, \vec{p}_\gamma)$, the above condition becomes $-m_e E + 0 \cdot \vec{p}_\gamma = 0$. Therefore, four-momentum conservation requires a photon of zero energy. Hence, the process is not possible.

1.5 Invariance of the phase space volume

We have seen that it is very convenient to work with quantities that we know are Lorentz invariant. Often individual quantities are components of four vectors, so they change if we change to a different frame, but appropriate combinations of such quantities are invariant because the individual changes cancel out. An important such quantity is the phase space volume $d^3r\, d^3p$: both the three-volume in space, d^3r, and the three-volume in momentum space, d^3p, change individually, but their product does not.

Consider first d^3r. Let us think of a particle in a volume $d^3r' = dx'dy'dz'$, where we have denoted the quantities in the particle rest frame, $\mathbf{K'}$, with primes. If we consider a frame that moves with respect to this frame along the common x-axes, the perpendicular y- and z-axes are unaffected by the Lorentz transformation, but there is a length contraction in x-direction: $dx = \gamma^{-1}dx'$. Therefore, the three-volumes are related by

$$d^3r = \frac{1}{\gamma}d^3r'. \tag{1.87}$$

However, the particle energy has also changed: in its rest frame, it was given by $E' = mc^2$; in the frame where the particle moves, it is $E = \gamma mc^2$, so we have $\gamma = E/E'$. If we insert this into Eq. (1.87), we have

$$Ed^3r = E'd^3r' = \text{Lorentz invariant.} \tag{1.88}$$

Let us now turn to the three-volume in momentum space. Consider a set of particles that are nearly at rest but have a small spread in their three-momenta, $\Delta p'^1$, $\Delta p'^2$, and $\Delta p'^3$. Then, to first order in $\Delta p'^i$, the four-momentum spread is $(\Delta p'^\mu) = (0, \Delta p'^1, \Delta p'^2, \Delta p'^3)$ with the energy component being zero, as it is quadratic in the momenta. The term $(\Delta p'^\mu)$ is a four-vector, so it transforms like

$$(\Delta p^\mu) = \begin{pmatrix} \Delta p^0 \\ \Delta p^1 \\ \Delta p^2 \\ \Delta p^3 \end{pmatrix} = \begin{pmatrix} \gamma & \gamma\beta & 0 & 0 \\ \gamma\beta & \gamma & 0 & 0 \\ 0 & 0 & 1 & 0 \\ 0 & 0 & 0 & 1 \end{pmatrix} \begin{pmatrix} 0 \\ \Delta p^1 \\ \Delta p^2 \\ \Delta p^3 \end{pmatrix}' = \begin{pmatrix} \gamma\beta\Delta p'^1 \\ \gamma\Delta p'^1 \\ \Delta p'^2 \\ \Delta p'^3 \end{pmatrix}, \tag{1.89}$$

and therefore

$$d^3p = \gamma d^3p'. \tag{1.90}$$

If we again express the Lorentz factor via the energy ratio as before, we find

$$\frac{d^3p'}{E'} = \frac{d^3p}{E} = \text{Lorentz invariant} \tag{1.91}$$

and obviously for the phase space volume

$$d^3r \, d^3p = \text{Lorentz invariant.} \tag{1.92}$$

This also has direct implications for the *phase-space distribution* f. The number of particles in a volume of phase space $d^3r \, d^3p$ is given by

$$dN = f d^3r \, d^3p. \tag{1.93}$$

Because this number is a countable quantity and therefore invariant and the phase-space volume is also Lorentz invariant, the same must also be true for the phase-space distribution f.

1.6   Relativistic electrodynamics

For completeness, we briefly give the main equations of special relativistic electrodynamics. A good overview over the topic can be found in J. D. Jackson's book, *Classical Electrodynamics*.

To formulate electrodynamics in a covariant way, we need to identify suitable four-vectors. As is shown in Chapter 3, the fields $\vec{E}$ and $\vec{B}$ can be expressed completely in terms of scalar potential, Φ, and a vector potential, $\vec{A}$:

$$\vec{B} = \vec{\nabla} \times \vec{A} \tag{1.94}$$

and

$$\vec{E} = -\vec{\nabla}\Phi - \frac{1}{c}\frac{\partial \vec{A}}{\partial t}. \tag{1.95}$$

It is important to realize that the continuity equation for the electric charge,

$$\frac{\partial \rho}{\partial t} + \vec{\nabla} \cdot \vec{j} = 0, \tag{1.96}$$

can be written in tensor form as

$$j^{\mu}_{,\mu} = 0, \tag{1.97}$$

where we have used the comma notation introduced in Eq. (1.64) and (j^{μ}) is the four-current

$$(j^{\mu}) = (\rho c, \vec{j}). \tag{1.98}$$

This is already our first four-vector needed to express Maxwell's equations in covariant form. In Lorentz gauge (see Chapter 3) the potentials Φ and $\vec{A}$ can be related to their sources ρ and $\vec{j}$ via

$$\vec{\nabla}^2 \Phi - \frac{1}{c^2} \frac{\partial^2 \Phi}{\partial t^2} = -4\pi\rho \qquad (1.99)$$

$$\vec{\nabla}^2 \vec{A} - \frac{1}{c^2} \frac{\partial^2 \vec{A}}{\partial t^2} = -\frac{4\pi}{c} \vec{j}. \qquad (1.100)$$

This motivates the introduction of the four-potential

$$(A^\mu) = (\Phi, \vec{A}). \qquad (1.101)$$

With this definition, the wave Eqs. (1.99) and (1.100) can be compactly written as

$$\Box A^\mu = -\frac{4\pi}{c} j^\mu, \qquad (1.102)$$

where $\Box$ is the Quabla- or D'Alembert operator given by $\partial^\alpha \partial_\alpha$ and $\partial_\alpha = (\partial/\partial ct, \partial/\partial x^i)$. Because the electric and magnetic fields are calculated as derivatives of the potentials (see Eqs. [1.94] and [1.95]), it is useful to define the anti-symmetric *electromagnetic field tensor*

$$F_{\mu\nu} \equiv A_{\nu,\mu} - A_{\mu,\nu}, \qquad (1.103)$$

which contains six independent components (if the entries in the matrix below the diagonal are known, the ones above the diagonal are obtained by a multiplication with -1; the diagonal only contains zeros) just like the physical fields $\vec{E}$ and $\vec{B}$. Explicitly (see Exercise 15) it reads

$$(F_{\mu\nu}) = \begin{pmatrix} 0 & -E_x & -E_y & -E_z \\ E_x & 0 & B_z & -B_y \\ E_y & -B_z & 0 & B_x \\ E_z & B_y & -B_x & 0 \end{pmatrix}. \qquad (1.104)$$

Because this object has by construction a well-defined behavior under a Lorentz transformation, it can be used to calculate the physical fields $\vec{E}'$ and $\vec{B}'$ in a different frame

$$F'_{\mu\nu} = \tilde{\Lambda}^\rho_\mu \tilde{\Lambda}^\lambda_\nu F_{\rho\lambda}, \qquad (1.105)$$

where $\tilde{\Lambda}^\nu_\mu$ is the matrix introduced in Eq. (1.46). The inhomogeneous (with source terms) Maxwell equations in covariant form are

$$\partial^\mu F_{\nu\mu} = \frac{4\pi}{c} j_\nu, \qquad (1.106)$$

and the homogeneous ones are then given as

$$\partial_\alpha F_{\beta\gamma} + \partial_\gamma F_{\alpha\beta} + \partial_\beta F_{\gamma\alpha} = 0. \qquad (1.107)$$

The forces on a particle of charge q moving in an electromagnetic field described by the tensor $F_{\alpha\beta}$ are written in covariant form as

$$\frac{dp_\alpha}{d\tau} = \frac{q}{c} F_{\alpha\beta} U^\beta. \tag{1.108}$$

This is the relativistic generalization of the Lorentz force.

1.7 Exercises

1.1 Cosmic rays
Cosmic rays that bombard the Earth's atmosphere produce muons as secondary particles (rest mass $mc^2 = 105$ MeV). They are typically produced at a height of 6 km and have a lifetime of 2200 ns but can still be detected on the ground. How is that possible? Give a lower limit on their energy.

1.2 Source size
"If a source varies substantially on a timescale δt, the source size must (from causality arguments) be smaller than $D = c \cdot \delta t$."
This often-used argument, which is valid for nonrelativistic speeds, has to be modified for relativistic sources. Consider an optically thick sphere that expands relativistically with a Lorentz factor of $\gamma \gg 1$. Show that for the radius of the source is $R < 2\gamma^2 \cdot c\delta t$. (This is very important in the context of active galactic nuclei and gamma-ray bursts.)

1.3 Relativistic beaming
Consider a light source that emits its photons isotropically in its own rest frame. Consider 12 such rays, and calculate how they appear to an observer moving relativistically toward the source.

1.4 Twin paradox
A spaceship sets off from Earth for a distant destination, traveling in a straight line at uniform speed $3c/5$. Ten years later, a second spaceship sets off in the same direction with a speed of $4c/5$. By the time that the second vessel catches up with the first, show that the first captain is older than the second by 4 years if the captains of the two vessels are twins.

1.5 Space travel 1
As a spaceship passes the Earth with a speed of $0.8c$, observers on this ship and on the Earth agree that the time is 12:00 in both places. Thirty minutes later (on the spaceship clock), the ship passes an interplanetary navigation station fixed relative to the Earth. The clock on this station reads Earth time.
(i) Show that, as the spaceship passes, the time on the station clock is 12:50.
(ii) Show that the station is 7.2×10^{13} cm from the Earth (as measured in the Earth frame).
(iii) As the ship passes the station, it reports back to Earth by radio. Show that (by Earth clock) the signal is received at 13:30.

(iv) There is an immediate reply from Earth. Show that (by the ship's clock) the reply is received at 16:30.

1.6 Two-body collision

Two subatomic particles approach each other with velocities (relative to an observer at rest in the laboratory) of $3c/5$ and $2c/5$. Show that their speed relative to each other is $25c/31$. Moreover, show that the observer would have to move with a velocity of $0.134c$ to measure their velocities as equal and opposite.

1.7 Space travel 2

A rocket moving away from the Earth with speed v emits light pulses once each second as measured by a clock on the rocket.

(i) Show that the rate at which the pulses are received on the Earth is

$$r = \sqrt{\frac{c - v}{c + v}}.$$
(1.109)

(ii) The rocket travels to a distant planet, and the signals are received on Earth both directly and by reflection from the planet. The pulse rates for the two signals are found to be in the ratio 1:2. Explain why this is so, and deduce the speed of the rocket.

(iii) If the rocket transmits only during its flight and the number of pulses received directly is 10^4, what is the distance of the planet from Earth?

1.8 Space travel 3

In a spaceship shuttle service from Earth to Mars, each spaceship is equipped with two identical lights, one at the front and one at the back. The spaceships usually travel at a speed v relative to the Earth, such that the headlight of a spaceship approaching the Earth appears green ($\lambda = 500$ nm) and the taillight of a departing spaceship appears red ($\lambda = 600$ nm).

(i) Show that $v/c = 1/11$.

(ii) One spaceship accelerates to overtake the spaceship ahead of it. Show that the overtaking spaceship has to travel with a speed of $0.18c$ relative to Earth so that the taillight of the Mars-bound spaceship ahead of it looks like a headlight (i.e., green).

1.9 Relativistic energy and momentum

A particle as observed in a certain reference frame has a total energy of 5 GeV and a momentum of 3 GeV/c.

(i) What is its mass in GeV/c^2?

(ii) What is its energy in a frame in which its momentum is equal to 4 GeV/c?

(iii) Use the velocity addition formulae, or the energy-momentum transformation, to find the relative speed of the two frames of reference, if the particles are moving in the same direction.

1.10 Supernova SN1987A

The supernova explosion SN1987A observed on January 23, 1987, in the Large Magellanic Cloud, a nearby galaxy only 160 000 light years from the Solar System, was the brightest and best studied supernova since 1604. A pulse of 20 neutrinos from SN1987A was detected by the IMB and Kamiokande experiments. All 20 neutrinos arrived within a time interval of 10 s, and their energies ranged from 7.5 to 40 MeV.

(i) If the neutrino has mass m_ν, derive an expression for the time it takes a neutrino of total energy E_ν to travel from the Large Magellanic Cloud to a detector on Earth.

(ii) From laboratory experiments based on studies of tritium β-decay, the neutrino mass is known to be very small: $m_\nu < 7\,\text{eV}/c^2$. Using the supernova data and assuming that the high-energy neutrinos (40 MeV) arrive first and the low-energy neutrinos (7.55 MeV) arrive last, show that the upper mass limit for neutrinos is $15.2\,\text{eV}/c^2$.

1.11 Tensor basics

(i) Show that $A^\mu B^\nu$ is a twice contravariant tensor.

(ii) Show that $A_\mu B^\nu$ is a mixed second rank tensor.

(iii) Show that the sum of two mixed second-rank tensors is again a mixed second-rank tensor: $C^\nu_\mu = A^\nu_\mu + B^\nu_\mu$.

1.12 Gradient of a tensor

Show that the gradient of a contravariant tensor A^μ, $\partial_\nu A^\mu$ transforms like a second-rank tensor of mixed type.

1.13 Inverse Compton scattering

A relativistic cosmic ray proton (momentum p, energy $E \gg m$, $c = 1$) has an encounter with a photon from the cosmic microwave background.

(i) Calculate the maximum energy that can be transferred to the photon.

As a guideline, you may want to proceed according to the following steps:

• write down four-momentum conservation,
• solve for the uninteresting momentum,
• square it,
• show that $E'_\gamma = [E_\gamma(E + pc)]/(2E_\gamma + E - pc)$, and
• show that it can be transformed to

$$E'_\gamma \approx \frac{E[1 - (m^2 c^4/4E^2)]}{1 + (m^2 c^4/4EE_\gamma)}. \tag{1.110}$$

(ii) Insert numbers for an ultra-high-energy proton with 10^{20} eV and a typical photon from the cosmic microwave background.

1.14 Resulting compound system

A particle of mass m_1 and velocity $\vec{v}_1$ collides with a particle at rest of mass m_2 and is absorbed. Show that the rest mass of the compound system is $m = (m_1^2 + m_2^2 + 2\gamma m_1 m_2)^{1/2}$ and its velocity is $\vec{v} = \frac{\vec{v}_1}{1 + (m_2/m_1\gamma)}$.

1.15 Electromagnetic field tensor

Calculate explicitly the components of the electromagnetic field tensor defined in Eq. (1.103).

1.16 Force on charged particle in electromagnetic field

Show that the covariant form of the force law, Eq. (1.108), reduces to the expressions known from elementary electrodynamics.

1.8   Further reading

Jackson, J. D. (1998). *Classical Electrodynamics*, 3rd edn. New York: Walter de Gruyter.

Misner, C. W., Thorne, K. S., and Wheeler, J. A. (1970). *Gravitation*. New York: W. H. Freeman.

Rybicki, G. B. and Lightman, A. P. (1979). *Radiative Processes in Astrophysics*. New York: Wiley.

Schutz, B. F. (1989). *A First Course in General Relativity*. Cambridge: Cambridge University Press.

2

Gas processes

In this chapter, we develop some elementary and fundamental concepts of gas and plasma physics that will be helpful for understanding later chapters of this book. An intuitive route to the understanding of gas dynamics starts at the microscopic level. First, we take a closer look at microscopic processes, in particular at collisions between gas particles, ionization, and transport processes. The bulk properties of fluids are described by the equations of fluid dynamics, which we introduce in Section 2.4. Subsequently, those equations will be applied to the theory of shock waves. Shock waves play an important role in many areas of high-energy astrophysics, as for example, in supernova explosions (Chapter 4), gamma-ray bursts (Chapter 7), or in accelerating particles to high energies, as explained in Section 2.8. Finally, we discuss some fluid instabilities that play a prominent role in astrophysics. All these topics are relevant to the following chapters on various astrophysical phenomena. Clearly, this introduction can neither be as complete nor as thorough as a pure physics textbook. The purpose of this chapter is to focus on key physical processes in gases starting from a basic knowledge of statistical physics and thermodynamics. For further reading, we give references at the end of this chapter.

2.1 Collisions

When is it justified to treat a collection of particles as a fluid? The answer is that these particles need to collide often enough so that they can be described by mean quantities. Grossly speaking, a macroscopic system of particles may be described as a fluid if the average distance between two subsequent collisions of the constituent particles is much less than the macroscopic scale that is of interest. Inevitably, the number of particles in such a system is large. On top of the random velocities of the particles, the fluid may possess a mean bulk velocity. Because of the frequent collisions, the random motions will never carry a particle very far from its neighbor, and the bulk velocity is not affected by the random motion.

The average distance between two subsequent collisions is called *mean free path*. The nature of the collisions between particles depends on the nature of the force between the particles. If the interacting particles are electrically charged, they interact via Coulomb forces, and if they are neutral, they interact only via short-range forces that act over microscopic length scales.

2.1.1   *Neutral particles*

If the particles interact only via short-range forces, as is the case for neutral atoms or molecules, the mean free path can be calculated as follows: imagine that all particles are solid spheres of radius R and that all particles are stationary except the test particle, which is traveling from left to right, as shown in Fig. 2.1. As the test particle travels along its trajectory, it sweeps out a cylinder of radius $2R$. The test particle can now collide with any other particle located within a distance $2R$ from its projected path. If we denote the mean free path by l, then a cylinder of radius $2R$ and length l contains exactly one particle. If we denote the cross section $\pi(2R)^2$ by σ, this implies that

$$n\sigma l = 1, \tag{2.1}$$

where n is the number density of the particles. This can be rearranged to give

$$l = \frac{1}{n\sigma}. \tag{2.2}$$

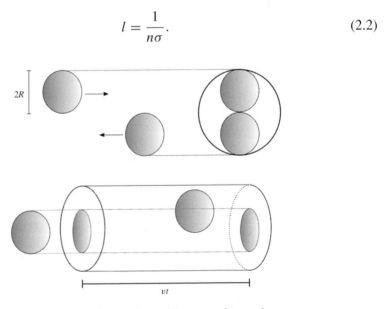

Figure 2.1   Visualization of the mean free path.

The cross section for neutral hydrogen atoms may be estimated as $\pi(2a_0)^2$, where $a_0 \approx \hbar^2/m_e e^2 \approx 0.5 \cdot 10^{-8}$ cm is the radius of the first Bohr orbit. Then, the mean free path for neutral hydrogen atoms is $l \sim 0.3 \cdot 10^{16}(n/\text{cm}^{-3})^{-1}$ cm. In passing, we note that the concept of mean free paths can also be applied to photons. This is discussed in Chapter 3.

In thermal equilibrium, the velocities of the constituent particles of a gas follow a Maxwell-Boltzmann distribution.[1] The root-mean-square velocity of this distribution is

$$v_{\text{rms}} = \left(\frac{3k_B T}{\mu m_H}\right)^{1/2}, \tag{2.3}$$

where k_B is Boltzmann's constant, T temperature, and μ the mean mass per particle in units of the mass of a hydrogen atom, m_H. The quantity μ is sometimes called the mean molecular weight. For neutral hydrogen, $\mu = 1$, whereas for fully ionized hydrogen $\mu = 0.5$ because the hydrogen atom has been split into two particles, an electron and a proton. Generally, one can show that for a fully ionized gas

$$\mu = \left[\sum_i \frac{X_i(Z_i + 1)}{A_i}\right]^{-1}, \tag{2.4}$$

where X_i is the relative mass fraction of atom i, with atomic number Z_i and atomic weight A_i.

By plugging the constants into Eq. (2.3), we find that $v_{\text{rms}} \approx 6.7 \cdot 10^5 \, (T/\text{K})^{1/2}$ cm s^{-1} for electrons and $\approx 1.6 \cdot 10^4 \, (T/\text{K})^{1/2}$ cm s^{-1} for protons.

2.1.2 Charged particles

Most fluids that we deal with in high-energy astrophysics are at least partially ionized and thus contain positively charged ions and negatively charged electrons. Charged particles interact via the Coulomb force. The magnitude of this force is given by[2]

$$F_{\text{Coul}} = \frac{q_1 q_2}{r^2}, \tag{2.5}$$

where q_1 and q_2 are the charges and r is their separation.

One can distinguish between two types of collisions. *Strong collisions* are collisions that lead to a drastic change in the directions and velocities of the particles involved in the collision. In contrast, *weak collisions* alter the paths and velocities

---

[1] The distribution of velocities, $f(v)$, in a classical gas is given by $f(v) = 4\pi(m/2\pi k_B T)^{3/2} v^2 \exp(-mv^2/2k_B T)$, where m is the mass of the gas particles, v the velocity and T the temperature.
[2] Keep in mind that we are using cgs units throughout the book.

only very slightly. Let us start with strong collisions. A condition for a strong collision could be that the magnitude of the potential energy at the point of closest approach between the two particles is larger than their kinetic energy. In this case, the Coulomb interaction is strong enough to affect the dynamics of the charges. If we consider the collision between an electron of velocity v and a stationary ion of charge $+Ze$, this condition can be expressed as

$$\frac{Ze^2}{r} > \frac{1}{2}m_e v^2. \tag{2.6}$$

In other words, we demand that the electrostatic energy at closest approach, Ze^2/r, is larger than the kinetic energy of the electron, $m_e v^2/2$.

   This can be arranged to yield the condition for a strong encounter

$$r < r_s = \frac{2Ze^2}{m_e v^2}. \tag{2.7}$$

The length r_s is sometimes called *strong collision radius*, and the cross section for such a strong collision is roughly πr_s^2. We can work out the time between two successive strong collisions using our expression for the mean free path given by Eq. (2.2):

$$t_{strong} = \frac{l}{v} = \frac{1}{v n \sigma} = \frac{m_e^2 v^3}{4\pi n Z^2 e^4}, \tag{2.8}$$

where n is the number density of charges.

   So much for the strong collisions. Next, we describe the dynamics of a weak encounter between two charged particles. Again, let us consider the deflection of an electron that passes a stationary ion with charge $+Ze$. We assume that the electron passes the ion at an impact parameter b (see Fig. 2.2) with a velocity v. As the electron passes the ion, it is deflected toward the ion and gains a velocity component, $v_\perp$, perpendicular to its undisturbed trajectory. Now we compute only the effect of weak encounters, in which the deflection angle and thus the ratio

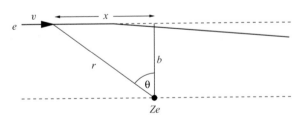

Figure 2.2   Geometry for weak collisions. The bold line traces the real trajectory of the electron, whereas the upper dashed line is the trajectory assumed in the weak collision approximation.

$v_\perp/v$ are small. This allows us to make two assumptions that greatly simplify the following calculation:

1. The electron passes the ion on a straight-line trajectory.
2. The ion is stationary during the encounter.

The deflecting force (i.e., the component of the force perpendicular to the undisturbed trajectory) is given by

$$F_\perp = \frac{Ze^2 \cos\theta}{b^2 + x^2}, \tag{2.9}$$

where x is the distance of the electron from its point of closest approach with the ion (see Fig. 2.2). Because $\cos\theta = b/r = b/(b^2 + x^2)^{1/2}$, we can write

$$F_\perp = \frac{Ze^2 b}{(b^2 + x^2)^{3/2}}. \tag{2.10}$$

Now, $x = vt$, and we can write

$$F_\perp = \frac{Ze^2}{b^2} \left[1 + \left(\frac{vt}{b}\right)^2 \right]^{-3/2}. \tag{2.11}$$

Using Newton's second law, $F = dp/dt$, we can work out the perpendicular momentum imparted on the electron during one encounter by integrating along the entire trajectory using $s = vt/b$

$$\Delta p_\perp \approx \int_{-\infty}^{\infty} F_\perp(t)\, dt = \frac{Ze^2}{bv} \int_{-\infty}^{\infty} (1 + s^2)^{-3/2}\, ds = \frac{Ze^2}{bv} \left[\frac{s}{\sqrt{1+s^2}} \right]_{-\infty}^{\infty} = \frac{2Ze^2}{bv}. \tag{2.12}$$

This results in a perpendicular velocity of

$$\Delta v_\perp = \frac{\Delta p_\perp}{m_e} \approx \frac{2Ze^2}{m_e bv}. \tag{2.13}$$

A quick way to get roughly the same result is by taking the perpendicular force at closest approach, which is

$$F_\perp = \frac{Ze^2}{b^2}, \tag{2.14}$$

and estimating that this force acts on the electron for a time $\Delta t \sim b/v$. Now, the change in the perpendicular momentum is $\Delta p_\perp \sim F_\perp \Delta t \sim Ze^2/bv$, which is only off by a factor of 2 from Eq. (2.12).

To obtain the cumulative effect of all weak collisions, we have to sum all $\Delta v_\perp$ over all impact parameters. Because $\Delta v_\perp$ can be positive or negative, we should sum its square.

In a time interval, dt, our test electron collides at impact parameter, b, with all particles in a cylindrical shell of radius b and thickness db. The length of this cylindrical shell is $v dt$ because this is the distance the electron covers in a time interval dt. If the number density of charges is n, the total number of charges in the shell is $2\pi b db\, v dt\, n$.

Thus, the total rate of change of $(\Delta v)^2$ is

$$\frac{d(\Delta v)^2}{dt} \approx nv \int_{b_{min}}^{b_{max}} \Delta v_\perp^2\, 2\pi b\, db = \frac{8\pi n Z^2 e^4}{m_e^2 v} \ln(b_{max}/b_{min}), \qquad (2.15)$$

where b_{min} and b_{max} are the smallest and largest values of the impact parameter that contribute to the integral. The integral in Eq. (2.15) is divergent as $b_{max} \to \infty$ and as $b_{min} \to 0$, so we need to find nonzero and finite values for b_{min} and b_{max}. For all our purposes, the ratio b_{max}/b_{min} appears only in the argument of the logarithm. The logarithm $\ln(b_{max}/b_{min}) \equiv \ln \Lambda$ is called the *Coulomb logarithm*. Because b_{min} and b_{max} appear only inside the logarithm, the Coulomb logarithm is not too sensitive to the exact values for b_{max} and b_{min}, and simple estimates for their values suffice. A bound on b_{min} comes from our assumption that the encounter should be weak. Our assumption that the electron remains on a straight-line trajectory during its passage breaks down when $b < r_s = 2Ze^2/m_e v^2$. In other words, our calculation is only valid for weak encounters, which occurs when $b > r_s$. Another restriction comes from quantum mechanics. The uncertainty principle states that we cannot know simultaneously the momentum and position of a particle with arbitrary precision. The uncertainties in momentum and position, Δp and Δx, are related via $\Delta p \Delta x \sim \hbar$. As a result, b_{min} may not be smaller than $\hbar/m_e v$.

The upper limit on b, b_{max}, is given by the maximum distance to which the Coulomb force extends. In free space this would be infinity, but in a plasma charges are *screened* by opposite charges that cluster around it. At distances larger than a length that is called *Debye length*, the Coulomb force is effectively shielded, and thus b_{max} may be set to the Debye length $b_{max} = (k_B T/4\pi n e^2)^{1/2}$. For all astrophysical plasmas, the Coulomb logarithm lies in the range $\sim$10–30.

So far, we have only computed the rate of change of $(\Delta v)^2$. We can convert this into a time for our electron to lose all memory of its original path, which happens after

$$t_{weak} = \frac{v^2}{d(\Delta v)^2/dt} = \frac{m_e^2 v^3}{8\pi n Z^2 e^4 \ln \Lambda}. \qquad (2.16)$$

This time can be compared directly to the time between two successive strong encounters. By comparing Eq. (2.8) and Eq. (2.16), we find

$$t_{weak} = \frac{t_{strong}}{2 \ln \Lambda}. \tag{2.17}$$

For all plasmas $\ln \Lambda \gg 1$, the time for a charge to be deflected appreciably via multiple weak encounters is much shorter than the time between two strong encounters. The ratio between these two timescales is given by the Coulomb logarithm. Hence, it turns out that for Coulomb collisions the cumulative effect of many weak collisions is larger than that of less frequent, strong encounters. This is an important result, and it implies that, for the sake of computing relaxation timescales, we can safely ignore strong collisions and consider weak ones only.

Finally, we can work out the distance a charge can travel until it is appreciably deflected from its original path. This distance is given by

$$l_{Coul} = v \frac{v^2}{d(\Delta v)^2/dt} = \frac{m_e^2 v^4}{8\pi n Z^2 e^4 \ln \Lambda}. \tag{2.18}$$

If we set the kinetic energy to the typical thermal energy, $m_e^2 v^4 \sim (k_B T)^2$, we can write Eq. (2.18) as

$$l_{Coul} \approx \frac{1.4 \cdot 10^4}{Z^2 \ln \Lambda} \left(\frac{T}{K}\right)^2 \left(\frac{n}{cm^{-3}}\right)^{-1} cm. \tag{2.19}$$

Interestingly, this is independent of the mass of the charge and therefore is the same for electrons and ions.

We conclude this section by considering what happens if charged particles move in a magnetic field. In the presence of magnetic fields, charged particles behave very differently. An electric charge that travels through a magnetic field experiences a Lorentz force

$$\vec{F}_{Lorentz} = \frac{q}{c}(\vec{v} \times \vec{B}), \tag{2.20}$$

where q is the charge, $\vec{v}$ the velocity of the charge, and $\vec{B}$ the magnetic field strength. This force is always perpendicular to $\vec{v}$ and $\vec{B}$, and consequently the charges spiral around the magnetic field lines. The radius of this spiralling motion, r_L, which is called *gyroradius* or *Larmor radius*, is found by equating the force from Eq. (2.20) to the centrifugal force, that is,

$$\frac{q}{c}vB = \frac{mv^2}{r_L}, \tag{2.21}$$

where m is the mass of the charged particle. Rearranging gives

$$r_L = \frac{mvc}{qB}. \tag{2.22}$$

The associated angular frequency is $\omega_L = 2\pi/T$, where T is the period, which is given by $T = 2\pi r_L/v$, assuming that the charges follow a circular orbit of radius r_L. This yields $\omega_L = v/r_L = qB/mc$ for the Larmor frequency. For typical field strengths in the cosmos and for practically all velocities, the gyroradius is much smaller than the dimensions of the field. As a result, the particles are effectively tied to the field lines and move along with the field. They are said to be frozen to the field, and this type of flow is called *frozen-in flow*. The only way they can be untied from "their" field line is through the collision with another particle. If such collisions occur frequently, the particles can effectively diffuse across the magnetic field.

The presence of magnetic fields can cause particles to behave like a fluid. There are many dilute gases in astrophysics, where the mean free path is very large. For example, in blast waves from stellar explosions, protons are accelerated to energies of 2 MeV. The typical stopping distance for such a proton would be $\sim 10^3$ pc, which is much larger than the size of the blast wave (see Chapter 4). According to our definition at the beginning of this chapter, this gas should not behave like a fluid. However, the gyroradius is much smaller and ties the particles together by preventing them from traveling large linear distances. As a result, the particles effectively behave like a fluid and can be described by macroscopic variables.

2.2   Ionization

Inside an atom, an electron can be promoted from its ground state to a state of higher energy. This can occur by the absorption of a photon or by collisions between atoms. If the absorbed energy is larger than the binding energy of a given electron, the electron is knocked out of the atom, leaving behind a positively charged ion. This process is called ionization.

The probability of an atom being in a particular energy state, E, is determined by the Boltzmann factor, $\exp(-E/k_B T)$. More than one state in an atom can have the same energy, in which case the states are called degenerate. The number of degenerate states is called the statistical weight, g_i. For a large number of atoms, the ratio of occupation probabilities must be the same as the ratio of numbers of atoms in the two energy levels. Hence the number density ratio of atoms in two different energy levels, E_1 and E_2, is given by

$$\frac{n_1}{n_2} = \frac{g_1}{g_2} \exp\left(-\frac{E_1 - E_2}{k_B T}\right), \tag{2.23}$$

which is the Boltzmann equation. An atom can change its state of ionization by capturing or releasing an electron. For example, iron has 26 electrons, and it has 26 different ions denoted by Fe I, Fe II, and up to Fe XXVI, corresponding to atoms with 26, 25, and down to 1 electrons, respectively.

Let us look at ionization reaction

$$X_i \leftrightarrow X_{i+1} + e^-, \tag{2.24}$$

where X_i denotes the ith state of ionization of atom X.

The energy difference between the ground state of a neutral atom and the ground state of the same atom with one electron removed is

$$\Delta E = U_{\text{ionis}} + \frac{1}{2} m_e v^2, \tag{2.25}$$

where the first term on the right-hand side is the ionization potential, U_{ionis} (the minimum amount of energy needed to remove an electron from its ground state), and the second term is the kinetic energy of the free electron, where v is the velocity of the electron, and we have assumed that the electron moves nonrelativistically. For energies comparable to atomic binding energies, $E \sim 10$ eV, this is a valid assumption because $E \ll m_e c^2 = 511$ keV. According to Boltzmann's law, the ratio between the population of this ion to that of its neutral atom is given by

$$\frac{dN_{i+1}(v)}{N_i} = \frac{g_{i+1} dg_e}{g_i} \exp\left[-\frac{(U_{\text{ionis}} + m_e v^2/2)}{k_B T}\right], \tag{2.26}$$

where $dN_{i+1}(v)$ is the differential number density of ions in ionization state $i + 1$ in the ground state with the free electron having a velocity between v and $v + dv$. Furthermore, N_i is the number density of ions in state i in the ground state, g_{i+1} and g_i are the statistical weights of these ions, respectively, in their ground states, and dg_e is the statistical weight of the electron in the phase space volume $d^3x \, d^3p$. The statistical weight denotes the number of particles that can have the same properties. We explain what "having the same properties" means. Every particle is characterized by its positions given by three spatial coordinates (e.g., $[x, y, z]$), and its momentum, which again has three components (e.g., $[p_x, p_y, p_z]$). The space spanned by these six coordinates is called *phase space*. Electrons are fermions, and according to Pauli's exclusion principle no two fermions may have the same properties. What is meant by "the same" is determined by Heisenberg's uncertainty principle, which states that you cannot distinguish two particles if their difference in momentum, Δp, multiplied by their difference in position, Δx, is less than Planck's constant, h. Hence, particles that occupy a cell in phase space with dimensions

$$\Delta x \, \Delta y \, \Delta z \, \Delta p_x \, \Delta p_y \, \Delta p_z = h^3 \tag{2.27}$$

are indistinguishable from each other. There is one more complication: electrons possess spin, which is quantized. For electrons, the corresponding spin quantum number can assume values of $+1/2$ and $-1/2$ (often labeled "up" and "down"). Separate spin quantum states are distinguishable, so that actually two electrons, one with spin $+1/2$ and one with spin $-1/2$, can occupy one single phase space cell. Therefore, each phase space cell is assigned a spin degeneracy factor, g_s, which for electrons is equal to 2.

The resulting statistical weight of an electron is given by

$$dg_e = \frac{g_s \, dx \, dy \, dz \, dp_x \, dp_y \, dp_z}{h^3}, \tag{2.28}$$

where h is Planck's constant. The volume $dx \, dy \, dz$ is the volume that a single electron occupies and therefore is equal to the inverse of the electron number density, n_e^{-1}. For an isotropic momentum distribution,

$$dp_x \, dp_y \, dp_z = 4\pi p^2 \, dp = 4\pi m_e^3 v^2 \, dv. \tag{2.29}$$

Substituting this together with $g_s = 2$ into Eq. (2.26) yields

$$\frac{dN_{i+1}(v)}{N_i} = \frac{8\pi m_e^3}{h^3 n_e} \frac{g_{i+1}}{g_i} \exp\left[-\frac{(U_{\text{ionis}} + m_e v^2/2)}{k_B T}\right] v^2 \, dv. \tag{2.30}$$

Integrating over v from 0 to ∞ yields

$$\frac{N_{i+1} n_e}{N_i} = \left(\frac{2\pi m_e k_B T}{h^2}\right)^{3/2} \frac{2g_{i+1}}{g_i} \exp\left(-\frac{U_{\text{ionis}}}{k_B T}\right). \tag{2.31}$$

Now remember that N_{i+1} and N_i denote the number densities of ions in their respective ionization states *in their ground states*. However, every ion can assume a whole range of energy states depending on what energy levels the bound electrons occupy within the ion. To relate the number density of ions in the ground state to the total number density of the respective ion, we invoke Boltzmann's law. Denoting the total number densities by small letters n, we have

$$\frac{N_i}{n_i} = \frac{g_i}{Z_i(T)} \tag{2.32}$$

and

$$\frac{N_{i+1}}{n_{i+1}} = \frac{g_{i+1}}{Z_{i+1}(T)}, \tag{2.33}$$

where $Z_i(T)$ and $Z_{i+1}(T)$ are the partition functions of the respective ionization states. The partition function is defined as $Z(T) = \sum g_i \exp(-E_i/k_B T)$. This will be used in the Box "Nuclear statistical equilibrium (NSE)" in Chapter 4.

Substituting this into Eq. (2.31) gives

$$\frac{n_{i+1}n_e}{n_i} = \left(\frac{2\pi m_e k_B T}{h^2}\right)^{3/2} \frac{2Z_{i+1}(T)}{Z_i(T)} \exp\left(-\frac{U_{\text{ionis}}}{k_B T}\right), \qquad (2.34)$$

which is called the *Saha equation* after Meghnad Saha,[3] who first derived it in 1920.

Let us proceed by applying this equation to the simplest atom, the hydrogen atom. First, we need to work out the partition functions. The partition function of the positive hydrogen ion is simply 1 because a hydrogen ion is just a bare proton and has no degeneracy. The energy of the first excited state of hydrogen is $E_2 - E_1 = 10.2$ eV above the ground state energy. Because 10.2 eV $\gg k_B T$ for temperatures $\ll 10^5$ K, the Boltzmann factors, $\exp[-(E_2 - E_1)/k_B T]$, in the partition function are much smaller than 1. Hence, virtually all neutral hydrogen atoms are in the ground state, and the partition function simplifies to $Z_i \approx g_1 = 2$. This means that the degeneracy of the hydrogen atom is essentially 2, which corresponds to the electron being in the ground state with its spin either up or down.

Let us define the degree of ionization as $\xi = n_{\text{H\,II}}/(n_{\text{H\,I}} + n_{\text{H\,II}})$. Electrical neutrality and conservation of nucleon number imply that the number density of positive ions must equal the number density of free electrons:

$$n_e = n_{\text{H\,II}}. \qquad (2.35)$$

The Saha equation can now be written as

$$\frac{\xi^2}{1 - \xi} = \frac{1}{n}\left(\frac{2\pi m_e k_B T}{h^2}\right)^{3/2} \exp\left(-\frac{U_{\text{ionis}}}{k_B T}\right), \qquad (2.36)$$

where $n = n_{\text{H\,I}} + n_{\text{H\,II}}$. One may verify the limits for ξ for small and large temperatures. As $T \to \infty$, $\xi \to 1$, and as $T \to 0$, $\xi \to 0$, as one would expect.

By substituting numerical values for hydrogen ($U_{\text{ionis}} = 13.6$ eV), we find

$$\frac{\xi^2}{1 - \xi} = \frac{4 \cdot 10^{-9} \text{g cm}^{-3}}{\rho} T^{3/2} \exp\left(-\frac{1.6 \cdot 10^5 \text{K}}{T}\right). \qquad (2.37)$$

The Saha equation predicts that hydrogen gas should be 50% ionized at a temperature of around $T \sim 10^4$ K.

Finally, we should mention some important limitations of Saha's equation. In its derivation, we assumed that the atoms are isolated. However, at high densities, the electrostatic potential wells of the atoms start to overlap, which effectively

---

[3] Meghnad N. Saha, Indian astrophysicist (1893–1956).

lowers the ionization energy because the electrons can escape more easily from the potential wells. This leads to easier ionization, which is called *pressure ionization*.

The characteristic separation a of atoms with a number density n can be worked out via

$$\frac{4}{3}\pi a^3 = \frac{1}{n}. \tag{2.38}$$

Pressure ionization happens when a becomes of the order of the Bohr radius of hydrogen, a_0, which corresponds to matter densities larger than $3\,\mathrm{g\,cm}^{-3}$. Then the Saha equation becomes invalid. An example for pressure ionization occurs at the center of the Sun. There, the temperature is $T \sim 1.5 \cdot 10^7$ K, and the hydrogen number density is $n_\mathrm{H} \sim 10^{26}\,\mathrm{cm}^{-3}$. According to Saha's equation, the solar interior would not be fully ionized even though helioseismology tells us that it is. This is the result of pressure ionization.

The Saha equation is also invalid if the gas is irradiated, say, by an external source of radiation. For example, gases in cold clouds around active galactic nuclei have a very different ionization structure than if they were not illuminated. Also, the Saha equation only applies to gases that are in thermal equilibrium because otherwise the Boltzmann equation that we used in its derivation is invalid. If the atoms and ions have not had enough time to collide and thus to *thermalize* the gas, the ionization fraction could have any value in principle. As we saw earlier, this is more likely in very dilute environments where the collision frequencies are smaller.

2.3 Transport processes

2.3.1 *Viscosity*

Viscosity in a fluid transports momentum.[4] Consider a simple, plane-parallel shear flow, in which the velocity in the x-direction, u_x, depends only on y as indicated in Fig. 2.3. On top of the bulk motion in the x-direction, the particles in the fluid perform thermal random motions with their thermal velocity v_rms. If there are n particles per unit volume, about $1/3$ have a velocity predominantly in the y-direction. Half of these (i.e., $1/6$) will travel in the positive y-direction and the other half in the negative y-direction. As a result, an equal and opposite flux of particles, $\sim n v_\mathrm{rms}/6$, crosses a plane $y = y_0$ (see Fig. 2.3). After traveling a mean distance l, these particles collide with other fluid particles and share their properties with the particles at their new location. As the particles with $y > y_0$ have a larger x-component of momentum than particles with $y < y_0$, the gas below y_0 will gain x-momentum, whereas gas above y_0 will lose x-momentum.

---

[4] Curiously, the word viscosity stems from the Latin word *viscum* for mistletoe. Mistletoe berries were used to make a viscous glue used for catching birds.

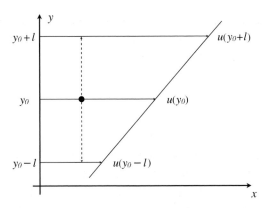

Figure 2.3   Geometry for calculation of viscosity. The particle marked by the black dot exchanges x-momentum with parts of the shear flow within a mean free path, l, of its position. See text for details.

We can attempt to quantify the resulting rate of change of momentum. Each particle carries an x-momentum of mu_x, where m is the mass of the particle. The mean x-component of momentum transported per unit time and area across a plane $y = y_0$ in the upward direction is $(1/6)\,nv_{\mathrm{rms}}mu_x(y_0 - l)$, where $u_x(y_0 - l)$ means u_x evaluated at $(y_0 - l)$. Conversely, the mean x-momentum carried downward is $(1/6)\,nv_{\mathrm{rms}}mu_x(y_0 + l)$. The net transport of momentum is found by subtracting one from the other. The stress is thus given by

$$F_{\mathrm{visc}} = \frac{1}{6}nmv_{\mathrm{rms}}[u_x(y_0 - l) - u_x(y_0 + l)]. \tag{2.39}$$

Assuming that the mean free path is smaller than the scale of variation of u_x, we can expand u_x around y_0 in a Taylor expansion:

$$u_x(y_0 \pm l) = u_x(y_0) \pm \left(\frac{\partial u_x}{\partial y}\right)_{y=y_0} l + O(l^2). \tag{2.40}$$

By substituting this into Eq. (2.39) and generalizing to any y, we obtain

$$F_{\mathrm{visc}} = \frac{1}{6}nmv_{\mathrm{rms}}l\left(-2\frac{\partial u_x}{\partial y}\right) \equiv -\eta\frac{\partial u_x}{\partial y}, \tag{2.41}$$

where

$$\eta = \frac{1}{3}nmv_{\mathrm{rms}}l \tag{2.42}$$

is the coefficient of molecular viscosity. By substituting $v_{rms} = (3k_B T/\mu m_H)^{1/2}$ from Eq. (2.3) and $l = 1/n\sigma$ from Eq. (2.2), we find

$$\eta = \frac{(mk_B T)^{1/2}}{\sqrt{3}\sigma} \propto T^{1/2}. \tag{2.43}$$

It is interesting to note that the coefficient of viscosity is independent of the number density and only appears to depend on temperature. At first, this may seem strange because if there are twice as many particles per unit volume, say, one would expect the transport of momentum to be twice as efficient. However, if the number density is doubled, the mean free path of the particles is halved, which leaves the efficiency for momentum transport unchanged.

For hydrogen atoms, this yields a numerical value of

$$\eta \approx 5.7 \cdot 10^{-5} T^{1/2} \, \mathrm{g\,cm^{-1}s^{-1}}, \tag{2.44}$$

where T is given in units of Kelvin. For a fully ionized gas using Eq. (2.19),

$$\eta \approx 2 \cdot 10^{-15} T^{5/2} (\ln \Lambda)^{-1} \, \mathrm{g\,cm^{-1}s^{-1}}, \tag{2.45}$$

where $\ln \Lambda$ is the Coulomb logarithm defined in Section 2.1. Sometimes the coefficient of *kinematic viscosity* is used, which is defined as

$$\nu \equiv \eta/\rho. \tag{2.46}$$

The ability of a fluid to damp out turbulence is quantified by a dimensionless parameter called *Reynolds number*. The Reynolds number[5] is the ratio of inertial forces to viscous forces and is used for determining whether a flow will be laminar or turbulent. It is defined as

$$\mathrm{Re} = \frac{Lv\rho}{\eta}, \tag{2.47}$$

where L is the size of a typical turbulent eddy, which has a turnover velocity, v. Flows with $\mathrm{Re} \gg 1$ will sustain turbulence, whereas flows with $\mathrm{Re} < 1$ damp out turbulent motions. Because $\nu \sim vl$, where l is the mean free path, the Reynolds number can be written as $\mathrm{Re} \sim (v/c_s)(L/l)$. Now for a fluid description to be at all valid, we need $L \gg l$, so the Reynolds number will typically be large.

2.4   Equations of fluid dynamics

Consider a fluid that is flowing in and out of a volume V that is bounded by a surface S. The rate of mass flowing through a surface element $d\vec{S}$ is $\rho\vec{v} \cdot d\vec{S}$,

---

[5] The Reynolds number is named after the British physicist Osborne Reynolds (1842–1912), who proposed it in 1883.

where ρ is mass density and $\vec{v}$ velocity. In the absence of any sinks or sources, the mass flowing through the whole surface must equal the change in mass within the volume, V. This can be expressed as

$$-\int \frac{\partial \rho}{\partial t} dV = \int \rho \vec{v} \cdot d\vec{S}, \qquad (2.48)$$

where $\vec{S}$ is the outward pointing surface vector. The derivatives denoted by $\partial/\partial t$ are taken at a fixed point in space and are called *Eulerian* derivatives. Alternatively, one can express derivatives in fluid mechanics also in a coordinate system moving with the fluid. This derivative is called *Lagrangian* derivative and is denoted here by d/dt. We can now work out how these two types of derivatives are related. If $\vec{x}$ and $\vec{x} + \vec{v}\delta t$ are the positions of a fluid element at times t and $t + \delta t$, then the Lagrangian time derivative of some fluid property $f(\vec{x}, t)$ is defined as

$$\frac{df}{dt} = \lim_{\delta t \to 0} \frac{f(\vec{x} + \vec{v}\delta t, t + \delta t) - f(\vec{x}, t)}{\delta t}. \qquad (2.49)$$

Keeping the first-order terms in the Taylor expansion, we have

$$f(\vec{x} + \vec{v}\delta t, t + \delta t) = f(\vec{x}, t) + \frac{\partial f}{\partial t}\delta t + \vec{v} \cdot \vec{\nabla} f \delta t + O(\delta t^2). \qquad (2.50)$$

Thus, we obtain the relation between the Lagrangian and the Eulerian derivatives:

$$\frac{df}{dt} = \frac{\partial f}{\partial t} + (\vec{v} \cdot \vec{\nabla})f. \qquad (2.51)$$

Back to Eq. (2.48). Using Gauss's theorem, we can convert the surface integral into a volume integral in the following way:

$$-\int \frac{\partial \rho}{\partial t} dV = \int \vec{\nabla} \cdot (\rho \vec{v}) dV. \qquad (2.52)$$

Because this must be true for any volume, we can write this in differential form

$$\frac{\partial \rho}{\partial t} = -\vec{\nabla} \cdot (\rho \vec{v}), \qquad (2.53)$$

which is called the *continuity equation* and expresses the conservation of mass. Similarly, we can consider the force exerted by the fluid on the surface, S. The pressure force is $P d\vec{S}$. The integral of the pressure over S can be rewritten using Gauss's law,

$$-\int P d\vec{S} = -\int \vec{\nabla} P dV. \qquad (2.54)$$

Hence, the pressure force on a volume element dV is $-\vec{\nabla}P \, dV$. In the presence of a gravitational field, ϕ, the gravitational force is $-\rho \vec{\nabla}\phi \, dV$. Newton's second law can now be written as

$$\rho \frac{d\vec{v}}{dt} = -\vec{\nabla}P - \rho\vec{\nabla}\phi, \tag{2.55}$$

which is called the *Euler equation*. Note that the right-hand side of this equation can be complemented by other external forces, such as forces exerted by magnetic fields or viscosity. In Eulerian coordinates (i.e., fixed in space), Eq. (2.55) can be rewritten as

$$\rho \frac{\partial \vec{v}}{\partial t} + \rho(\vec{v} \cdot \vec{\nabla})\vec{v} = -\vec{\nabla}P - \rho\vec{\nabla}\phi. \tag{2.56}$$

In many cases in astrophysics, such as in stars, gas remains in a fairly steady state for very long times. Here, gravity and pressure forces balance to form a state that is called *hydrostatic equilibrium*. We can find the condition for hydrostatic equilibrium by setting $d\vec{v}/dt = 0$ and finding

$$\vec{\nabla}P = -\rho\vec{\nabla}\phi. \tag{2.57}$$

The equation of hydrostatic equilibrium is very important for the structure of stars, in which the fluid is approximately in hydrostatic equilibrium. The final fundamental equation of fluid dynamics is the energy equation. The total energy density of a fluid element consists of its kinetic energy density, $\rho v^2/2$, and its thermal energy per unit volume, ρe, where e is the internal energy per unit mass. For a monatomic gas

$$e = \frac{3k_{\mathrm{B}}T}{2\mu m_{\mathrm{H}}}, \tag{2.58}$$

where T is temperature and k_{B} Boltzmann's constant. The energy equation states that the change in total energy in a volume is equal to the flux of total energy through its surface plus the work done by pressure forces acting on it, plus energy gains or losses by other processes. This can be written as

$$\frac{\partial}{\partial t}\left(\frac{1}{2}\rho v^2 + \rho e\right) + \vec{\nabla} \cdot \left[\left(\frac{1}{2}\rho v^2 + \rho e + P\right)\vec{v}\right] = -\rho\vec{v} \cdot \vec{\nabla}\phi + \mathcal{L}. \tag{2.59}$$

For a detailed derivation, consult any fluid mechanics textbook. The term $\mathcal{L}$ denotes the rate at which energy is being lost or gained by radiation, thermal conduction, or other processes. Its dimension is energy per time and per volume.

When the gas can be approximated as ideal, pressure, density, and temperature are related via

$$P = nk_{\mathrm{B}}T = \frac{\rho k_{\mathrm{B}}T}{\mu m_{\mathrm{H}}} = \frac{2}{3}\rho e, \tag{2.60}$$

where n is the number density and μ the mean molecular weight. However, this pressure is only the pressure contributed by particles. There are other sources of pressure, such as radiation pressure and magnetic pressure. These concepts are introduced later.

2.5 Equation of state

The previous section concluded with the ideal gas law, which is a relation among pressure, density, and temperature in an ideal gas. Often in high-energy astrophysics, the ideal gas description is a poor approximation for the behavior of matter. For example, in the centers of very massive stars, the densities become so large that the forces between the constituent particles can no longer be neglected, and the particles move with relativistic speeds. Then, we have to employ more sophisticated equations of state. In the following, we briefly revise how the equation of state derives from the microscopic properties of the matter.

An equation of state can be regarded as a link between the macroscopic thermodynamic properties of the stellar material and its microphysics. An important concept in this context is the *phase space distribution* function, $\tilde{f}$, which denotes the number of particles per phase space volume, $d^3x\, d^3p$. The phase space distribution can be written as $\tilde{f} = \frac{g_s}{h^3} f$, where $g_s = 2S + 1$ and S is the spin of the particle.

Once $\tilde{f}$ is known, the macroscopic thermodynamic quantities are obtained by integrating over the appropriate section of phase space. For example, we obtain the space number density of a gas of particles, n, by integrating over momentum space,

$$n = \int \tilde{f} d^3 p. \tag{2.61}$$

The pressure is given by

$$P = \frac{1}{3} \int p v \tilde{f} d^3 p, \tag{2.62}$$

and the energy density as

$$u = \int E\, dn = \int E \tilde{f} d^3 p. \tag{2.63}$$

Here p is the momentum, v the velocity, and E the energy of a particle.

As long as particles are close enough to "know" about their quantum nature, we have to distinguish between fermions (particles whose spin is a multiple of 1/2) and bosons (with an integer spin):

$$f(E) = \frac{1}{\exp[(E - \mu)/k_B T] \pm 1},$$
(2.64)

where the upper sign refers to fermions (or the Fermi–Dirac statistics), the lower sign refers to to bosons (Bose–Einstein statistics), μ is the chemical potential of the particles, k_B is the Boltzmann constant, and T is the gas temperature.

When quantum-mechanical effects are irrelevant (i.e., if the gas is hot and has a low density), this expression reduces to

$$f(E) \approx \exp\left(\frac{\mu - E}{k_B T}\right),$$
(2.65)

the distribution function of a classical, nonrelativistic Maxwell–Boltzmann gas. Using the distribution function of Eq. (2.65) in the relations (2.61), (2.62), and (2.63), we recover the relations for the ideal gas:

$$n = g_s \left(\frac{m k_B T}{2\pi \hbar^2}\right)^{3/2} \exp\left(\frac{\mu - mc^2}{k_B T}\right),$$
(2.66)

$$P = n k_B T,$$
(2.67)

and

$$u = nmc^2 + \frac{3}{2} n k_B T.$$
(2.68)

This description is appropriate for most low-density gases.

At higher densities, however, the quantum nature of the particles does matter. In a white dwarf, for example, the electrons are so close to each other that they dominate the pressure, and it is a quantum effect, the degeneracy pressure, that determines (together with gravity) the structure of a white dwarf. Consider a "cold" (i.e., $\mu/k_B T \gg 1$) gas of fermions. In this case, the distribution function (2.64) becomes a step function (the chemical potential μ at zero temperature is called Fermi energy, E_F),

$$f(E) = \begin{cases} 1 & \text{for } E \leq E_F \\ 0 & \text{else.} \end{cases}$$

Defining the momentum that corresponds to the Fermi energy E_F as p_F, we have

$$E_F = \left(p_F^2 c^2 + m_e^2 c^4\right)^{1/2}.$$
(2.69)

In the considered situation, the equation for the number density, Eq. (2.61), can be easily integrated to yield

$$n_e = \frac{2}{h^3} \int_0^{p_F} 4\pi p^2 dp = \frac{8\pi}{3h^3} p_F^3, \qquad (2.70)$$

or

$$p_F = \left(\frac{3h^3}{8\pi} n_e \right)^{1/3}, \qquad (2.71)$$

which gives the Fermi momentum as a function of the density. Inserting the relativistically correct velocity $v = pc^2/E$ (see Eq. [1.75]) and the Fermi distribution of Eq. (2.64) into Eq. (2.62) yields the general expression for the pressure. However, this is rather complicated even in our $T = 0$ limit (see Section 5.6.1). For our purposes it is enough to consider only the two limiting cases, where the electrons are either nonrelativistic (i.e., $m_e c^4 \gg pc^2$) or extremely relativistic, $pc^2 \gg m_e c^4$. In these cases, the equation of state reduces to the simple form

$$P = K\rho^\Gamma, \qquad (2.72)$$

which is called *polytropic* equation of state or simply a *polytrope*. Equation (2.71) shows that, even at zero temperature, the electrons provide a pressure, the so-called *degeneracy pressure*, that results from their fermionic quantum nature.

The quantity

$$\Gamma = \frac{\partial \ln(P)}{\partial \ln(\rho)} \bigg|_s \qquad (2.73)$$

is called the *adiabatic exponent*.

At densities less than 10^6 g cm^{-3}, electrons are nonrelativistic (from Eq. (2.71), and the adiabatic exponent is

$$\Gamma_{nr} = \frac{5}{3}, \qquad (2.74)$$

whereas the extremely relativistic case, $\rho \gg 10^6$ g cm^{-3}, yields

$$\Gamma_{er} = \frac{4}{3}. \qquad (2.75)$$

A stability analysis shows that a gaseous sphere obeying a polytropic equation of state becomes dynamically unstable (in other words, it will collapse) once $\Gamma \le 4/3$. This result has important consequences for white dwarfs and supernovae (see Chapter 4). These results are derived explicitly in Section 5.6.1.

2.6   Bernoulli's equation

Bernoulli's equation applies to steady flows, for which $\partial v / \partial t = 0$ (this should not be confused with hydrostatic equilibrium, for which $dv/dt = 0$). Euler's equation then becomes

$$(\vec{v} \cdot \vec{\nabla})\vec{v} = -\vec{\nabla}\phi - \frac{\vec{\nabla}P}{\rho}. \tag{2.76}$$

By using the vector identity $(\vec{v} \cdot \vec{\nabla})\vec{v} = (1/2)\vec{\nabla}(v^2) - \vec{v} \times (\vec{\nabla} \times \vec{v})$, this can be rearranged to give

$$\frac{1}{2}\vec{\nabla}(v^2) + (\vec{\nabla} \times \vec{v}) \times \vec{v} + \vec{\nabla}\phi + \frac{\vec{\nabla}P}{\rho} = 0. \tag{2.77}$$

Now in certain cases, for example, in adiabatic flows, Eq. (2.77) can be put into a useful form. The total enthalpy H is defined to be $H = U + PV$, where U is the internal energy. The derivative of H is $dH = dU + P\,dV + V\,dP = T\,dS + V\,dP$, where in the last equation we used the first law of thermodynamics: $dU = T\,dS - P\,dV$. Dividing by the mass gives $dh = T\,ds + dP/\rho$, where the small letters denote specific quantities (i.e., quantities per unit mass).

Thus, for adiabatic flows, $ds = 0$, and hence $\vec{\nabla}P/\rho = \vec{\nabla}h$. If we take the scalar product of Eq. (2.77) with the unit vector $\hat{v}$, the term involving $\vec{\nabla} \times \vec{v}$ drops out. Moreover, the term $\hat{v} \cdot \vec{\nabla}$ is the derivative along the direction of motion (i.e., the streamline of the flow). Because we can write this derivative as d/dl, we have

$$\frac{d}{dl}\left(\frac{1}{2}v^2 + h + \phi\right) = 0. \tag{2.78}$$

This means that the expression in the bracket is conserved along streamlines. Now the specific enthalpy is just proportional to the pressure, which means that if we neglect the effect of gravity, an increase in fluid velocity must be accompanied by a drop in the pressure.

This effect is called the *Bernoulli effect*. Because the flux of gas is fixed, this requires that the gas velocity scales inversely with the cross-sectional area, resulting in a drop in pressure in the nozzle.

2.7   Shock waves

In an ideal gas, small disturbances in pressure cause linear compression waves, or sound waves. The equations that govern the propagation of these small disturbances can be derived by considering small perturbations of the pressure and density about some constant background value. If we linearize the continuity and the Euler equations in the small perturbations and assume that the disturbances are adiabatic, we end up with a wave equation that gives for the wave speed

$$c_s = \left[\left(\frac{\partial P}{\partial \rho} \right)_s \right]^{1/2}, \tag{2.79}$$

where P is pressure, ρ is density, and the derivative is taken at constant entropy, s. For a polytropic, ideal gas, this is

$$c_s = \left(\frac{\Gamma P}{\rho} \right)^{1/2}, \tag{2.80}$$

where Γ is the ratio of specific heats, which in the following we assume to be the value for a monatomic ideal gas, 5/3. This shows that variations in pressure imply variations in the sound speed. Thus each point on the waveform propagates with its local sound speed, which is greater at the peaks, where the density is higher, than in the troughs. With time, the waveform steepens and eventually would break as waves on an ocean. This would produce multiple values for the state variables of the gas at a given point in space, which is not permitted. Instead nature inserts a shock front just before the wave breaks, and the flow variables remain single valued. How does this happen? On a microscopic level, large gradients in temperature and velocity at the front of the steepening wave cause the irreversible processes of heat conduction and viscous stress to dominate in a region with a width equal to a few collision mean free paths and to counteract the self-steepening process so that a single-valued shock front forms. The net effect on a macroscopic level is that mass, momentum, and energy are conserved across the shock front, but entropy is not. The entropy increases as macroscopic kinetic energy is dissipated into heat through atomic or molecular collisions.

The basic definition of a shock is that it is an irreversible (entropy-generating) wave that causes a transition from supersonic to subsonic flow. Because the shock thickness is of the order of the mean free path of the gas particles, it is usually much smaller than the length scales of gradients in the gas on either side of the shock, and in the following, we regard it as infinitely thin.

Across an idealized discontinuity, the relations between the thermodynamic variables can be found using the laws of conservation of mass, energy, and momentum. As a reference frame, we chose a frame in which the shock front is stationary. Gas flows into the shock with density, velocity, and pressure (ρ_1, v_1, P_1) and flows out of the shock wave with the corresponding values ρ_2, v_2, and P_2. Let x denote the distance from the shock.

Conservation of mass flux across the discontinuity yields

$$\rho_1 v_1 = \rho_2 v_2 = \mathcal{F}_{mass}, \tag{2.81}$$

where $\mathcal{F}_{\text{mass}}$ is the mass flux. This equation merely states that no mass is lost or generated in the flow.

Similarly, from Euler's equation, Eq. (2.56), by neglecting gravity and looking for a steady solution (i.e., $\partial \vec{v}/\partial t = 0$) we find

$$\rho v \frac{dv}{dx} + \frac{dP}{dx} = 0 \tag{2.82}$$

or

$$\frac{d}{dx}(P + \rho v^2) = 0, \tag{2.83}$$

where we have used $\rho v = $ constant. Integrating this across the shock, we find

$$P_1 + \rho_1 v_1^2 = P_2 + \rho_2 v_2^2 = \mathcal{F}_{\text{mom}}, \tag{2.84}$$

where $\mathcal{F}_{\text{mom}}$ is called the *momentum flux*.

Finally, we demand the conservation of energy. If we neglect radiative losses and thermal conduction across the shock, the shock is often called *adiabatic*. This is a misleading term because, by its very nature, a shock transition increases the specific entropy of the flow. This contradicts the definition of the term *adiabatic*, which usually denotes processes in which the entropy does not change. A better term is *nonradiative shock*. From the energy equation, Eq. (2.59), we find

$$\frac{d}{dx}\left[\left(\frac{1}{2}\rho v^2 + \rho e + P \right) v \right] = 0. \tag{2.85}$$

Here we sketch the derivation of shock relations for a gas with an adiabatic exponent $\Gamma = 5/3$. The relations for a general Γ will be given at the end of this section.

Generally, the relation between the specific energy and pressure and density is given by $e = P/(\Gamma - 1)\rho = 3P/2\rho$. Using this together with $\rho v = $ const. we can rewrite Eq. (2.85) as

$$\rho v \frac{d}{dx}\left(\frac{1}{2}v^2 + \frac{5P}{2\rho} \right) = 0. \tag{2.86}$$

Again, integrating across the shock yields

$$\frac{1}{2}v_1^2 + \frac{5}{2}\frac{P_1}{\rho_1} = \frac{1}{2}v_2^2 + \frac{5}{2}\frac{P_2}{\rho_2} = E, \tag{2.87}$$

where $E = (1/2)v^2 + (5/2)(P/\rho)$ is the total specific energy. Relations (2.81), (2.84) and (2.87) are called *Rankine–Hugoniot* jump conditions. They determine the jumps in ρ, P, and v across the shock. We now show how, with the help of a little algebra, these conditions can be transformed into relations between the up- and downstream variables of the flow. First, it is easy to verify from Eq. (2.81) and

Eq. (2.84) that

$$\frac{\mathcal{F}_{\text{mom}}}{\mathcal{F}_{\text{mass}} v} = \frac{P}{\rho v^2} + 1. \tag{2.88}$$

Next, let us define the Mach number, $\mathcal{M}$, via

$$\mathcal{M}^2 = \frac{v^2}{c_s^2} = \frac{3}{5} \frac{\rho v^2}{P}, \tag{2.89}$$

where we have used Eq. (2.79). Now, we can write Eq. (2.88) and Eq. (2.87) as

$$\frac{\mathcal{F}_{\text{mom}}}{\mathcal{F}_{\text{mass}} v} = \frac{3}{5\mathcal{M}^2} + 1 \tag{2.90}$$

and

$$E = \frac{1}{2} v^2 + \frac{5}{2} \frac{P}{\rho} = \frac{1}{2} v^2 + \frac{5}{2} \left(\frac{\mathcal{F}_{\text{mom}} v}{\mathcal{F}_{\text{mass}}} - v^2 \right) \tag{2.91}$$

or

$$v^2 - \frac{5}{4} \frac{\mathcal{F}_{\text{mom}} v}{\mathcal{F}_{\text{mass}}} + \frac{1}{2} E = 0. \tag{2.92}$$

This is a quadratic equation for v in terms of the conserved quantities $\mathcal{F}_{\text{mass}}$, $\mathcal{F}_{\text{mom}}$, and E. This equation has two roots, which must correspond to the velocities up- and downstream from the shock, v_1 and v_2. By solving this quadratic equation, we find that their sum is

$$v_1 + v_2 = \frac{5}{4} \frac{\mathcal{F}_{\text{mom}}}{\mathcal{F}_{\text{mass}}}, \tag{2.93}$$

and consequently

$$\frac{v_2}{v_1} = \frac{5}{4} \frac{\mathcal{F}_{\text{mom}}}{\mathcal{F}_{\text{mass}} v_1} - 1 = \frac{5}{4} \left(\frac{3}{5\mathcal{M}_1^2} + 1 \right) - 1, \tag{2.94}$$

where we have used Eq. (2.90) and where $\mathcal{M}_1$ is the upstream Mach number.

In the limiting case of a strong shock, $\mathcal{M}_1 \gg 1$, we find

$$\frac{v_2}{v_1} = \frac{1}{4}, \tag{2.95}$$

and by using Eq. (2.81),

$$\frac{\rho_2}{\rho_1} = 4. \tag{2.96}$$

These relations can be generalized for any value of Γ:

$$\frac{\rho_1}{\rho_2} = \frac{(\Gamma + 1)P_1 + (\Gamma - 1)P_2}{(\Gamma - 1)P_1 + (\Gamma + 1)P_2}. \tag{2.97}$$

For the ratio of the temperatures on the two sides of the discontinuity, we find from the equation of state for a perfect gas $T_2/T_1 = P_2\rho_1/P_1\rho_2$ that

$$\frac{T_2}{T_1} = \frac{P_2}{P_1}\left[\frac{(\Gamma+1)P_1 + (\Gamma-1)P_2}{(\Gamma-1)P_1 + (\Gamma+1)P_2}\right]. \tag{2.98}$$

For the mass flux, we obtain

$$\mathcal{F}^2_{\text{mass}} = \rho_1\frac{(\Gamma-1)P_1 + (\Gamma+1)P_2}{2}. \tag{2.99}$$

Analogously, one may derive some additional, useful formulae that express the ratios of densities, pressures, and temperatures in a shock wave in terms of the Mach number $\mathcal{M}_1$:

$$\frac{\rho_2}{\rho_1} = \frac{v_1}{v_2} = \frac{(\Gamma+1)\mathcal{M}_1^2}{(\Gamma-1)\mathcal{M}_1^2 + 2} \tag{2.100}$$

and

$$\frac{P_2}{P_1} = \frac{2\Gamma\mathcal{M}_1^2}{\Gamma+1} - \frac{\Gamma-1}{\Gamma+1}. \tag{2.101}$$

The Mach number $\mathcal{M}_2$ is given in terms of $\mathcal{M}_1$ by

$$\mathcal{M}_2^2 = \frac{2 + (\Gamma-1)\mathcal{M}_1^2}{2\Gamma\mathcal{M}_1^2 - (\Gamma-1)}. \tag{2.102}$$

This is symmetrical in $\mathcal{M}_1$ and $\mathcal{M}_2$ and may be rearranged to give

$$2\Gamma\mathcal{M}_1^2\mathcal{M}_2^2 - (\Gamma-1)\left(\mathcal{M}_1^2 + \mathcal{M}_2^2\right) = 2. \tag{2.103}$$

On physical grounds, we expect $\mathcal{M}_1 > 1$ because otherwise a shock would not have formed in the first place. By writing Eq. (2.100) as

$$\frac{\rho_2}{\rho_1} = \frac{\Gamma+1}{(\Gamma-1) + 2\mathcal{M}_1^{-2}}, \tag{2.104}$$

we see that $\rho_2/\rho_1 > 1$ for $\mathcal{M}_1 > 1$ and increases with an increase in $\mathcal{M}_1$. This means (i) that the gas behind a shock is always compressed and (ii) a shock involving stronger compression has to move faster. When $\mathcal{M}_1$ tends to 1, ρ_2/ρ_1 also goes to 1, and the shock will disappear. Note that the Rankine–Hugoniot relations formally also allow rarefaction shocks, in which $\mathcal{M}_1 < 1$ and $\mathcal{M}_2 > 1$. The jump conditions alone do not prevent a reversal of roles for the regions labeled by our subscripts 1 and 2. However, rarefaction shocks, in which subsonically moving hot gas suddenly expands and accelerates to supersonic speeds, are unphysical. The specific entropy jump is positive for a compressive shock and negative for a rarefaction shock. The second law of thermodynamics disallows internal processes by which a gas can spontaneously lower its specific entropy. In other words, viscosity can transform the energy of bulk motion into heat, but it cannot do the reverse. For these reasons, the term shock always refers to a compression shock.

For the sake of completeness, we can also state the limiting results for strong shock waves for general Γ, which are (straight from Eq. [2.104])

$$\frac{\rho_1}{\rho_2} = \frac{\Gamma - 1}{\Gamma + 1}, \tag{2.105}$$

$$\frac{T_2}{T_1} = \frac{(\Gamma - 1)P_2}{(\Gamma + 1)P_1}, \tag{2.106}$$

and from Eq. (2.101),

$$\frac{P_2}{P_1} = \frac{2\Gamma \mathcal{M}_1^2}{(\Gamma + 1)}. \tag{2.107}$$

The ratio T_2/T_1 increases to infinity with P_2/P_1, which means that the temperature discontinuity in a shock can be arbitrarily large. In reality, at some temperature, cooling will become important to prevent this. The density ratio, however, tends to a constant limit.

In plasmas, the physics of shocks can be much more complicated than presented here because there can be various typical speeds, not just the sound speed, by which information is propagated in the fluid. Also, many space plasmas are *collisionless*, which means that they are so rarefied that Coulomb collisions between the constituent particles happen very infrequently. In this case, the charged particles interact via collective electromagnetic effects. In very dilute conditions, where collisions between particles are rare, the magnetic fields mediate the pressure. In the presence of magnetic fields, the energy and pressure balance needs to be complemented by the respective magnetic contributions. The resulting shocks are called hydromagnetic shocks. The speed at which pressure differences are communicated throughout the plasma is no longer the sound speed but the *Alfvén speed*,[6] which is given by $v_A = (B^2/4\pi\rho)^{1/2}$.

2.8 Shock acceleration

Shocks can change the velocity distribution of a gas by accelerating a fraction of the particles to very high energies. This has been observed in astrophysical shocks on a wide range of scales. Examples are the shock waves that form when the solar wind collides with planetary magnetospheres, shocks surrounding supernova

---

[6] Named after the Swedish plasma physicist Hannes Olof Gösta Alfvén (1908–1995).

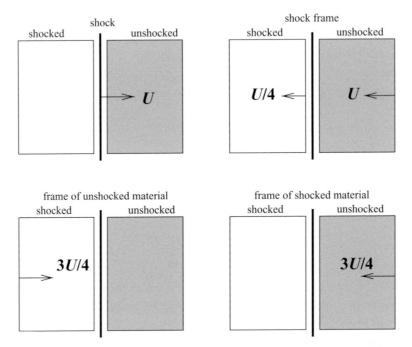

Figure 2.4   Illustration of first-order Fermi acceleration. A strong shock wave propagates at a supersonic velocity, U, through a stationary medium. In the shock frame, the ratio of the velocities in front of and behind the shock is $v_1/v_2 = 4$. The lower panels illustrate the flow in a frame in which the upstream gas is stationary (left) and where the downstream gas is stationary (right).

remnants in the interstellar medium, and shocks in active, extragalactic sources such as quasars and radio galaxies. The resulting particle energies reach up to about 100 keV in the shock in the Earth's magnetosphere and up to about 100 TeV in supernova remnants. Next we demonstrate how this works.

As before, we denote the variables in the upstream region by ρ_1, v_1, and P_1 and in the downstream region by ρ_2, v_2, and P_2. Furthermore, let the shock front propagate with a velocity U, let us assume that the gas is ideal with a ratio of specific heats $\Gamma = 5/3$, and let us assume that the shock is adiabatic in the sense that radiative losses and so forth are unimportant. Finally, we assume that the shock is strong such that Eqs. (2.95) and (2.96) apply.

As a result, the gas flows into the shock with a velocity $v_1 = U$ and flows out of the shock with a velocity $v_2 = U/4$. Viewed from the rest frame of the unshocked material, the shocked material moves toward it with a velocity $v \equiv |v_2 - v_1| = 3U/4$ (see Fig. 2.4). Now, picture a fast gas particle from the tail of the Maxwell–Boltzmann distribution in the unshocked region that crosses the shock front into the shocked region. Let the particle be relativistic such that its momentum is given by $p_x = E/c$, where E is its energy and c the speed of light. When our fast particle crosses the shock front, it collides with the downstream particles and thus is accelerated to a new energy

$$E' = E + p_x v, \qquad (2.108)$$

where we assume that the shock is nonrelativistic, that is, that the Lorentz factor of the shock front is ≈ 1. This gives

$$\Delta E / E = \frac{v}{c}, \qquad (2.109)$$

where $\Delta E = E' - E$.

Next, consider the same process from the point of view of the shocked region. In this frame, the upstream gas flows into the shock with a speed $v \equiv |v_1 - v_2| = 3U/4$. Now a particle that diffuses downstream from the shock is accelerated in the same way, leading to the same fractional energy increase. Thus, high-energy particles gain energy whenever they cross the shock front in either direction. Scattering processes make the velocity distribution isotropic with respect to the local rest frame on either side of the shock front. Consequently, there is a certain probability that a particle that has just gone through the shock and been accelerated will diffuse back through the shock to where it came from. Every time the particle crosses the shock it will gain energy; it will never lose energy. Moreover, the increase in energy is the same no matter from which direction the particle crosses the shock. Particles passing back and forth through the shock can thus attain very high energies. This process is called first-order Fermi (Fermi-I) acceleration after the Italian physicist Enrico Fermi (first order because it is proportional to the first power of v/c). Fermi-I acceleration can, of course, alter the velocity spectrum of the gas away from the thermal Maxwell–Boltzmann distribution.

We can work out the spectrum of particles that results from multiple crossings of the shock front.

Every time the particle passes through a shock, its energy increases by a factor $\beta = 1 + v/c$. Thus, after j crossings, a particle with initial energy E_0 will have an energy $E = E_0 \beta^j$. There is also the possibility that the particle gets away from the shock region altogether and will no longer get accelerated. Let us denote the probability of remaining in the shock region by $\mathcal{P}$. Then, after j crossings, there are $N = N_0 \mathcal{P}^j$ particles left in the acceleration region, where N_0 is the original number of particles. We can now eliminate j by writing

$$\frac{\log(N/N_0)}{\log(E/E_0)} = \frac{\log \mathcal{P}}{\log \beta}, \qquad (2.110)$$

resulting in

$$\frac{N}{N_0} = \left(\frac{E}{E_0}\right)^{\log \mathcal{P}/\log \beta}.$$

(2.111)

In differential form, we find

$$n(E) \propto E^{(\log \mathcal{P}/\log \beta)-1} \propto E^{-k} dE,$$

(2.112)

where $k = 1 - (\log \mathcal{P}/\log \beta)$ is the so-called power-law index. Such distributions are referred to as *nonthermal* because with their power-law spectrum they differ significantly from the exponential drop-off of a Maxwell–Boltzmann distribution. Finally, we can determine the probability to remain in the accelerating region, $\mathcal{P}$, as follows: the average number of particles that cross the shock with relativistic velocities is $nc/4$, where n is the number density of particles. This result comes about as follows: the flux of particles that crosses a surface is given by

$$\mathcal{F} = \int_{v_x>0} f(v)v \cos\theta \, d^3v,$$

(2.113)

where $f(v)$ is the phase space density of particles with velocities between v and $v + dv$. Evaluating the integrals

$$\mathcal{F} = \int_0^\infty f(v)v^3 \, dv \int_0^{\pi/2} \cos\theta \sin\theta \, d\theta \int_0^{2\pi} d\phi$$

(2.114)

yields

$$\mathcal{F} = \pi \int_0^\infty f(v)v^3 \, dv.$$

(2.115)

Note that the θ-integral only extends over half of the velocity space, from 0 to $\pi/2$, because we wish to consider only those particles that cross the surface in one direction. We can express this flux in terms of the average velocity in the gas, which is defined as

$$\langle v \rangle = \frac{1}{n} \int f(v)v \, d^3v.$$

(2.116)

Again evaluating the integrals gives

$$\langle v \rangle = \frac{4\pi}{n} \int_0^\infty f(v)v^3 \, dv.$$

(2.117)

Thus, we find

$$\mathcal{F} = \frac{n\langle v \rangle}{4},$$

(2.118)

which we have used previously. If we now assume relativistic velocities, $v \approx c$, we obtain the result that $\mathcal{F} = nc/4$.

Downstream from the shock, the particles are advected away from the shock with a bulk velocity $U/4$. Consequently, the fraction of particles lost per unit time from the shock front is $(nU/4)/(nc/4) = U/c$. Thus, the probability of remaining in the acceleration region is $\mathcal{P} = 1 - U/c$. Because $U \ll c$, we can write

$$\log \mathcal{P} = \log \left(1 - \frac{U}{c} \right) \approx -\frac{U}{c}. \tag{2.119}$$

To calculate β, we should take into account all particles that pass the shock with a velocity that makes an angle θ with the shock normal. Thus, the fractional gain in energy should really read

$$\frac{\Delta E}{E} = \frac{v}{c} \cos \theta. \tag{2.120}$$

Now, we need to use a standard result from kinetic theory. The number of particles that cross the shock front under angles $[\theta, \theta + d\theta]$ is proportional to $\sin \theta$. The rate at which the particles reach the shock is proportional to the component of their velocity parallel to the shock normal, so it is proportional to $\cos \theta$. To normalize the energy gain per shock crossing, we have to integrate

$$\left\langle \frac{\Delta E}{E} \right\rangle = \frac{v}{c} \int_0^{\pi/2} 2 \cos^2 \theta \sin \theta \, d\theta = \frac{2}{3} \frac{v}{c}. \tag{2.121}$$

A round trip produces twice this, so the fractional gain per round trip is

$$\beta = \frac{E}{E_0} = 1 + \frac{4}{3} \frac{v}{c}, \tag{2.122}$$

and consequently, because $v = 3U/4$,

$$\log \beta = \log \left(1 + \frac{U}{c} \right) \approx \frac{U}{c}. \tag{2.123}$$

This results in a power-law index

$$k = 1 - \frac{\log \mathcal{P}}{\log \beta} \approx 2, \tag{2.124}$$

which yields a spectrum of the form $n(E)dE \propto E^{-2}dE$. In other words, shock acceleration tends to produce a particle spectrum with a power-law index of 2. Now, many sources in high-energy astrophysics show a power-law energy spectrum for particles. For example, supernova blast waves, active galactic nuclei, and so forth all show nonthermal spectra with power-law indices in the range $k \sim 2.2$–3.0, as we discuss in Chapters 4 and 8. Now, this is a bit higher than the 2 that we worked out before. However, higher spectral indices can be obtained if some of our simplifying assumptions are relaxed. For example, we ignored the back reaction of the accelerated particles on the shock structure. As the accelerated particle population contains a mixture of relativistic and nonrelativistic particles, the ratio of specific heats will be somewhere between 5/3 and 4/3, and no simple equation of state exists. So, as the fraction of relativistic particles increases, the ratio of specific heats will change, which in turn will change the compression ratio of the shock and thus the acceleration efficiency. Moreover, many shocks radiate strongly and are no longer *adiabatic*. The radiative losses allow for higher compression ratios than the 4 assumed before. Finally, in many cases, for example in extragalactic jets, the shock itself is relativistic. In that case, the Rankine–Hugoniot relations derived in Section 2.7 are no longer valid, and higher compression ratios are possible. In summary, it seems that Fermi-I acceleration is a simple model that predicts a power-law particle spectrum with a relatively unique spectral index for a wide range of objects. However, the whole issue of shock acceleration in such dilute plasmas is very hard to test and difficult to model because the conditions are impossible to reproduce in laboratories on Earth.

2.9   Fluid instabilities

2.9.1 *Convective instability*

Convection is a ubiquitous physical phenomenon. It occurs in a teapot where hot water that is heated from below develops overturning motions to exchange heat with the cold air above the water. Convection is the process that distributes heat from a radiator in our living rooms. Convection occurs in the Earth's atmosphere and oceans and is important in astrophysics. In stars, the energy is usually created in a small region in the center where temperatures and densities are sufficiently high for nuclear reactions to take place. Now, this energy needs to be transported to the surface of the star where the photons are created that make the star shine. A very efficient mechanism to transport heat in a stratified medium is convection. Convection, and here we use it synonymously with thermal convection although there are also other forms of convection, is the motion of fluid that is driven by thermal gradients. Here, we derive the conditions under which a stratified medium will become convectively unstable.

Consider a blob of gas that is in equilibrium with its ambient medium. This means that it has the same pressure, temperature, density, and velocity as its surroundings. Now we investigate what happens when this blob is displaced from its equilibrium position. Will the surrounding fluid push it back into its original position, in which case the fluid is stable, or will the blob accelerate further away, in which case the fluid is unstable? This type of analysis is called local stability analysis. We consider a blob that is sufficiently small to maintain pressure equilibrium with its surroundings. Thus, when the blob is moved to a position with lower ambient pressure, it will expand adiabatically, in other words at constant entropy, to assume the same pressure as the surrounding fluid. Similarly, when it is moved to a position of higher pressure, it will compress adiabatically. After its displacement, the new matter density in the blob is related to the change in pressure via

$$d\rho_{\text{blob}} = \left(\frac{\partial \rho}{\partial P}\right)_{s} dP. \tag{2.125}$$

Meanwhile, the density of the ambient medium at the new position is given by

$$d\rho_{\text{amb}} = \left(\frac{\partial \rho}{\partial P}\right)_{s} dP + \left(\frac{\partial \rho}{\partial s}\right)_{P} ds, \tag{2.126}$$

where dP and ds are the small changes in pressure and specific entropy at the new position as compared with original position. Now, buoyancy occurs whenever the density at constant pressure differs from the ambient density. The fluid continues to rise when its density is lower than the ambient density, that is, when $d\rho_{\text{blob}} < d\rho_{\text{amb}}$ (or continues to sink if its density is higher than the ambient density). Because we assumed the blob to be so small that it always adjusts to the external pressure, the dP's in the two previous equations are the same. Hence, the condition for convective instability (along a negative pressure gradient or against gravity, for a gravitationally stratified fluid) can be obtained from Eq. (2.126) with $dP = 0$ and $d\rho_{\text{amb}} < 0$, which yields

$$\left(\frac{\partial \rho}{\partial s}\right)_{P} ds < 0. \tag{2.127}$$

Using the Maxwell relation

$$\left[\frac{\partial(\rho^{-1})}{\partial s}\right]_{P} = \left(\frac{\partial T}{\partial P}\right)_{s}, \tag{2.128}$$

we can substitute the partial derivative in Eq. (2.127). Now, all thermodynamically stable fluids have temperatures that increase upon adiabatic compression. This means that they have $(\partial T/\partial P)_s > 0$, which implies that all fluids obey

$$\left(\frac{\partial \rho}{\partial s}\right)_{P} < 0. \tag{2.129}$$

Using this information in Eq. (2.127) yields the *Schwarzschild criterion* for convective instability, which is simply

$$ds > 0 \qquad\qquad (2.130)$$

in the direction of gravity. To remember this result, associate specific entropy with buoyancy. If the entropy of a star or a planetary atmosphere increases inward, then more buoyant material underlies less buoyant material, and the medium has a tendency to overturn. The overturning mixes entropy, with high entropy (i.e., hotter than average material) rising and low entropy material (colder than average) sinking, so that there is a net convective transport of heat outward. The effect of this mixing must be to produce a more nearly uniform entropy. In other words, if heat transport by radiation plus conduction alone requires a superadiabatic temperature gradient, that is, a gradient that corresponds to entropy increasing inward, then convection would set in that tends to erode this gradient.

We have derived a simple condition for the onset of the convective instability. What really happens, once convection sets in, is very hard to model. Sophisticated computer simulations such as the one shown in Fig. 2.5 help to predict the properties of convection.

2.9.2   Rayleigh–Taylor instability

The Rayleigh–Taylor instability is an instability that can occur at the interface of two fluids of different densities. If the two fluids are accelerated such that the denser fluid lies on top of the lighter fluid, that is, if the acceleration vector points from the denser to the lighter fluid, the configuration will become unstable. This is because it is energetically favorable if the lighter fluid lies on top of the denser fluid. Here it is irrelevant what causes the acceleration: it could be gravity or an explosion (resulting in a blast wave). At first, when the initial perturbations to the interface between the fluids are small, the flow can be described by the linearized equations of fluid dynamics. Later on, the interface evolves into lighter bubbles and spikes of denser fluid penetrating the opposed fluid. In the absence of surface tension or viscosity, the growth rate increases indefinitely with the wave number. This means that disturbances of smaller wavelength grow faster. The presence of viscosity or surface tension changes this behavior, and the growth rate peaks at some critical wave number.

2.9.3   Kelvin–Helmholtz instability

The Kelvin–Helmholtz or shear instability occurs when two fluids slip past one another at their interface. This instability occurs, for example, when wind blows

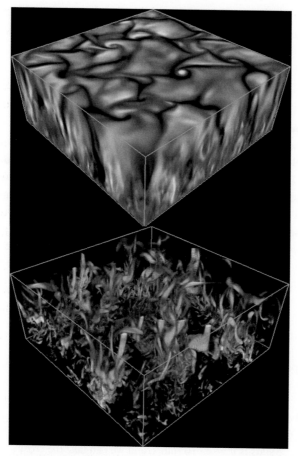

Figure 2.5   Three-dimensional simulations of compressible convection. The top panel shows a volume rendering of vertical velocity. The bottom panel shows enstrophy. Figure courtesy of Nic Brummell.

over water and produces waves. To understand this instability, consider an interface between two fluids that move at different speeds. Now, imagine perturbing the interface very slightly about its original position, causing it to undulate slightly about its equilibrium position. At those points where the interface is no longer perfectly flat, the fluids have to flow around another, which will lead to a slight centrifugal force. As a result, the pressure in concavities is higher than pressure in convexities. This causes the amplitude of the oscillation to grow, which marks the onset of the Kelvin–Helmholtz instability.

In the nonlinear stages of the Kelvin–Helmholtz instability, cyclonic structures resemble a breaking wave or a "cat-eye," as these structures are sometimes called. When combined with the Rayleigh–Taylor instability, the Kelvin–Helmholtz instability tends to transform the "fingers" of the Rayleigh–Taylor instability

into mushroom-shaped structures. As in the case of the Rayleigh–Taylor instability, magnetic fields, viscosity, or surface tension can suppress and alter this
instability.

2.10   Exercises

2.1   Mean free path

The elastic scattering between photons and charged particles is called Thomson scattering (see Section 3.7.8). For electrons, the cross section for Thomson scattering is
$\sigma_T = 6.65 \cdot 10^{-25}$ cm^2. Calculate the mean free path of photons in the core of the Sun,
given that the density in the center of the Sun is $\rho = 150$ g cm^{-3}. Assume that the core
of the Sun is fully ionized and consists of 25% helium and 75% hydrogen (by mass).

In the case of Thomson scattering, the dimensionless quantity τ, defined by $d\tau = n_e \sigma_T ds$, where n_e is the electron number density and s is the distance along the photon
path, is called the *optical depth* (see Chapter 3). It is a measure of the transparency of a
medium to radiation. Show that the equation of hydrostatic equilibrium may be written
as

$$\frac{dP}{d\tau} = \frac{g\mu_e}{\sigma_T}, \tag{2.131}$$

where P is pressure, g the gravitational acceleration, and μ_e the mean molecular mass
per electron.

2.2   Collisions I

The number density of stars in the Sun's vicinity is $n \sim 10^{-57}$ cm^{-3}. The Sun's velocity
relative to these stars is $v \sim 2 \cdot 10^6$ cm/s, and we can take the cross section for collision
with another star to be $\sigma \sim 5 \cdot 10^{22}$ cm^2. In the Jeans theory of the birth of the Solar
System, such an encounter is considered responsible for the formation of planets. How
probable is it that the Sun would have formed planets in $5 \cdot 10^9$ years? How many
planetary systems would we expect altogether in the Galaxy if there are 10^{11} stars and
if the Sun were representative? (Problem 4–6 from M. Harwit, *Astrophysical Concepts*,
1998.)

2.3   Collisions II

As a star of mass M moves with velocity v through an assembly of stars of the same
mass with a stellar number density n, it suffers many weak encounters through the
gravitational interaction with the other stars. Show that the relaxation time for such a
star through weak encounters is

$$t_{\text{relax}} \approx \frac{v^3}{8\pi G^2 M^2 n \ln \Lambda_g}, \tag{2.132}$$

where $\Lambda_g = b_{max}/b_{min}$ is the ratio of the maximum to the minimum impact parameter.
How would one choose appropriate values for b_{max} and b_{min}? Show that the relaxation

time can be expressed as

$$t_{\text{relax}} \approx \frac{2 \cdot 10^9 \text{ yrs}}{\ln \Lambda_g} \left(\frac{v}{10 \text{ km s}^{-1}} \right)^3 \left(\frac{M}{M_\odot} \right)^{-2} \left(\frac{n}{10^3 \text{ pc}^{-3}} \right)^{-1}. \tag{2.133}$$

2.4 Shocks I

Show that the velocities of propagation of the shock relative to the downstream and upstream gas are given by

$$u_1^2 = \frac{1}{2\rho_1} [(\Gamma - 1)P_1 + (\Gamma + 1)P_2] \tag{2.134}$$

and

$$u_2^2 = \frac{1}{2\rho_1} \frac{[(\Gamma - 1)P_1 + (\Gamma - 1)P_2]^2}{(\Gamma - 1)P_1 + (\Gamma + 1)P_2}, \tag{2.135}$$

their difference being

$$u_1 - u_2 = \left[\frac{2(P_2 - P_1)}{\rho_1[(\Gamma - 1)P_1 + (\Gamma + 1)P_2]} \right]^{1/2}. \tag{2.136}$$

2.5 Shocks II

If cooling behind a shock front is very effective, the gas temperature returns quickly to its preshock value. If we are not interested in the details of the cooling region, we can regard it and the shock itself as a single discontinuity, usually called an isothermal shock, and can treat it in the manner used for adiabatic shocks. Thus, we can retain the mass and momentum jump conditions but replace the energy condition by

$$T_1 = T_2, \tag{2.137}$$

where T_1 are T_2 are the gas temperatures on each side of the discontinuity, or equivalently the equality of sound speeds:

$$c_1 = c_2 = c, \tag{2.138}$$

where $c_1 = (P_1/\rho_1)^{1/2}$ and so forth.
Show that

$$v_2 = \frac{c}{\mathcal{M}_\infty}, \tag{2.139}$$

$$\rho_2 = \mathcal{M}_\infty^2 \rho_1, \tag{2.140}$$

and

$$P_2 = \rho_1 v_1^2, \tag{2.141}$$

where $\mathcal{M}_\infty$ is the preshock Mach number. Strong isothermal shocks result in very large compressions, and in all cases the postshock gas pressure is equal to the preshock ram pressure. (From Frank, King, & Raine, *Accretion Power in Astrophysics*, 2002.)

2.6 Viscosity

Gas flows between two smooth plates at $z = \pm a$. It has uniform density ρ and bulk velocity $\vec{u} = (V'z, 0, 0)$, where V' is a constant. The gas molecules have a mean free path l and a mean speed c. By considering the flux of gas molecules across a unit area in the $z = 0$ plane, show that the gas in the volume $z > 0$ exerts a force f_x per unit area in the x-direction on the gas in the volume $z < 0$ given by

$$f_x = k l \rho c V', \tag{2.142}$$

where k is a constant to be determined. Describe qualitatively the effect that this force has on the bulk fluid motion.

2.7 Ionization

Negative hydrogen ions H^- have a dissociation energy of 0.754 eV and play an important role in stars. Assuming an electron density of $n_e = 10^{17}$ cm^{-3}, estimate the temperature at which the number densities of H and H^- are equal.

2.8 Hydrostatic equilibrium I

Derive the pressure, density, and temperature profiles of an adiabatically stratified plane-parallel atmosphere under constant gravitational acceleration g. Assume that the atmosphere consists of an ideal gas of molecular weight μ.

Assume that the Earth's atmosphere is adiabatically stratified and plane parallel and that it consists of an ideal gas of molecular weight $\mu = 14$. If the gravitational acceleration is $g = 981$ cm s^{-2} and the temperature at sea level is 300 K, what is the temperature at the summit of Mt. Everest at a height of 8848 m? What is the pressure on the summit compared to the pressure at sea level?

2.9 Hydrostatic equilibrium II

Write down the equations of hydrostatic equilibrium for gas of density ρ and pressure P. Show that if $P = \alpha\rho^2$, these can be combined to yield

$$\nabla^2\rho + \frac{2\pi G}{\alpha}\rho = 0. \tag{2.143}$$

Show that this equation admits spherically symmetric solutions of the form $\rho(r) = A\sin(kr)/r$, if $r < \pi/k$ and $\rho = 0$ otherwise, where $k = (2\pi G/\alpha)^{1/2}$ and A is an arbitrary constant (which is fixed once the total mass is specified).

2.11   Further reading

Batchelor, G. (1967). *An Introduction to Fluid Dynamics*. Cambridge: Cambridge University Press.

Chandrasekhar, S. (1961). *Hydrodynamic and Hydromagnetic Stability*. International Series of Monographs on Physics. Oxford: Clarendon.

Frank, J., King, A., and Raine, D. J. (2002). *Accretion Power in Astrophysics*, 3rd edn. Cambridge: Cambridge University Press.

Harwit, M. (1988). *Astrophysical Concepts*, 2nd edn. New York: Springer-Verlag.

Longair, M. S. (1994). *High Energy Astrophysics*, 2nd edn. Cambridge: Cambridge University Press.

Shu, F. H. (1992). *Gas Dynamics. Vol. II of Physics of Astrophysics*. Mill Valley, CA: University Science Books.

3

Radiation processes

3.1 Introduction

The first observations of celestial objects were done with the naked eye and then, since the early seventeenth century, with the aid of telescopes. Naturally, such observations were restricted to optical wavelengths. They showed a rather quiet universe, mainly consisting of planets and peacefully burning stars. Over the last decades of the twentieth century, several other electromagnetic wavelength bands have been explored, ranging from the long radio waves over ultraviolet to the short-wavelength X and gamma rays. These observations revealed a different, much more violent, and until then hidden side of the universe. For example, they showed relativistic radio jets resulting from accretion onto supermassive black holes or extremely bright explosions called gamma-ray bursts, to name just two examples.

To interpret electromagnetic observations, we must understand both the processes that produce the radiation and how the radiation is altered on its way to the observer. Such changes could come, for example, from the interaction with the interstellar or intergalactic medium or from light-bending effects that are caused by the curvature of space-time.

Our electromagnetic picture of the universe is complemented by the observation of particles that are produced in space. Neutrinos have allowed us a glimpse at the deep interiors of stars that are inaccessible to electromagnetic observations. For example, the detection of a burst of neutrinos from the supernova SN 1987A in our neighboring galaxy, the Large Magellanic Cloud, allowed us to take a look into the deep interior of a dying star. The observations of neutrinos from the Sun have confirmed our ideas about the solar interior. The same observations have also revealed that neutrinos are *not* massless as was thought for a long time. Moreover, Earth is continuously bombarded by energetic cosmic-ray particles, which can tell us about supernovae and even more violent cosmic catastrophes.

A completely new window to the universe could be opened up by the detection of gravitational waves, disturbances of space-time that travel at the speed of light. Their existence has been proven indirectly by the orbital decay of a binary system consisting of two neutron stars; see Chapter 6. In the past few years, tremendous efforts have been undertaken to build ground-based gravitational-wave detectors, some of which have recently started taking data. Such detections could, again, allow us to see a so far unknown face of the universe. In this chapter, we deal with the physics of electromagnetic radiation. We outline basics of radiative transfer, discuss the properties of blackbody radiation, summarize the radiation from accelerated charges, and then discuss the specifics of particularly important radiation processes. Some of the derivations have to remain fragmentary, as the techniques from quantum electrodynamics that are required are beyond the scope of this book. A good reference is the book *Radiative Processes in Astrophysics* by Rybicki and Lightman at which parts of this chapter are oriented.

3.2   Basic photon properties and spectrum

Massless particles called photons are the constituents of electromagnetic radiation. Photons have an energy, E, that is related to their wavelength, λ, or equivalently to their frequency, ν. These quantities are related via

$$c = \lambda \cdot \nu \quad \text{and} \quad E = h \cdot \nu, \tag{3.1}$$

where the constants are the speed of light in vacuum,[1]

$$c = 2.998 \cdot 10^{10} \, \text{cm s}^{-1}, \tag{3.2}$$

and Planck's constant

$$h = 6.626 \cdot 10^{-27} \, \text{erg} \cdot \text{s}. \tag{3.3}$$

In many cases the photon energy is closely related to the temperature of the emitter. Therefore, the photon energy is often expressed in units of temperature:

$$T = \frac{E}{k_B}, \tag{3.4}$$

with k_B being Boltzmann's constant

$$k_B = 1.381 \cdot 10^{-16} \, \text{erg K}^{-1}. \tag{3.5}$$

It is often convenient to express photon energies in units of *electron volts*, where one electron volt (eV) is the kinetic energy that an electron attains in vacuum

[1] As usual in astrophysics, we are using cgs units throughout the book.

after passing through an electrostatic potential difference of one volt. This unit is particularly useful as the energy levels in an atom are of this order. The ground state of hydrogen, for example, has an energy of -13.6 eV. Useful rules of thumb for the conversion between energy and temperature units are

$$10\,\mathrm{eV} \leftrightarrow 10^5\,\mathrm{K} \qquad \text{and} \qquad 1\,\mathrm{MeV} \leftrightarrow 10^{10}\,\mathrm{K}. \tag{3.6}$$

The frequencies that are usually dealt with in astronomy span a very large range. The different frequency bands are characterized, via Eq. (3.4), by the following temperatures:

- gamma rays, $T \gtrsim 10^9$ K;
- X-rays, $10^9 \gtrsim T \gtrsim 10^6$ K;
- ultraviolet, $10^6 \gtrsim T \gtrsim 10^5$ K;
- optical, $T \sim 4 \cdot 10^4$ K;
- infrared, $10^4 \gtrsim T \gtrsim 10^2$ K; and
- radio, $10^1 \gtrsim T \gtrsim 10^{-1}$ K.

A more complete overview over the electromagnetic spectrum is given in Fig. 3.1.

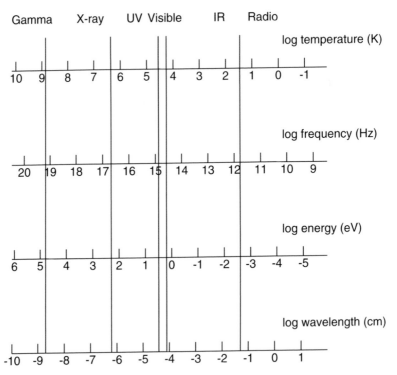

Figure 3.1   Overview over the electromagnetic spectrum.

3.3   Basic definitions

If the scales of the system of interest are far greater than the wavelength of the considered radiation, the quantum nature of the photons may be ignored, and the radiation can be described via rays of light. In the following, we give several definitions that are used to characterize radiation in such a macroscopic description.

3.3.1   Radiative energy flux

Imagine radiation passing through a surface dA for a time interval dt. The corresponding energy is then

$$dE = F\, dA\, dt, \tag{3.7}$$

where F is called the radiative *energy flux*. Note that in general F depends on the orientation of the surface. Its units are erg s^{-1} cm^{-2}.

   If one is interested in the energy flux *per frequency*, one can define the *monochromatic energy flux* as the energy passing per time dt through a surface dA in the frequency interval $[\nu, \nu + d\nu]$:

$$dE = F_\nu\, dA\, dt\, d\nu. \tag{3.8}$$

The total energy flux F can then be obtained by integration over the frequencies, $F = \int_0^\infty F_\nu\, d\nu$.

   Now consider a source that emits isotropically, which means that it radiates the same amount of energy in each direction. An example of such an isotropic source is a spherical, isolated star. Because no energy is gained or lost, energy conservation determines the dependence of the flux on the distance r to the source:

$$F \propto \frac{1}{r^2}. \tag{3.9}$$

This is often referred to as the inverse square law.

3.3.2   Specific intensity

The radiative flux is a measure of the energy carried by *all* photons that pass through a surface element dA. A more subtle measure is the *specific monochromatic intensity* I_ν. Consider a surface element, dA, that is perpendicular to the direction of a particular ray, given by the direction $\vec{n}$ as shown in Fig. 3.2. Then the energy in the frequency range $d\nu$ transported within a time dt through the surface dA by all those rays that lie within a solid angle $d\Omega$ around $\vec{n}$ is given by

$$dE = I_\nu(\Omega)\, dt\, dA\, d\nu\, d\Omega. \tag{3.10}$$

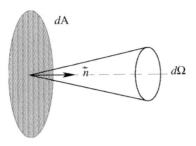

Figure 3.2   Geometry for the definition of the specific intensity, I_ν.

Note that I_ν in general depends on the location with respect to the source and the frequency of the radiation. Its unit is erg s^{-1} cm^{-2} Hz^{-1} ster^{-1}. In a sloppy way, one can remember I_ν as "energy per everything."

If one is not interested in the angular dependence of I_ν, one can use an angle-averaged quantity, the so-called *mean intensity*

$$J_\nu = \frac{1}{4\pi} \int I_\nu(\Omega) \, d\Omega, \qquad (3.11)$$

where the normalization constant is determined by the integral over the full solid angle, $\int d\Omega = \int \sin\theta \, d\theta \, d\varphi = 4\pi$.

For a surface of uniform brightness $B = \int I_\nu \, d\nu$, the flux and the brightness at the surface are simply related by a factor of π:

$$F = \pi \cdot B. \qquad (3.12)$$

For a derivation of this relation, see Exercise 3.3.

3.3.3   *Radiative energy density*

The *specific radiative energy density*, u_ν, is defined as

$$u_\nu(\Omega) = \frac{\text{energy}}{\text{volume} \cdot \text{frequency} \cdot \text{solid angle}}. \qquad (3.13)$$

Consider a cylinder of length dl in direction Ω and cross-section area dA, as sketched in Fig. 3.3. The energy contained in the cylinder is given by

$$dE = u_\nu(\Omega) \, dl \, dA \, d\nu \, d\Omega. \qquad (3.14)$$

When we substitute dl by the length traveled by light in a time dt, which is $c \, dt$, we obtain

$$dE = u_\nu(\Omega) \, c \, dt \, dA \, d\nu \, d\Omega. \qquad (3.15)$$

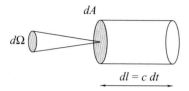

Figure 3.3 Energy contained in a cylinder of length $c\,dt$ to relate the specific energy density $u_\nu(\Omega)$ to the specific intensity I_ν.

Comparing this expression with Eq. (3.10), we see that the specific radiative energy density is equal to the specific intensity multiplied by the speed of light:

$$I_\nu(\Omega) = u_\nu(\Omega)\,c. \tag{3.16}$$

By integrating over the full solid angle, one obtains

$$u_\nu = \int u_\nu(\Omega)\,d\Omega = \frac{1}{c}\int I_\nu(\Omega)\,d\Omega = \frac{4\pi}{c}J_\nu, \tag{3.17}$$

where we have used the mean intensity introduced in Eq. (3.11). By integrating once more over the frequencies, we obtain the total energy density (energy per volume),

$$u = \int_0^\infty u_\nu\,d\nu = \frac{4\pi}{c}\int_0^\infty J_\nu\,d\nu. \tag{3.18}$$

The relative velocity between the frames in which radiation is emitted and observed can be very large. An example is a relativistic jet emerging from a supermassive black hole in the center of a galaxy, see Chapter 8. Therefore, one has to know how the specific intensity behaves under a Lorentz transformation. As we had seen in Chapter 1, Eq. (1.93), the phase space distribution function, f is Lorentz invariant. We can use it to express the energy per volume

$$du = f(\vec{r}, \vec{p})\,E\,d^3p = f(\vec{r}, \vec{p})\,E\,p^2 dp\,d\Omega. \tag{3.19}$$

For photons, we can use $p = E/c = h\nu/c$ and $dp = (h/c)d\nu$, so that

$$du = u_\nu(\Omega)d\nu\,d\Omega = \frac{h^4}{c^3}f(\vec{r}, \vec{p})\nu^3 d\nu\,d\Omega, \tag{3.20}$$

yielding

$$u_\nu(\Omega) = \frac{h^4}{c^3}f(\vec{r}, \vec{p})\nu^3 \tag{3.21}$$

or, by using Eq. (3.16),

$$\frac{I_\nu}{\nu^3} = \frac{h^4}{c^2}f(\vec{r}, \vec{p}) = \text{Lorentz invariant}, \tag{3.22}$$

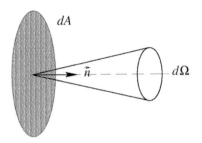

Figure 3.4   Geometry for the definition of the specific intensity, I_ν.

because f is Lorentz invariant. Equation (3.22) becomes important again in Chapter 8.

3.3.4   Net and momentum flux

Imagine a surface element that has an arbitrary angle, θ, with respect to the surface normal, $\vec{n}$, as shown in Fig. 3.4. The area of this surface element projected onto a plane perpendicular to $\vec{n}$ is then $dA_\perp = dA \cos \theta$. Therefore, the differential monochromatic flux per unit area is

$$dF_\nu = I_\nu(\Omega) \cos \theta \, d\Omega. \tag{3.23}$$

Clearly, if $\theta = 90°$, the rays in this direction do not contribute to dF_ν; this is accounted for by the factor $\cos \theta$. The *net flux* is obtained by integration over the solid angle:

$$F_\nu = \int dF_\nu = \int I_\nu(\Omega) \cos \theta \, d\Omega. \tag{3.24}$$

If the radiation is *isotropic*, or, in other words, if I_ν does not depend on the angles, the net flux is zero as $\int \cos \theta \, d\Omega = 2\pi \int_0^\pi \sin \theta \cos \theta \, d\theta = 0$. That means that as much radiation passes the surface in a positive direction, $\vec{n}$, and in a negative direction, $-\vec{n}$.

In radio astronomy, the flux F_ν is often measured in the unit Jansky,[2] where

$$1 \, \text{Jy} = 10^{-23} \, \frac{\text{erg}}{\text{s} \cdot \text{cm}^2 \cdot \text{Hz}}. \tag{3.25}$$

Each individual photon of energy E carries a momentum $\vec{p} = E/c \cdot \hat{e}_\gamma$, where $\hat{e}_\gamma$ is the unit vector in the direction of the photon. The component of the photon momentum perpendicular to the surface is $p_\perp = |\vec{p}| \cdot \cos \theta$ and hence the differential momentum flux is $(1/c)dF_\nu \cos \theta$. By integrating over all directions, we find the

---

[2] After the American physicist and radio engineer Karl Guthe Jansky (1905–1950).

monochromomatic momentum flux as

$$P_\nu = \frac{1}{c} \int I_\nu(\Omega) \cos^2 \theta \, d\Omega. \tag{3.26}$$

Multiplying an angle-dependent function with the nth power of $\cos \theta$ and then integrating over the solid angle is often called *taking the nth moment* of this function. In this sense, the energy density, u_ν, is the *0th moment* of the specific intensity, I_ν, Eq. (3.17),

$$u_\nu = \frac{1}{c} \int I_\nu(\Omega) \, d\Omega; \tag{3.27}$$

the energy flux F_ν, Eq. (3.24), is the first moment,

$$F_\nu = \int I_\nu(\Omega) \cos \theta \, d\Omega; \tag{3.28}$$

and the momentum flux, Eq. (3.26), is the second moment of the specific intensity I_ν,

$$P_\nu = \frac{1}{c} \int I_\nu(\Omega) \cos^2 \theta \, d\Omega. \tag{3.29}$$

3.3.5 Radiation pressure

Now consider an isotropic radiation field enclosed in a cavity with reflecting walls. What pressure does the field exert on the walls? When a ball bounces off a wall, the wall takes up twice the momentum component perpendicular to its normal direction. This is simply a result of momentum conservation. The same is true for radiation. The rate at which momentum is transferred to the wall per unit frequency is

$$P_\nu = \frac{2}{c} \int_{\cos \theta > 0} I_\nu(\Omega) \cos^2 \theta \, d\Omega, \tag{3.30}$$

where we integrate only over those photons that move toward the wall, that is, those that have $\cos \theta > 0$. Note that this is consistent with our definition in Eq. (3.29), as the integration here runs over only 2π, contrary to before where it ran over 4π. As P_ν is the momentum transferred per time, area, and frequency, the radiation pressure (momentum transferred per time and area) exerted is

$$P = \int_0^\infty P_\nu \, d\nu = \frac{2}{c} \int_0^\infty \int_{\cos \theta > 0} I_\nu(\Omega) \cos^2 \theta \, d\Omega \, d\nu, \tag{3.31}$$

or, as we have assumed isotropy, $I_\nu = J_\nu$ (see Eq. [3.11]),

$$P = \frac{2}{c} \int_0^\infty J_\nu \, d\nu \int_{\cos\theta > 0} \cos^2\theta \, d\Omega = \frac{2}{c} \cdot \frac{cu}{4\pi} \int_0^\pi d\varphi \int_0^\pi \sin\theta \cos^2\theta \, d\theta = \frac{u}{3},$$

(3.32)

where we have used Eq. (3.18). Thus, we have found that for isotropic radiation, the radiation pressure is equal to one third of the radiative energy density.

3.4 Basics of radiative transfer

Now that we have introduced some basic definitions and terms, we can work out how the intensity of a wave, I_ν, changes as it travels through some medium.

3.4.1 *Free space: Intensity along a given ray is constant*

Let us first discuss a ray traveling through vacuum, where no photons are absorbed, emitted, or scattered out of or into the ray path. Consider two circular surfaces that are oriented perpendicular to a particular ray and that have a separation of R, as shown in Fig. 3.5.

How much energy is contained in those rays that pass through both surfaces? According to the definition given in Eq. (3.10), the energies passing through the disks are

$$dE_1 = I_{\nu 1} \, dA_1 \, dt \, d\nu_1 \, d\Omega_1 \quad \text{and} \quad dE_2 = I_{\nu 2} \, dA_2 \, dt \, d\nu_2 \, d\Omega_2,$$

(3.33)

where $d\Omega_1$ is the solid angle subtended by dA_2 at dA_1 and vice versa. As energy is conserved, we have $dE_1 = dE_2$. Because we consider vacuum only, where nothing changes the photon frequencies, we also have $d\nu_1 = d\nu_2$. The solid angles involved in this equation are related to the surfaces by $d\Omega_1 = dA_2/R^2$ and $d\Omega_2 = dA_1/R^2$. When we insert everything in this equation, we find $I_{\nu 1} = I_{\nu 2}$, which means that the specific intensity is conserved along a ray in free space. If the traveled path increment is called ds, the constancy of the specific intensity along a ray can be

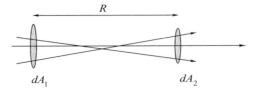

Figure 3.5 Geometry for the proof that I_ν is constant along a given ray.

expressed in differential form as

$$\frac{dI_\nu}{ds} = 0. \tag{3.34}$$

3.4.2   *Emission*

Now consider a ray that passes through matter that itself emits radiation and thus adds to the intensity of the beam. If the matter is contained in a volume dV, it will add the energy

$$dE = j_\nu(\Omega)\, dV\, dt\, d\nu\, d\Omega, \tag{3.35}$$

where the *monochromatic emission coefficient*, j_ν, is generally dependent on the direction into which emission takes place. As usual, there is also a frequency-integrated emission coefficient, $j = \int_0^\infty j_\nu\, d\nu$.

Consider now a ray of cross section dA and length ds, similar to the cylinder sketched in Fig. 3.3. The energy contained in this cylinder is

$$dE = I_\nu(\Omega)\, dt\, dA\, d\nu\, d\Omega + j_\nu(\Omega)\, ds\, dt\, dA\, d\nu\, d\Omega = (I_\nu + j_\nu\, ds)\, dt\, dA\, d\nu\, d\Omega. \tag{3.36}$$

Thus the specific intensity is incremented by the second term in the brackets on the right-hand side, so

$$dI_\nu = j_\nu\, ds. \tag{3.37}$$

3.4.3   *Absorption*

If the beam passes through absorbing material, the change in specific intensity is proportional to the incident intensity and the length element ds:

$$dI_\nu = -\alpha_\nu I_\nu\, ds, \tag{3.38}$$

where the absorption constant, α_ν, has the dimension of an inverse length.

To find a microscopic interpretation for α_ν, one can adopt the following simple model.[3] Think of a set of spheres with an effective absorbing area (cross section) σ_ν randomly distributed at a number density n. If we use again a cylindrical volume element, $dV = dA\, ds$, it will contain a number of $dV n$ spheres that will "block" a surface of $dV n\, \sigma_\nu = dA\, ds\, n\, \sigma_\nu$, if we assume that the spheres are placed sufficiently sparse, so that there are no spheres shadowing each other. The energy absorbed out of the beam is then $-d I_\nu\, dA\, d\Omega\, dt\, d\nu = I_\nu\, (n\, \sigma_\nu\, dA\, ds)\, d\Omega\, dt\, d\nu$, and

---

[3] This is similar to Section 2.1.1.

therefore the intensity is reduced by

$$dI_\nu = -n\sigma_\nu I_\nu \, ds, \tag{3.39}$$

which implies that

$$\alpha_\nu = n\sigma_\nu. \tag{3.40}$$

The quantity

$$l = \frac{1}{\alpha_\nu} = \frac{1}{n\sigma_\nu} \tag{3.41}$$

is called the *local mean free path* as it corresponds to the average distance traveled by a photon without being absorbed. The same relation was derived for a gas in Chapter 2.

Often the absorption coefficient α_ν is written as $\alpha_\nu = \rho\kappa_\nu$, where the quantity κ_ν with the dimension of a surface *per mass* is called the *opacity coefficient*.

3.4.4 The equation of radiative transfer

Combining the effects of emission and absorption, Eqs. (3.37) and (3.38), one arrives at the *radiative transfer equation*

$$\frac{dI_\nu}{ds} = -\alpha_\nu I_\nu + j_\nu. \tag{3.42}$$

To solve this equation, we first have to identify the important physical processes that contribute to emission and absorption. Then α_ν and j_ν can be calculated. Although this equation looks straightforward to solve, its solution can become quite complicated in reality as the emission and absorption coefficients themselves can depend on I_ν. Moreover, photons can be scattered into and out of the rays so that different rays couple, and it is not enough to solve the equation just along a ray. In most cases, this equation has to be solved numerically.

The special cases of pure emission and absorption, however, are easy to solve. First consider the case of pure emission, $\alpha_\nu = 0$. In this case, Eq. (3.42) can be integrated to give

$$I_\nu(s) = I_\nu(s_0) + \int_{s_0}^{s} j_\nu(s') \, ds', \tag{3.43}$$

which implies that the intensity at the source, $I_\nu(s_0)$, is incremented by emission along the line of sight.

For pure absorption, $j_\nu = 0$, the integration of Eq. (3.42) yields

$$I_\nu(s) = I_\nu(s_0) \exp\left\{ -\int_{s_0}^{s} \alpha_\nu(s')\, ds' \right\}. \tag{3.44}$$

Hence, the initial brightness decreases exponentially with the integral of the absorption coefficient along the line of sight. The decrease in brightness is determined by the quantity

$$\tau_\nu \equiv \int_{s_0}^{s} \alpha_\nu(s')\, ds', \tag{3.45}$$

which is a dimensionless measure of how much absorption occurs. The quantity τ_ν is called *optical depth*: if a medium has $\tau_\nu \ll 1$, it is called *optically thin* or *transparent*, if $\tau_\nu \gg 1$ it is called *optically thick* or *opaque*. In an optically thin medium, a photon can transverse matter more or less unhindered; in an optically thick medium, photons are doomed to be absorbed. Generally, as indicated by the subscript ν, the optical depth depends on the photon frequency. It is entirely possible that a source is transparent in one frequency regime and opaque in another.

The optical depth can be related to the local microscopic cross section by using our previous model, Eq. (3.40),

$$d\tau_\nu = n\sigma_\nu\, ds = \frac{ds}{l}. \tag{3.46}$$

Often, the radiative transfer equation is expressed in terms of the dimensionless optical depth τ_ν instead of the true physical path s. Divide Eq. (3.42) by α_ν to obtain

$$\frac{1}{\alpha_\nu} \frac{dI_\nu}{ds} = \frac{dI_\nu}{d\tau_\nu} = \frac{j_\nu}{\alpha_\nu} - I_\nu \equiv S_\nu - I_\nu, \tag{3.47}$$

where we have introduced the *source function* $S_\nu = j_\nu/\alpha_\nu$. This equation is formally solved by

$$I_\nu(\tau_\nu) = I_{\nu,0} \cdot e^{-\tau_\nu} + \int_0^{\tau_\nu} e^{-(\tau_\nu - \tau_\nu')} S_\nu(\tau_\nu')\, d\tau_\nu'. \tag{3.48}$$

Because the source function itself is an integral over intensity, this solution is of only limited use. If the source function S_ν is a constant, the solution reads

$$I_\nu(\tau_\nu) = I_{\nu,0} \cdot e^{-\tau_\nu} + S_\nu(1 - e^{-\tau_\nu}). \tag{3.49}$$

Note that for $\tau_\nu \to \infty$, the observed intensity approaches the source function, $I_\nu \to S_\nu$, so the source function is the function that the specific intensity tries to reach.

3.5 Summary of blackbody radiation

Because blackbody radiation is part of essentially every undergraduate physics course, we do not repeat the details of the derivation here, but instead restrict ourselves to summarizing the main assumptions and properties.

3.5.1 *Thermal versus nonthermal radiation*

Generally, one distinguishes between thermal and nonthermal radiation. Thermal radiation is emitted by a body because of its thermal energy, for example, by a hot plate, and the emission is determined by the body's temperature. Nonthermal radiation is not a direct result of the thermal motions in the source. The most common astrophysical example is synchrotron radiation emitted by an electron spiraling around a magnetic field line. This is discussed in some detail in Section 3.8.2. Thermal and nonthermal radiation differ clearly in their spectra. Thermal radiation shows a characteristic exponential drop-off at high energies, that is, relatively few high-energy photons are emitted. Nonthermal radiation often exhibits "high-energy tails" described by power laws of the form $I(E) \propto E^{-p}$, where p is the "characteristic," "power-law," or "spectral" index of the distribution. In a log-log plot, such a power-law distribution is a straight line with a slope of $-p$. Qualitative examples of typical spectral shapes are shown in Fig. 3.6.

3.5.2 *Thermodynamics of blackbody radiation and the Stefan–Boltzmann law*

Consider a box at a given temperature, and do not let radiation in or out of the box until equilibrium has been reached. Once the photons inside the box are emitted

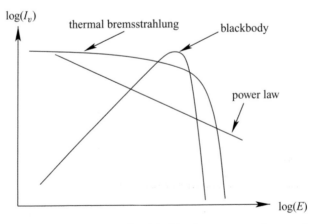

Figure 3.6 Qualitative shapes of a blackbody, a power law, and a thermal bremsstrahlung spectrum are shown.

and absorbed by the walls at equal rates, they are in equilibrium. The photons can be considered to constitute a photon gas. Now open a hole that is small enough not to disturb the state inside the box. The radiation measured at this opening is called "blackbody radiation," and its specific intensity is denoted B_ν and entirely determined by the temperature of the box.

Being in thermodynamic equilibrium, the photon gas can be treated with the methods of thermodynamics, in particular, the first and second law of thermodynamics can be applied. If blackbody radiation is isotropic, we have $B_\nu = J_\nu$, where J_ν is the mean intensity defined in Eq. (3.11). As B_ν depends on only temperature, we find via Eq. (3.18) that the energy density,

$$u = \frac{4\pi}{c} \int_0^\infty J_\nu \, d\nu = \frac{4\pi}{c} \int_0^\infty B_\nu(T) \, d\nu = u(T), \qquad (3.50)$$

is also only a function of the temperature. In addition, if we use that $P = u/3$, Eq. (3.32), together with the first and second laws of thermodynamics, we find that

$$u(T) = aT^4, \qquad (3.51)$$

which is the well-known *Stefan–Boltzmann law*.[4] By using $u(T) = (4\pi/c)B(T)$ (see Eq. [3.18]), one finds the frequency integrated Planck function

$$B(T) = \frac{c}{4\pi} aT^4. \qquad (3.52)$$

As we had seen earlier (see Eq. [3.12] and Exercise 3.3), the flux from an isotropically emitting surface (such as blackbody radiation) is just π times the brightness; therefore

$$F = \int_0^\infty F_\nu \, d\nu = \pi \int_0^\infty B_\nu \, d\nu = \frac{c}{4} aT^4 \equiv \sigma_{SB} T^4, \qquad (3.53)$$

which is another form of the Stefan–Boltzmann law.

3.5.3   *Specific intensity of blackbody radiation: Planck law*

The derivation of the specific intensity of blackbody radiation, called the *Planck law*,[5] can be found in most textbooks of undergraduate physics courses. The strategy for deriving the Planck spectrum includes two steps: first one derives the density of photon states and then the average energy in each state. The combination of both

---

[4] Found experimentally by the Slovene-Austrian physicist, mathematician, and poet Joseph Stephan (1835–1893) and later derived theoretically by the Austrian physicist Ludwig Boltzmann (1844–1906).
[5] After the German physicist Max Planck (1858–1947), who won the Nobel Prize in physics 1918 for his discovery that photon energies are quantized.

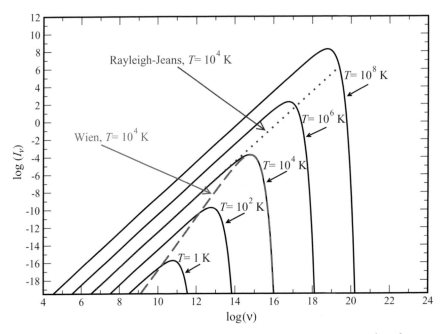

Figure 3.7   The solid curves are Planck spectra for $T = 1, 100, 10^4, 10^6$, and 10^8 K. For the $T = 10^4$ K case, the Rayleigh–Jeans (dots) and the Wien approximation (dashed) are overlaid. Frequency units are Hz; I_ν is measured in erg s^{-1} cm^{-2} Hz^{-1} ster^{-1}.

finally yields the desired specific intensity of blackbody radiation:

$$B_\nu(T) \equiv I_\nu^{\mathrm{BB}}(T) = \frac{2h\nu^3/c^2}{\exp(h\nu/k_\mathrm{B}T) - 1}. \tag{3.54}$$

Examples of Planck spectra for different temperatures are displayed in Fig. 3.7.

With the explicit form of the Planck spectrum, we can now express the radiation constants a and σ_{SB} that occurred in the context of the Stefan–Boltzmann law (see Eq. [3.51] and [3.53]) in terms of fundamental constants via $B = \int_0^\infty B_\nu(T)\,d\nu = (c/4\pi)aT^4$:

$$a = \frac{8\pi^5 k_\mathrm{B}^4}{15c^3h^3} = 7.56 \cdot 10^{-15}\mathrm{erg\,cm^{-3}\,K^{-4}} \tag{3.55}$$

and

$$\sigma_{\mathrm{SB}} = \frac{c}{4}a = \frac{2\pi^5 k_\mathrm{B}^4}{15c^2h^3} = 5.67 \cdot 10^{-5}\mathrm{erg\,cm^{-2}\,K^{-4}\,s^{-1}}. \tag{3.56}$$

One distinguishes between thermal and blackbody radiation. For thermal radiation, the source function S_ν (see Eq. [3.47]), is equal to the blackbody intensity B_ν

$$S_\nu = B_\nu(T). \tag{3.57}$$

This relation is known as *Kirchhoff's law*. In other words, the emission and absorption coefficients for thermal radiation are related by the blackbody intensity:

$$j_\nu = \alpha_\nu B_\nu(T) \tag{3.58}$$

(see Eq. [3.47]). Blackbody radiation, in contrast, is defined by $I_\nu = B_\nu$. So from Eq. (3.49) one concludes that thermal radiation only becomes blackbody radiation for optically thick media.

3.5.4   *Properties of blackbody radiation*

Historically, some properties of blackbody radiation were known before Planck came up with his distribution. In particular, the behavior of the spectrum at high and low frequencies was known empirically. In the following, we discuss these two limiting cases.

The Rayleigh–Jeans limit

For energies much lower than the typical thermal energy, $h\nu \ll k_B T$, the exponential in Eq. (3.54) can be expanded in a series such that

$$\exp\left(\frac{h\nu}{k_B T}\right) - 1 = \frac{h\nu}{k_B T} + O\left[\left(\frac{h\nu}{k_B T}\right)^2\right] \cdots \tag{3.59}$$

This yields the low-energy or *Rayleigh–Jeans limit* of the Planck distribution,

$$I_\nu^{RJ}(T) = \frac{2\nu^2}{c^2} k_B T, \tag{3.60}$$

which is often applicable to a good approximation in radio astronomy. Interestingly, the quantum-mechanical constant h has dropped out in this limit. The typical photon energy in this limit is just $k_B T$, the classical equipartition value for the energy of an electromagnetic wave. It is obvious that the Rayleigh–Jeans law cannot describe the full spectrum: integrating over all frequencies, $\int_0^\infty d\nu\, \nu^2 \ldots$, shows that the energy content at finite temperature is infinitely large. This failure to describe the high-energy part of the spectrum has been coined the *ultraviolet catastrophe*. The reason for this failure is that the discrete nature of the photons for $h\nu \gg k_B T$ is not accounted for. The Rayleigh–Jeans law is, however, a good approximation to the linear part of the spectrum in a $\log(I_\nu) - \log(\nu)$ plot, as one can see in Fig. 3.7.

The Wien limit

For energies much larger than the characteristic thermal temperature, $h\nu \gg k_B T$, the -1 in the denominator is a negligible correction. Thus, we can write

$$\exp\left(\frac{h\nu}{k_B T}\right) - 1 \approx \exp\left(\frac{h\nu}{k_B T}\right), \tag{3.61}$$

so that in this limit

$$I_\nu^W(T) = \frac{2h\nu^3}{c^2} \exp\left(-\frac{h\nu}{k_B T}\right). \tag{3.62}$$

This limit does contain the Planck constant, and it accounts properly for the quantum nature of the photons. This approximation is called the *Wien limit*,[6] and it describes properly the exponential drop-off of the spectrum at high energies, as is shown in Fig. 3.7.

Monotonicity

The derivative $\partial B_\nu(T)/\partial T$ is always, which means for all frequencies, larger than zero (also see Exercise 3.4). This implies that the higher temperature curves lie always entirely above the low-temperature curves; compare to Fig. 3.7. This property is called the *monotonicity of the Planck spectrum with temperature*.

Wien's displacement law

The frequency at which the maximum of the blackbody spectrum occurs is determined by the temperature. The maximum of B_ν is found by setting

$$\left.\frac{\partial B_\nu}{\partial \nu}\right|_{\nu=\nu_{max}} = 0. \tag{3.63}$$

This yields

$$\left.\frac{\partial}{\partial \nu}\left\{\nu^3\left[\exp\left(\frac{h\nu}{k_B T}\right) - 1\right]^{-1}\right\}\right|_{\nu_{max}} = 0, \tag{3.64}$$

which becomes

$$3\left[1 - \exp\left(-\frac{h\nu_{max}}{k_B T}\right)\right] = \frac{h\nu_{max}}{k_B T}. \tag{3.65}$$

This is a transcendental equation for $h\nu_{max}/k_B T$ that cannot be solved analytically. However, a first guess is that $h\nu_{max}/k_B T \approx 3$. If we substitute this into the left-hand side, we find a new value for $h\nu_{max}/k_B T$. After a few steps, the solution converges

---

[6] After Wilhelm Wien (1864–1928), German physicist.

to $h\nu_{max}/k_B T \approx 2.82$ or

$$E_{peak} = h\nu_{max} \approx 2.82 \cdot k_B T. \tag{3.66}$$

The photon energy at the peak of the spectrum is of order $k_B T$ with higher temperatures shifting the peak to higher frequencies:

$$\nu_{max} = 5.88 \cdot 10^{10}\, \mathrm{Hz\,K^{-1}} \cdot T. \tag{3.67}$$

This is *Wien's displacement law*.[7] It can also be expressed in terms of wavelengths

$$\lambda_{max} \cdot T = 0.290\,\mathrm{cm\,K}. \tag{3.68}$$

Note that λ_{max} and ν_{max} are *not* related by c; in other words $\lambda_{max} \cdot \nu_{max} \neq c$.

The frequency of the maximum, ν_{max}, can also be used to distinguish the different limits: for $\nu \ll \nu_{max}$ the spectrum is accurately described by the Rayleigh–Jeans law; for $\nu \gg \nu_{max}$ Wien's law is applicable.

3.5.5   *Temperature definitions*

Brightness temperature

To characterize the brightness of an object *at a given frequency*, one can equate the measured brightness to the brightness of a blackbody at the brightness temperature T_b:

$$I_\nu = B_\nu(T_b). \tag{3.69}$$

The main advantage of this definition is that the brightness is measured simply in K instead of $\mathrm{erg\,s^{-1}\,cm^{-2}\,Hz^{-1}\,ster^{-1}}$, and this is related to the physical temperature of the emitting object. In particular, in radio astronomy, where the Rayleigh–Jeans law, Eq. (3.60), is often applicable, this provides the simple relation

$$T_b = I_\nu \frac{c^2}{2k_B \nu^2}. \tag{3.70}$$

Effective temperature

The effective temperature, T_{eff}, characterizes the emitting surface of a radiating object. It is defined as the temperature of a blackbody that emits the same (frequency-integrated) flux as the observed object:

$$F_{obs} \equiv \sigma_{SB} T_{eff}^4. \tag{3.71}$$

---

[7] Not to be confused with the Wien limit of blackbody radiation.

For example, using the energy flux of the Sun, one finds an effective temperature of $T_{\text{eff}} \approx 5800$ K.

3.6 Radiative transitions

Atoms and ions can absorb or emit photons by exciting or de-exciting electrons in their shells. These transitions are called radiative transitions and give rise to absorption and emission lines, which are invaluable tools to determine conditions in astrophysical environments.

There are three kinds of electronic transitions that lead to important radiative processes:

- *Bound–bound*: electrons move between bound states in an atom or ion. A photon is either emitted or absorbed.
- *Bound–free*: electron transitions between bound and unbound states. If the electron changes from a bound to an unbound state, this transition is called ionization; if an initially unbound electron is captured, this is called recombination.
- *Free–free*: free electron gains energy by absorbing a photon as it passes an ion or loses energy by emitting a photon. The latter emission process is called *bremsstrahlung* and is discussed later in this chapter; see Section 3.8.1.

Electron transitions are possible only between specific energy levels: they must obey so-called *selection rules*. Selection rules arise because photons carry angular momentum, which must be conserved in any emission or absorption process: *permitted transitions* are those allowed by selection rules. So-called *forbidden transitions* may be

- genuinely impossible. For example, an electron in the spherically symmetric $n = 2$ state of hydrogen (2s) cannot decay to the ground state (1s) by single-photon emission. Only collisions or two-photon emission allow this to happen.
- forbidden only in some approximate description of the transition. In this case, the transition can occur, but the rate is low and the states are long-lived.

To introduce some quantitative concepts of radiative transitions, let us consider transitions between two energy levels. An electron in the lower of two states can absorb a photon and jump to the upper energy level. In the upper state, it can either spontaneously emit a photon and return to the lower state or it can be induced to emit a photon through stimulation by radiation of the appropriate frequency. In statistical equilibrium, the sum of induced and spontaneous downward transitions must equal the number of upward transitions.

The probability that an atom or ion will undergo a radiative transition from a state of higher energy, E_i, to a state of lower energy, E_j, is given by

$$P_{ij} = A_{ij} + B_{ij}u_\nu, \tag{3.72}$$

where A_{ij} is the coefficient for spontaneous transition from level i to level j and B_{ij} is the coefficient for stimulated emission induced by radiation of energy density $u_\nu\, d\nu$ in the frequency range ν to $\nu + d\nu$. Albert Einstein introduced the concepts of spontaneous and stimulated emission in 1917. Hence, the coefficients A and B are called *Einstein coefficients*. The process of stimulated emission is sometimes explained by regarding the atom as an oscillator and the radiation as a driving force. An oscillator absorbs or emits energy depending on its phase with respect to its driving force. So, Einstein argued, the radiation field causes a downward transition of an electron at a rate proportional to the radiation energy density. Thus, stimulated emission is the process by which an electron is induced to jump from a higher energy level to a lower one by the presence of electromagnetic radiation at (or near) the frequency of the transition.

The probability for absorption of photons leading to an excitation of an electron from level j to level i is

$$P_{ji} = B_{ji}u_\nu, \tag{3.73}$$

where B_{ji} is the coefficient for absorption of a photon. The energy density of blackbody radiation at temperature T in the frequency range ν to $\nu + d\nu$ is

$$u_\nu = \frac{8\pi h\nu^3}{c^3}\left[\exp\left(\frac{h\nu}{k_{\rm B}T}\right) - 1\right]^{-1}, \tag{3.74}$$

where we have used Eq. (3.17) and Eq. (3.54).

In statistical equilibrium, the number of de-excitations must equal the number of excitations. Moreover, in equilibrium, the population of each state is determined by the Boltzmann distribution. These two conditions yield the equation

$$g_j \exp\left(-\frac{E_j}{k_{\rm B}T}\right) B_{ji}u_\nu = g_i \exp\left(-\frac{E_i}{k_{\rm B}T}\right)(A_{ij} + B_{ij}u_\nu), \tag{3.75}$$

where g_i and g_j are the statistical weights of level i and j, respectively. By setting $E_i - E_j = h\nu$, we find

$$A_{ij} = \frac{8\pi h\nu^3}{c^3}B_{ij} \tag{3.76}$$

and

$$g_i B_{ij} = g_j B_{ji}. \tag{3.77}$$

The coefficient A_{ij} is the probability per second that an atom with an electron in level i will spontaneously emit a photon of energy $h\nu_{ij} = E_i - E_j$. The energy emitted per unit time is hence $h\nu_{ij}A_{ij}$. Therefore, the energy emitted per unit time, per unit volume, and per unit solid angle by spontaneous bound–bound transitions is

$$j = n_i A_{ij} \frac{h\nu_{ij}}{4\pi}, \tag{3.78}$$

where n_i is the number density of ions in level i. The quantity j is called line emissivity and has units of erg s^{-1}cm^{-3} ster^{-1}. This is an example of an emission coefficient as defined in Section 3.4.2. A database on atomic transition probabilities is available on the World Wide Web at http://physics.nist.gov/.

3.7   Radiation from moving charges

3.7.1   *Why do accelerated charges radiate?*

Consider a charge q that is initially at rest and then, over a short time Δt, is accelerated to a velocity Δv. The situation after a time t is sketched in Fig. 3.8. Outside a radius $R_2 = ct$, the field lines are still pointing radially toward the original position because the time was too short for the information to arrive. Inside the radius $R_1 = c(t - \Delta t)$, the new velocity is already known; the field lines are pointing radially to the current position of the charge. In the shell of width $c\Delta t$ between R_1 and R_2, there must be a tangential component of the field (radiation) that propagates at the speed of light. The energy for this radiation pulse comes from the force that accelerated the charge. In the direction of acceleration, no such tangential field exists. As is shown later, this field component drops off with only

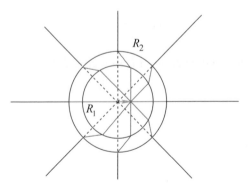

Figure 3.8   Sketch to qualitatively show why radiation occurs from accelerated charges; see text for details.

$1/r$, whereas the radial component behaves like $1/r^2$. At distances far from the charge, the radiation part dominates.

3.7.2   *Maxwell's equations and Poynting's theorem*

Maxwell's equations describe how the electric and the magnetic fields are related to each other and to their sources, the charge density ρ and the current density $\vec{j}$[8]:

$$\vec{\nabla} \cdot \vec{D} = 4\pi\rho \quad \text{and} \quad \vec{\nabla} \cdot \vec{B} = 0, \tag{3.79}$$

$$\vec{\nabla} \times \vec{E} = -\frac{1}{c}\partial_t \vec{B} \quad \text{and} \quad \vec{\nabla} \times \vec{H} = \frac{4\pi}{c}\vec{j} + \frac{1}{c}\partial_t \vec{D}. \tag{3.80}$$

The electric displacement field $\vec{D}$ and the magnetic field $\vec{H}$ can often be linearly related to the electric field $\vec{E}$ and the magnetic flux density $\vec{B}$ by means of the dielectric constant ϵ and the magnetic permeability μ: $\vec{D} = \epsilon\vec{E}$ and $\vec{B} = \mu\vec{H}$. In the following and unless otherwise stated, we assume that no dielectric or permeable media are present and that the distinction between $\vec{D}$ and $\vec{E}$ and $\vec{B}$ and $\vec{H}$, respectively, can be ignored; that is, we assume $\epsilon = \mu = 1$.

An immediate consequence of Maxwell's equations is the *conservation of electric charge*: taking the divergence of the $\vec{\nabla} \times \vec{H}$ equation yields the charge continuity equation

$$\partial_t\rho + \vec{\nabla} \cdot \vec{j} = 0. \tag{3.81}$$

Of course, energy is also conserved. The energy conservation law of electrodynamics, also known as *Poynting's theorem*, has to take into account that the electromagnetic field, and the charges can mutually do work on each other. In differential form, Poynting's theorem can be written as

$$\partial_t\left(\frac{E^2 + B^2}{8\pi}\right) + \vec{j} \cdot \vec{E} = -\vec{\nabla} \cdot \vec{S}, \tag{3.82}$$

where the first term represents the energy density of the electromagnetic field, $u_{em} = (E^2 + B^2)/(8\pi)$. The second term, $\vec{j} \cdot \vec{E}$, is the rate of change of the mechanical energy per volume due to the action of the electric field on the charges. The vector $\vec{S}$ is called the *Poynting vector* and is given by

$$\vec{S} = \frac{c}{4\pi}\vec{E} \times \vec{B}. \tag{3.83}$$

It describes the electromagnetic energy flux (energy per time and surface) and its magnitude can thus also be written as $S = dW/(dt \cdot dA)$.[9] Integrating Eq. (3.82)

---

[8] Throughout this book, we use Gaussian cgs units and often write ∂_t instead of $\frac{\partial}{\partial t}$.
[9] We use here W rather than E for the energy to avoid possible confusion with the electric field $\vec{E}$.

over the volume yields

$$\frac{d}{dt} \int_V \left(\frac{E^2 + B^2}{8\pi} \right) dV + \int_V \vec{j} \cdot \vec{E} dV = -\int_\Sigma \vec{S} \cdot d\vec{A}, \tag{3.84}$$

where Σ denotes the surface enclosing the volume V. On the right-hand side we have used Gauss's divergence theorem, $\int_V (\vec{\nabla} \cdot \vec{X}) dV = \int_\Sigma \vec{X} d\vec{A}$, to transform the right-hand side into a surface integral. If we extend the integrals to infinity, the right-hand side measures the energy per time that flows out to infinity. As the surface element scales like r^2, energy can only flow to infinity if S does not drop faster than r^{-2}. This becomes important in the following discussion of radiation: the radiative parts of $\vec{E}$ and $\vec{B}$ both behave like $\propto 1/r$ and therefore $S \propto 1/r^2$. Therefore, a finite amount of radiative energy can reach infinity.

3.7.3 *Electromagnetic potentials*

The electromagnetic fields $\vec{E}$ and $\vec{B}$ can be expressed in terms of a scalar potential Φ and a vector potential $\vec{A}$. One might wonder what the advantage is of expressing two quantities by two other quantities, but using potentials instead of physical fields has definite advantages. First of all, instead of dealing with six numbers, the components of $\vec{E}$ and $\vec{B}$, one only has to deal with four numbers, Φ, A_1, A_2, and A_3. Second, the equations determining Φ and $\vec{A}$ are simpler than Maxwell's equations. And, third, as briefly outlined in Chapter 1, Φ and $\vec{A}$ can be combined into a four-vector, which is used for a covariant description of electrodynamics.

The *no-monopoles condition* of the Maxwell equations, $\vec{\nabla} \cdot \vec{B} = 0$, allows us to write the magnetic field as the curl of a vector potential,

$$\vec{B} = \vec{\nabla} \times \vec{A}, \tag{3.85}$$

because $\vec{\nabla} \cdot (\vec{\nabla} \times \vec{A})$ vanishes identically. Using Eq. (3.85) and the $\vec{\nabla} \times \vec{E}$ Maxwell equation, Eq. (3.80), yields

$$\vec{\nabla} \times \left(\vec{E} + \frac{1}{c} \partial_t \vec{A} \right) = 0. \tag{3.86}$$

As $\vec{\nabla} \times (\vec{\nabla} \Phi) \equiv 0$, we can express the term in brackets as the (negative) gradient of a scalar potential, $-\vec{\nabla}\Phi = \vec{E} + (1/c)\partial_t \vec{A}$. Hence, once we know the potentials Φ and $\vec{A}$, the physical fields can be calculated from Eq. (3.85) and from

$$\vec{E} = -\vec{\nabla}\Phi - \frac{1}{c} \partial_t \vec{A}. \tag{3.87}$$

This leaves us with the question of how to obtain the potentials from the distributions of the physical sources of the fields, the charge density ρ, and the currents $\vec{j}$. The strategy is to find wave-like equations containing the sources for both Φ and $\vec{A}$ that can be solved via a so-called Green's function approach. If we use the $\vec{\nabla} \cdot \vec{E}$ Maxwell equation, Eq. (3.79), and insert Eq. (3.87), we can write

$$\vec{\nabla}^2\Phi - \frac{1}{c^2}\partial_t^2\Phi + \frac{1}{c}\partial_t\left(\vec{\nabla}\cdot\vec{A} + \frac{1}{c}\partial_t\Phi\right) = -4\pi\rho. \tag{3.88}$$

One may wonder why the time derivatives of Φ (that cancel out) have been inserted. This is to keep the equation similar to the vector potential equation that follows.

By using vector identities and Eq. (3.85), the $\vec{\nabla} \times \vec{B}$ equation can be transformed to

$$\vec{\nabla}^2\vec{A} - \frac{1}{c^2}\partial_t^2\vec{A} - \vec{\nabla}\left(\vec{\nabla}\cdot\vec{A} + \frac{1}{c}\partial_t\Phi\right) = -\frac{4\pi}{c}\vec{j}; \tag{3.89}$$

see Exercise 3.7.

As we have seen, the physical fields can be calculated by differentiating the potentials as defined in Eqs. (3.85) and (3.87). However, this implies that the potentials are not uniquely determined. We have some freedom to choose the additive constants in a way that simplifies our equations. For example, when we add the gradient of an arbitrary scalar function Ψ to the vector potential, $\vec{A} \to \vec{A} + \vec{\nabla}\Psi$, we obtain exactly the same magnetic field. If at the same time we also change the scalar potential, $\Phi \to \Phi - (1/c)\partial_t\Psi$, the electric field is also unchanged. This freedom to change the (auxiliary) quantities in a way that the physical fields remain untouched is called *gauge freedom*. The corresponding changes of the potentials are called *gauge transformations*. We can choose Ψ such that

$$\vec{\nabla}\cdot\vec{A} + \frac{1}{c}\partial_t\Phi = 0, \tag{3.90}$$

which brings our Eqs. (3.88) and (3.89) into a symmetric and particularly simple form:

$$\Box\Phi \equiv \vec{\nabla}^2\Phi - \frac{1}{c^2}\partial_t^2\Phi = -4\pi\rho, \tag{3.91}$$

$$\Box\vec{A} \equiv \vec{\nabla}^2\vec{A} - \frac{1}{c^2}\partial_t^2\vec{A} = -\frac{4\pi}{c}\vec{j}, \tag{3.92}$$

where we have used the so-called D'Alembert operator $\Box \equiv \vec{\nabla}^2 - (1/c^2)\partial_t^2$ that we encountered in Chapter 1. The choice of Ψ, Eq. (3.90), is called the *Lorentz gauge*. Equations (3.91) and (3.92) were used in Chapter 1 to motivate the introduction of the four-potential; see Section 1.6.

Equations such as (3.91) and (3.92) can be solved using the method of Green's functions. The result is[10]

$$\Phi(\vec{r}, t) = \int \frac{[\rho] \, d^3 r'}{|\vec{r} - \vec{r}'|} \tag{3.93}$$

$$\vec{A}(\vec{r}, t) = \frac{1}{c} \int \frac{[\vec{j}] \, d^3 r'}{|\vec{r} - \vec{r}'|}, \tag{3.94}$$

where the bracket notation means $[X] \equiv X(\vec{r}', t - |\vec{r} - \vec{r}'|/c)$. This means that the potentials at point $\vec{r}$ are determined by the sources *as they were* a time $|\vec{r} - \vec{r}'|/c$ *earlier* than t. The interpretation is that the information can propagate only at a finite speed, c, from $\vec{r}'$ to $\vec{r}$. Therefore, potentials (3.93) and (3.94) are called *retarded potentials*.[11] At first sight, the distinction between different times may seem somewhat sophisticated, but it has very important implications. As we see later, this allows for physical fields that fall off with $1/r$ rather than with $1/r^2$, as for example for the Coulomb field. Such $1/r$ fields are called *radiation fields*.

The strategy is now the following: i) given the distribution of the sources ρ and $\vec{j}$, find the retarded potentials via integration (see Eqs. [3.93] and [3.94]) and ii) calculate the physical fields using Eqs. (3.85) and (3.87).

3.7.4   *Liénard–Wiechert potentials*

The retarded potentials can be calculated for the special case of a single moving charge, in which case they are called *Liénard–Wiechert potentials*. Consider a particle of charge q that moves on a trajectory given by $\vec{r}_0(t)$ and whose velocity is

$$\vec{u} = \frac{d\vec{r}_0(t)}{dt}, \tag{3.95}$$

as sketched in Fig. 3.9.

In this case, the sources are

$$\rho(\vec{r}, t) = q \, \delta[\vec{r} - \vec{r}_0(t)] \tag{3.96}$$

and

$$\vec{j}(\vec{r}, t) = q\vec{u} \, \delta[\vec{r} - \vec{r}_0(t)], \tag{3.97}$$

where $\delta(\vec{r} - \vec{r}_0)$ is the delta distribution that localizes the particle at $\vec{r}_0$. With these sources known, one can perform the integrations in Eqs. (3.93) and (3.94) using

---

[10] See Jackson, *Classical Electrodynamics* (1998) for details.
[11] Mathematically there are also solutions with advanced potentials that are not considered here.

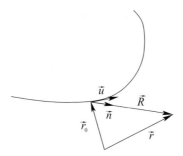

Figure 3.9   Liénard–Wiechert potentials: single charge moving along a trajectory $\vec{r}_0(t)$; r is the position we are interested in, R points from the particle's current position to r.

the properties of the delta distribution and suitable substitutions.[12] As a result one finds

$$\Phi(\vec{r}, t) = \left[\frac{q}{\kappa R}\right] \qquad \text{and} \qquad \vec{A}(\vec{r}, t) = \left[\frac{q\vec{u}}{c\kappa R}\right], \tag{3.98}$$

where the brackets indicate again that the quantities are to be taken at *retarded times* $t_{\text{ret}} = t - |\vec{R}|/c$, where $\vec{R} = \vec{r} - \vec{r}_0(t_{\text{ret}})$. The function κ is given by

$$\kappa(t) = 1 - \frac{1}{c}\vec{n}(t) \cdot \vec{u}(t), \tag{3.99}$$

where $\vec{n}$ is the unit vector pointing from the position at the retarded time to the position of interest at $\vec{r}$, $\vec{n} = \vec{R}/|\vec{R}|$; see Fig. 3.9.

There are two important differences to the potentials you may have encountered in electrostatic theory. The first comes from the function κ. For small velocities $\vec{u}$, it simply approaches unity, and no effect is noticeable. For large velocities, however, it leads to *relativistic beaming*, an effect closely related to the Lorentz transformations that were discussed in some detail in Chapter 1. The second effect comes from taking the arguments at the *retarded times* t_{ret}. This has the consequence that *radiative solutions* exist, that is, the physical fields $\vec{E}$ and $\vec{B}$ fall off $\propto 1/r$ rather than $\propto 1/r^2$. This allows radiative energy to flow off to infinity. See also the discussion after Eq. (3.84).

3.7.5   *Velocity and acceleration field*

Using the Liénard–Wiechert potentials, one can calculate $\vec{E}$ and $\vec{B}$ via Eqs. (3.85) and (3.87). In principle, this is straightforward, but in practice this turns out to be

---

[12] For details of the calculation see Jackson, *Classical Electrodynamics* (1998).

a rather lengthy calculation.[13] The resulting physical fields are

$$\vec{E}(\vec{r}, t) = q \left[\frac{(\vec{n} - \vec{\beta})(1 - \beta^2)}{\kappa^3 R^2} \right] + \frac{q}{c} \left[\frac{\vec{n}}{\kappa^3 R} \times \{(\vec{n} - \vec{\beta}) \times \dot{\vec{\beta}}\} \right], \quad (3.100)$$

$$\vec{B}(\vec{r}, t) = [\vec{n} \times \vec{E}], \quad (3.101)$$

where $\vec{\beta} = \vec{u}/c$ and, like before, $\kappa = 1 - \vec{n} \cdot \vec{\beta}$ has been used. It is worth highlighting a few points:

- The electric field has two contributions: the first falls off $\propto 1/R^2$ and depends via $\vec{\beta}$ on the particle velocity. The second contribution falls off $\propto 1/R$ and depends, apart from $\vec{\beta}$, also on the acceleration $\dot{\vec{\beta}}$. The first contribution is called the *velocity field*, the second one the *acceleration field*.
- In the low-velocity limit without acceleration, $\beta \to 0$, $\kappa \to 1$, and $\dot{\vec{\beta}} = 0$, and the electric field becomes

$$\vec{E}(\vec{r}, t) = q \frac{\vec{r} - \vec{r}_0}{|\vec{r} - \vec{r}_0|^3}, \quad (3.102)$$

  that is, the *Coulomb law* is reproduced by the velocity field.
- The *acceleration field* is only present for $\dot{\vec{\beta}} \neq 0$.
- At time t, the particle already has advanced further along its path, but the fields are still determined from the conditions at the previous, *retarded time*.
- The magnetic field is always perpendicular to both $\vec{E}$ and $\vec{n}$, see Eq. (3.101).
- The $1/r$ contributions are called the *radiation field* and are given by

$$\vec{E}_{\mathrm{rad}}(\vec{r}, t) = \frac{q}{c} \left[\frac{\vec{n}}{\kappa^3 R} \times \{(\vec{n} - \vec{\beta}) \times \dot{\vec{\beta}}\} \right], \quad (3.103)$$

$$\vec{B}_{\mathrm{rad}}(\vec{r}, t) = [\vec{n} \times \vec{E}_{\mathrm{rad}}(\vec{r}, t)]. \quad (3.104)$$

3.7.6   *The nonrelativistic limit: Larmor's formula and the dipole approximation*

The expressions for $\vec{E}_{\mathrm{rad}}$ and $\vec{B}_{\mathrm{rad}}$ that we derived before are valid even in the relativistic case. However, in many cases a simpler, nonrelativistic treatment is good enough, and this is what we are going to do first.

Let us first get some feeling for the orders of magnitude of the involved terms. Roughly speaking, the acceleration term in Eq. (3.100) behaves like $E_{\mathrm{rad}} \sim (q/c^2)(\dot{u}/\kappa^3 R)$ and the velocity term like $E_{\mathrm{vel}} \sim (q/\kappa^3 R^2)$, hence

$$\frac{E_{\mathrm{rad}}}{E_{\mathrm{vel}}} \sim \frac{\dot{u} R}{c^2}. \quad (3.105)$$

---

[13] The calculation can be found in Jackson, *Classical Electrodynamics* (1998).

Now, when we attribute the radiation to an oscillation of a charge distribution with a typical oscillation period T or frequency ν, we can write $\dot{u} \sim u/T = u \cdot \nu$. By using $\nu/c = 1/\lambda$, Eq. (3.105) becomes

$$\frac{E_{\text{rad}}}{E_{\text{vel}}} \sim \left(\frac{R}{\lambda}\right)\left(\frac{u}{c}\right). \tag{3.106}$$

Thus, we can distinguish two regions depending on the distance from the source compared to the typical wavelength λ: in the *near zone* at $R \sim \lambda$, the velocity field dominates over the radiation field by c/u, and the *far zone*, $R \gg \lambda$, where the radiation field dominates, because it falls off more slowly.

Let us now have a more careful look at the case of low velocities, $\beta \ll 1$. We have $\vec{\dot{\beta}} = \dot{\vec{u}}/c$ and, in this limit, $\vec{n} - \vec{\beta} \to \vec{n}$, $\kappa^3 \to 1$, so the acceleration field becomes

$$\vec{E}_{\text{rad}} = \frac{q}{Rc^2}\left[\vec{n} \times (\vec{n} \times \dot{\vec{u}})\right]. \tag{3.107}$$

Because of Eq. (3.104), we can write

$$|\vec{E}_{\text{rad}}| = |\vec{B}_{\text{rad}}| = \frac{q\dot{u}}{Rc^2}\sin\theta, \tag{3.108}$$

where θ is the angle between $\vec{n}$ and $\dot{\vec{u}}$. Therefore, the Poynting vector has a magnitude of

$$S = |\vec{S}| = \frac{c}{4\pi}|\vec{E}_{\text{rad}} \times \vec{B}_{\text{rad}}| = \frac{q^2\dot{u}^2}{4\pi R^2 c^3} \cdot \sin^2\theta. \tag{3.109}$$

If we keep in mind that $S = dW/(dt \cdot dA)$ and $dA = R^2 d\Omega$, we can write the emitted energy per time and solid angle

$$\frac{dW}{dt\,d\Omega} = R^2 S = \frac{q^2\dot{u}^2}{4\pi c^3} \cdot \sin^2\theta. \tag{3.110}$$

From this expression, we can, via integration over the solid angle, calculate the total emitted power

$$\frac{dW}{dt} = \int \frac{dW}{dt\,d\Omega}d\Omega = \frac{q^2\dot{u}^2}{4\pi c^3}\int \sin^2\theta\,d\Omega, \tag{3.111}$$

or, with $\int \sin^2\theta\,d\Omega = 8\pi/3$,

$$\frac{dW}{dt} = \frac{2q^2\dot{u}^2}{3c^3}. \tag{3.112}$$

This is *Larmor's formula*, an extremely useful relation for the power emitted by nonrelativistic, accelerated charge. The following points are important to note:

- The radiated power is proportional to the square of the charge q and to the square of the acceleration $\dot{u}$.
- The differential radiation power $\frac{dW}{dt\,d\Omega}$, is proportional to $\sin^2\theta$; compare to Eq. (3.110), which is characteristic for dipole radiation. This means that *no radiation is emitted in the acceleration direction*, but the emission is maximal in the direction perpendicular to it.
- The direction of the electric field vector of the radiation is determined by $\dot{u}$ and $\vec{n}$; see Eq. (3.107). If the acceleration occurs along one space-fixed line, for example, if the charge is accelerated back and forth along the x-axis, the direction of the $\vec{E}$-field is fixed in space; that is, the radiation is 100% polarized in the $\dot{u}$-$\vec{n}$-plane.

So far, we have considered only individual, accelerated charges. Let us now consider a cloud of charged particles that move, accelerate, or decelerate each other and, consequently, radiate. In principle, to calculate the radiation at a particular point, the retarded time of each particle has to be considered. This means that one has to do the bookkeeping for the phase relations of every single particle. Now, consider a cloud of size L that changes on a typical timescale τ. If the light travel time across the system is much shorter than the timescale on which the cloud as a whole changes, $L/c \ll \tau$, we can ignore the tiny changes in the retarded times and drop the brackets in Eqs. (3.103) and (3.104). Because $\nu \sim 1/\tau$, $L/c \ll \tau$ also means $c/\nu \gg L$, or $\lambda \gg L$. *In words, if the wavelength of the radiation largely exceeds the source size, differences resulting from the retarded times can be safely ignored.*

Now consider a set of nonrelativistic particles whose electric radiation field in the nonretarded form is given by Eq. (3.107) as

$$\vec{E}_{\text{rad}} = \sum_i \frac{q_i}{c^2} \frac{\vec{n} \times (\vec{n} \times \dot{\vec{u}}_i)}{R_i}. \tag{3.113}$$

If we are far away from the source in the sense that the source extension is negligible in comparison with the distance between us and the source, R_0, this can be written as

$$\vec{E}_{\text{rad}} \approx \sum_i \frac{q_i}{c^2} \frac{\vec{n} \times (\vec{n} \times \dot{\vec{u}}_i)}{R_0} = \frac{\vec{n} \times (\vec{n} \times \ddot{\vec{d}})}{c^2 R_0}, \tag{3.114}$$

where in the last step we have inserted the dipole moment of the distribution

$$\vec{d} = \sum_i q_i \vec{r}_i. \tag{3.115}$$

As the radiation field is determined by the (second time derivative of the) dipole moment, this approximation is usually referred to as the *dipole approximation*.

Completely analogously to the Larmor formula, we can write

$$\frac{dW}{dt\,d\Omega} = \frac{\ddot{d}^2}{4\pi c^3} \cdot \sin^2\theta \tag{3.116}$$

and

$$\left(\frac{dW}{dt}\right)_{dip} = \frac{2\ddot{d}^2}{3c^3}. \tag{3.117}$$

3.7.7 *The spectrum of dipole radiation*

The time and the frequency domain are related by Fourier transforms. The Fourier transform of a function $E(t)$ is defined as

$$E(\omega) = \frac{1}{2\pi}\int_{-\infty}^{+\infty} E(t)e^{i\omega t}\,dt \tag{3.118}$$

and the inverse as

$$E(t) = \int_{-\infty}^{+\infty} E(\omega)e^{-i\omega t}\,d\omega. \tag{3.119}$$

Parseval's theorem states that integrating the square of the absolute value over either the whole time or the whole frequency domain should give the same result up to a factor of 2π [14]:

$$\int_{-\infty}^{+\infty} |E(t)|^2\,dt = 2\pi \int_{-\infty}^{+\infty} |E(\omega)|^2\,d\omega. \tag{3.120}$$

This has useful applications in the context of spectra. We have seen before that the absolute value of the Poynting vector can be written as

$$S = \frac{dW}{dt\,dA} = \frac{c}{4\pi}E(t)^2 \tag{3.121}$$

and hence, by integrating over time,

$$\frac{dW}{dA} = \frac{c}{4\pi}\int_{-\infty}^{+\infty} |E(t)|^2\,dt = \frac{c}{2}\int_{-\infty}^{+\infty} |E(\omega)|^2\,d\omega = c\int_{0}^{+\infty} |E(\omega)|^2\,d\omega, \tag{3.122}$$

where we have used Parseval's relation and $|E(\omega)|^2 = |E(-\omega)|^2$. To obtain the emitted energy per surface area and *frequency*, we need the Fourier transform of

---

[14] Note that the factor 2π appearing here is related to the normalization of the Fourier transform that we have chosen. If both forward and backward Fourier transform are normalized with the same factor $(2\pi)^{-1/2}$, the factor 2π does not occur in the Parseval relation.

the electric field:

$$\frac{dW}{dA\,d\omega} = c|E(\omega)|^2. \tag{3.123}$$

In the case of dipole radiation, we first write the magnitude of the electric field, given by Eq. (3.114), as

$$E(t) = \ddot{d}(t)\frac{\sin\theta}{c^2 R_0}, \tag{3.124}$$

where we have assumed that the direction of $\vec{d}$ remains fixed. We then have to find its Fourier transform, which is simply

$$E(\omega) = -\frac{1}{c^2 R_0}\omega^2 d(\omega)\sin\theta; \tag{3.125}$$

see Exercise 3.9. When we again use $dA = R_0^2 d\Omega$ and $\int \sin^2\theta\,d\Omega = 8\pi/3$, we have

$$\frac{dW}{d\omega\,d\Omega} = \frac{\omega^4}{c^3}|d(\omega)|^2 \sin^2\theta \tag{3.126}$$

and

$$\frac{dW}{d\omega} = \frac{8\pi\omega^4}{3c^3}|d(\omega)|^2. \tag{3.127}$$

Hence, the frequency of dipole radiation is directly related to the oscillation frequency of the source.

3.7.8 *Application of the dipole approximation: Thomson scattering*

An important application of the dipole approximation is the scattering of an electromagnetic wave off a free electron. This process is known as *Thomson scattering*.

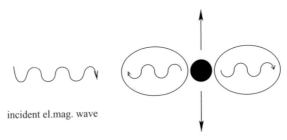

incident el.mag. wave

Figure 3.10   Sketch of the situation in Thomson scattering: an incident electromagnetic wave forces an electron to oscillate. The accelerated electron emits radiation in the characteristic dipole pattern perpendicular to its acceleration direction.

Thomson scattering is valid for the case where the electron has little energy or is at rest and the incident photons have an energy much smaller than the electron rest mass, $h\nu \ll mc^2$, so that quantum-particle electrodynamical effects can be ignored. Consider a free particle of mass m and charge q and a linearly polarized, incident electromagnetic wave, whose electric field vector is given by

$$\vec{E} = E_0 \sin \omega t \cdot \hat{e}. \tag{3.128}$$

This electric field exerts a force on the charge, $\vec{F} = q\vec{E}$, and accelerates it. To calculate the power emitted by the accelerated charge via the dipole formula (see Eq. [3.116]), we need the second time derivative of the dipole moment, $\vec{d} = q\vec{r}$, which is easily calculated as

$$\ddot{\vec{d}} = \frac{q^2 E_0}{m} \sin \omega t \cdot \hat{e}. \tag{3.129}$$

When θ denotes the angle between the direction of interest and the $\vec{E}$-field direction $\hat{e}$, Eq. (3.116) yields

$$\frac{dW}{dt\,d\Omega} = \frac{q^4 E_0^2}{8\pi m^2 c^3} \sin^2 \theta. \tag{3.130}$$

Note that we have not inserted the instantaneous second derivative of $\vec{d}$. To get rid of the time dependence, we have taken the average over one period of $\ddot{\vec{d}}^2$, that is,

$$\frac{1}{T} \int_0^T \sin^2 \omega t \, dt = \frac{1}{2T} \left[t - \frac{1}{2\omega} \sin(2\omega t) \right]_0^T = \frac{1}{2}. \tag{3.131}$$

Hence, the total power is given by

$$\frac{dW}{dt} = \int \frac{dW}{dt\,d\Omega} d\Omega = \frac{q^4 E_0^2}{3m^2 c^3}, \tag{3.132}$$

where we have used $\int \sin^2 \theta \, d\Omega = 8\pi/3$. The incident energy flux is given by the Poynting vector S and, because $|\vec{E}| = |\vec{B}|$, the time average over one period ($T = 2\pi/\omega$) is

$$\langle S \rangle = \frac{1}{T} \int_0^T \left(\frac{c}{4\pi} |\vec{E}_{\text{rad}} \times \vec{B}_{\text{rad}}| \right) dt = \frac{c E_0^2}{4\pi T} \int_0^T \sin^2 \omega t \, dt = \frac{c E_0^2}{8\pi}. \tag{3.133}$$

When we now define the *differential cross section* $\frac{d\sigma}{d\Omega}$ as

$$\frac{dW}{dt\,d\Omega} = \langle S \rangle \frac{d\sigma}{d\Omega}, \tag{3.134}$$

we have

$$\frac{d\sigma}{d\Omega} = \frac{q^4 \sin^2 \theta}{m^2 c^4}. \tag{3.135}$$

For the case of electrons $q = e$, this can be written as

$$\frac{d\sigma}{d\Omega} = r_0^2 \sin^2 \theta. \tag{3.136}$$

Here, we have introduced a quantity that has the dimension of a length,

$$r_0 = \frac{e^2}{m_e c^2}, \tag{3.137}$$

and is historically called *the classical electron radius*. This, however, has nothing to do with the physical radius of the elementary particle electron. It is merely a quantity with the dimension of a length, constructed from the electron mass and charge.

The *total cross section* is then found via integration over the solid angle

$$\sigma_T \equiv \int \frac{d\sigma}{d\Omega} d\Omega = \frac{8\pi}{3} r_0^2 \equiv \sigma_T = 6.65 \cdot 10^{-25} \text{cm}^2 \tag{3.138}$$

and is called the *Thomson cross section*. Even though we have derived this total cross section for polarized incident radiation, it can be shown that the total cross section for unpolarized radiation is exactly the same. It is also worth pointing out that the cross section does *not* depend on wavelength, which means that high- and low-energy photons are scattered equally efficiently.

However, we should remember the assumptions that we made in the beginning: we have applied the nonrelativistic dipole limit; therefore the incident field must not be so strong as to accelerate the electron to relativistic velocities. Moreover, if the energy of the incident photons becomes comparable to the electron rest mass, $m_e c^2 = 511$ keV, the process must be treated in a quantum-mechanical approach rather than the classical approach that we have used here.

3.8 Individual radiation processes

We now discuss individual radiation processes that are particularly important in high-energy astrophysics, namely *bremsstrahlung*, synchrotron emission, and Compton scattering. Because our discussion is entirely based on classical physics, we can derive only relations that are of pedagogical value. The proper treatment is often rather involved and requires the use of quantum electrodynamics (QED). Nevertheless, the classical approach provides some insight into the functional dependencies of the radiation on physical parameters such as densities or temperatures.

Figure 3.11   Orion nebula. Photograph courtesy of Robert Gendler.

3.8.1   Bremsstrahlung

Bremsstrahlung[15] is emitted during encounters of charged particles in a hot, ionized plasma. The radiating particle is free, both before and after the deflection. Hence, also the term *free–free radiation* is commonly used in astronomical contexts. As the changes in the energy of the charged particle due to the collision are continuous, *bremsstrahlung* is characterized by a continuous spectrum over a very wide range, that, for typical plasma temperatures, extends into X-rays. By contrast, transitions inside an atom (see Section 3.6) or cyclotron processes (nonrelativistic electron in orbit around a magnetic field line) produce line emission.

Bremsstrahlung is important for several environments, for example, for the emission from the ionized gas clouds around young and hot stars, so-called *HII regions*. An example of such an HII region, the Orion Nebula, is shown in Fig. 3.11. Further examples include supernova remnants (see Chapter 4), or massive clusters of galaxies, where *bremsstrahlung* produces a diffuse X-ray emission with $k_B T \approx 10$ keV.

While cyclotron and synchrotron radiation can be treated to a good approximation in a classical picture, *bremsstrahlung* under most circumstances requires an interpretation in the framework of quantum electrodynamics. Only in the low-frequency limit is a classical description appropriate. Most plasmas, however, can be treated in a semiclassical model, where one starts out from a classical

---

[15] From the German *bremsen*, to brake, and *Strahlung*, radiation, thus "braking radiation."

description and then encapsulates the quantum-mechanical effects into a correction factor, the so-called *Gaunt factor*.

Radiation from individual collisions

Bremsstrahlung arises predominantly from collisions between electrons and ions. Collisions between particles of the same type, or more accurately of the same charge-to-mass ratio, do not produce *bremsstrahlung*, at least not in the dipole approximation that we are using here. The reason is that in this case the second time derivative of the dipole moment has to vanish because of momentum conservation:

$$\ddot{\vec{d}} = \sum_i q_i \ddot{\vec{r}}_i = \sum_i \frac{q_i}{m_i} m_i \ddot{\vec{r}}_i = \frac{q}{m} \sum_i \dot{\vec{p}}_i = 0. \tag{3.139}$$

Because of the large mass ratio between the electron and the ion (keep in mind a proton is 1836 times more massive) it is mainly the electron that is accelerated; the ion can be considered to be immobile.

As we have seen in Chapter 2, strong collisions, where the electron is substantially deflected from its original path, are relatively rare: the bulk of encounters are so-called weak collisions. Consider now an individual electron with velocity v approaching an immobile ion with an impact parameter b, as shown in Fig. 2.2. The duration for the deflection is characterized by the ratio of the typical length and velocity, $\tau = b/v$, and the typical frequency is $\omega_c = 1/\tau$. Little radiation is emitted at frequencies above ω_c.

Because we are interested in the energy emitted per unit frequency, we take the Fourier transform of the dipole moment; see Eq. (3.127). Straight from its definition, Eq. (3.118), we see that the Fourier transform of $\ddot{\vec{d}}$ is just $-\omega^2 d(\omega)$. Moreover, because $\vec{d} = -e\vec{r}$, we find

$$\ddot{\vec{d}} = -e\dot{\vec{v}}, \tag{3.140}$$

so that

$$-\omega^2 d(\omega) = -\frac{e}{2\pi} \int_{-\infty}^{+\infty} \dot{\vec{v}}(t) e^{i\omega t} \, dt. \tag{3.141}$$

The integral can be simplified substantially: for frequencies above ω_c, the exponential oscillates rapidly, which makes the integral small. For frequencies well below ω_c, the exponential is essentially 1, and the integral $\int \dot{\vec{v}}(t) dt = \Delta \vec{v}$ is just the velocity change due to the collision with the ion. In Chapter 2 we found in Eq. (2.13) that $\Delta v = (2Ze^2)/(mbv)$ and thus

$$d(\omega) \sim \begin{cases} \frac{1}{\omega^2} \frac{e}{2\pi} \Delta v & \omega \ll \omega_c \\ 0 & \omega \gg \omega_c \end{cases} . \tag{3.142}$$

Using Eq. (3.127), the energy emission per frequency from a single collision with velocity v and impact parameter b is given by

$$\frac{dW(b)}{d\omega} = \begin{cases} \frac{8Z^2e^6}{3\pi c^3 m^2} \frac{1}{b^2 v^2} & \omega \ll \omega_c \\ 0 & \omega \gg \omega_c \end{cases} . \tag{3.143}$$

This means that the radiation spectrum is essentially flat at a value $\propto 1/b^2 v^2$ up to a cutoff frequency ω_c. This was qualitatively shown in Fig. 3.6. The total energy released in such a collision is $\Delta W \sim \frac{dW}{d\omega} \cdot \omega_c \propto (b^3 v)^{-1}$. This inverse proportionality to the velocity comes from the slow particles being accelerated longer by the ion than the fast particles.

Radiation from a stream of single-speed electrons

Consider a distribution of ions fixed in space with a number density n_{ion} and a stream of incident electrons with a number density n_e and a single velocity v. Focus first on a single electron. In a time dt it will encounter collisions with an impact parameter between b and $b + db$ with all those ions that lie in a cylinder wall of length $v\,dt$ and a thickness db with a volume of $dV = (2\pi b\,db) \cdot (v\,dt)$. The number of ions in this volume is $n_{\text{ion}} \cdot dV$, and so the number of collisions per time of a single electron is $\frac{dN}{dt} = 2\pi b v n_{\text{ion}}\,db$. By multiplying with the electron number density, we get the rate per volume, $\frac{dN}{dt\,dV} = 2\pi b v n_{\text{ion}} n_e\,db$. Finally, the emitted energy per time, frequency, and volume is

$$\frac{dW}{d\omega\,dV\,dt} = 2\pi n_e n_{\text{ion}} v \int_{b_1}^{b_2} \frac{dW(b)}{d\omega} b\,db, \tag{3.144}$$

where finite limits b_1 and b_2 were chosen to avoid the divergence of the integral. By substituting the low-frequency result for $dW/d\omega$, Eq. (3.142), we find

$$\frac{dW}{d\omega\,dV\,dt} = \frac{16e^6 Z^2}{3c^3 m^2} \frac{n_e n_{\text{ion}}}{v} \ln\left(\frac{b_2}{b_1}\right). \tag{3.145}$$

Still, we need to specify the integration limits b_1 and b_2. They are difficult to determine exactly. However, as they occur inside the logarithm, the result is not terribly sensitive to these values. The lower limit could, for example, be determined from quantum mechanics. In the derivation we assumed that the electron is a point particle, but because of the uncertainty principle an electron of momentum $p = mv$ can be localized only to within an uncertainty of $\Delta x \sim h/(mv)$. For impact parameters b smaller than $\sim h/(mv)$, our classical picture breaks down. One possibility is therefore $b_1 = h/(mv)$. For b_2 we can simply take $b_2 = v/\omega$.

Because we have made some crude approximations in our derivation so far, we do not want to elaborate further on the choice of the integration limits. We merely quote the result of a more exact calculation that looks very similar to this result but incorporates quantum-mechanical effects in the so-called *Gaunt factor* g_{ff}

$$\frac{dW(v, \omega)}{d\omega \, dV \, dt} = \frac{16e^6 Z^2 \pi}{3\sqrt{3}c^3 m^2} \frac{n_e n_{ion}}{v} g_{ff}(v, \omega).$$ (3.146)

The $\pi/\sqrt{3}$ occurs because of a convention in the definition of the Gaunt factor. Extensive tables of Gaunt factors exist in the literature.[16]

Radiation from a distribution of electrons: Thermal bremsstrahlung

So far, we have only considered single-speed electrons. However, astrophysical plasmas contain electrons with a distribution of velocities. To obtain the emission properties of a thermal plasma, we have to average the single-speed results over a velocity distribution of the electrons. Assuming a Maxwellian velocity distribution, the probability dP to find an electron in a velocity range d^3v is

$$dP \propto e^{-(E/k_B T)} d^3 v = e^{-(mv^2/2k_B T)} 4\pi v^2 dv,$$ (3.147)

where in rewriting the volume element of the velocity space we have assumed that the distribution is isotropic. To find averages of our single-speed formula, we have to give each velocity a weight of $v^2 e^{-(mv^2/2k_B T)}$. In choosing the integration limits, we have to keep in mind that the energy for the emitted photon comes at the expense of the kinetic energy of the electrons, so no photon can be emitted with an energy in excess of the electron kinetic energy:

$$h\nu \leq \frac{1}{2}mv^2.$$ (3.148)

Hence, the minimum possible velocity to emit a photon of energy $h\nu$ is $v_{min} = \sqrt{(2h\nu/m)}$. This is sometimes called the *photon discreteness effect*. By generalizing Eq. (3.146) to a thermal distribution of electrons, we write

$$\frac{dW(T, \omega)}{dV dt \, d\omega} = \frac{\int_{v_{min}}^{\infty} (dW(v, \omega)/d\omega \, dV \, dt) \, v^2 \exp(-mv^2/2k_B T) \, dv}{\int_0^{\infty} v^2 \exp(-mv^2/2k_B T) \, dv}.$$ (3.149)

By using $d\omega = 2\pi d\nu$, we find for the free–free emission of a thermal plasma (erg s^{-1} cm^{-3} Hz^{-1})

$$\frac{dW}{dV \, dt \, d\nu} = \frac{2^5 \pi e^6}{3mc^3} \left(\frac{2\pi}{3k_B m}\right)^{1/2} Z^2 n_e n_{ion} T^{-1/2} e^{-(h\nu/k_B T)} \bar{g}_\nu^{ff},$$ (3.150)

where $\bar{g}_\nu^{ff}$ is the velocity-averaged Gaunt factor.

---

[16] For example, Brussaard and van de Hulst (1962), Karzas and Latter (1961), and Sutherland (1998).

To find the total power emitted by thermal *bremsstrahlung*, we integrate Eq. (3.150) over frequencies to obtain

$$\frac{dW}{dV\,dt} = \frac{2^5 \pi e^6}{3hmc^3} \left(\frac{2\pi k_B}{3m}\right)^{1/2} Z^2 n_e n_{ion} T^{1/2} \bar{g}^{ff}, \tag{3.151}$$

where $\bar{g}^{ff}$ is now the frequency average over the velocity-averaged Gaunt factor.

Also the inverse process to *bremsstrahlung* can occur: in a plasma of ions and electrons, the electrons can absorb photons. This process is important, for example, in stellar interiors, where the absorption of photons by fully ionized hydrogen and helium is the main source of opacity. If we use Kirchhoff's law, $\alpha_\nu = j_\nu / B_\nu(T)$, Eq. (3.58), we see that the absorption is strongly frequency dependent. For low frequencies, where the Rayleigh–Jeans limit applies, we have $\alpha_\nu \propto \nu^{-2}$. This implies that absorption is most effective for low frequencies.

Summary

Let us summarize the main results:

- *Bremsstrahlung* is produced by charged particle collisions in ionized plasmas. Electrons dominate the emission; emission from collisions between particles of the same type is suppressed.
- The spectrum is rather flat up to a cutoff frequency.
- The total emission (erg s^{-1} cm^{-3}) is proportional to the product of the electron and ion number densities and the square root of the plasma temperature $\epsilon^{ff} \propto n_e n_{ion} T^{1/2}$.
- At very low frequencies, self-absorption occurs with $\alpha_\nu \propto \nu^{-2}$.

3.8.2 *Synchrotron radiation*

Synchrotron emission is one of the most important nonthermal radiation processes in astrophysics. It is produced by highly *relativistic electrons* that spiral around magnetic field lines. To stay on their spiral path, the electrons experience a constant acceleration and thus emit electromagnetic radiation. Synchrotron emission occurs in a large variety of astrophysical environments: in the ionized gas around hot stars (HII regions), in the intergalactic medium, in the remnants of supernovae, near neutron stars like the Crab Pulsar (Chapter 5), in Active Galactic Nuclei (Chapter 8); it is responsible for the radio emission of our own Galaxy and is thought to be the main emission process for the nonthermal component in gamma-ray bursts (Chapter 7).

Typical observational signatures of synchrotron emission are a smooth spectrum over a large range of wavelengths without emission lines. The spectra can be approximated by power laws that usually turn over at low and high frequencies. The emission shows strong linear polarization of up to $\sim 70\%$.

There is no *fundamental* difference between cyclotron and synchrotron radiation: both are produced as charged particles spiral around magnetic field lines. The difference comes from the fact that synchrotron radiation comes from highly relativistic electrons.

To summarize the main differences right from the beginning:

- Cyclotron emission comes from nonrelativistic electrons, that is, $\gamma \approx 1$; synchrotron emission comes from ultrarelativistic electrons.
- The frequency of cyclotron radiation is essentially the gyro or Larmor frequency, $\omega_L = (qB)/(mc)$, of the orbiting electron; the typical frequency of synchrotron radiation is $\gamma^2 \omega_L$ and therefore *much* higher.

Synchrotron power of a single electron

The rigorous derivation of results for synchrotron radiation is rather involved. Therefore, we present only a simplified derivation of the most important results here.

Let us begin by considering the motion of a single electron around a magnetic field line. Previously, we had derived the *nonrelativistic* Larmor formula (see Eq. [3.112]), that gives the emitted radiation power of an accelerated charge. The relativistic version of Larmor's formula[17] is

$$\frac{dW}{dt} = \frac{2q^2}{3c^3}\gamma^4(a_\perp^2 + \gamma^2 a_\parallel^2), \tag{3.152}$$

where $a_\perp$ and $a_\parallel$ are the accelerations perpendicular and parallel to the velocity, respectively. This means that once we know the acceleration of the electron, we can calculate the total emitted power. The relativistic equations of motion for a particle of rest mass m and charge q moving with velocity $\vec{v}$ in a pure magnetic field[18] are

$$\frac{d\vec{p}}{dt} = \frac{d(\gamma m\vec{v})}{dt} = q\frac{\vec{v}}{c} \times \vec{B} \tag{3.153}$$

$$\frac{d(\gamma mc^2)}{dt} = q\vec{v} \cdot \vec{E} = 0. \tag{3.154}$$

The first equation reads like the usual Lorentz force equation but with a relativistically increased particle mass by a factor of γ. The second equation states that the electric field, which is assumed to be zero, does not do any work on the charged

---

[17] For a derivation see, for example, Rybicki and Lightman (1979) or Jackson (1998).
[18] They follow from the covariant expression Eq. (1.105).

particle.[19] This implies that the Lorentz factor and therefore the particle velocity $|\vec{v}|$ remains constant. Thus, the first equation simplifies to

$$\gamma m \frac{d\vec{v}}{dt} = q \frac{\vec{v}}{c} \times \vec{B}. \tag{3.155}$$

We can split up the velocity into components parallel and perpendicular to the magnetic field, $\vec{v}_\parallel$ and $\vec{v}_\perp$. From Eq. (3.155) we see that there is no change in in the parallel component,

$$\frac{d\vec{v}_\parallel}{dt} = 0 \quad \text{and} \quad \frac{d\vec{v}_\perp}{dt} = \frac{q}{\gamma m} \frac{\vec{v}_\perp}{c} \times \vec{B}, \tag{3.156}$$

but as $|\vec{v}| = \text{constant}$ and $\vec{v}_\parallel = \text{constant}$, also $|\vec{v}_\perp|$ must be constant. This implies that the motion of the particle is helical. In other words, the electron spirals around the field line while its original velocity component along the magnetic field direction stays constant. The radius R of this circular motion is given by the balance between relativistic centrifugal force and Lorentz force,

$$\frac{(\gamma m) v_\perp^2}{R} = \frac{q v_\perp B}{c}. \tag{3.157}$$

Consequently, the frequency of rotation is

$$\omega_B = \frac{v_\perp}{R} = \frac{1}{\gamma} \cdot \left(\frac{q B}{mc} \right) = \frac{1}{\gamma} \cdot \omega_L, \tag{3.158}$$

where ω_L is the usual Larmor frequency of a nonrelativistic particle. The relativistic increase of the mass in Eq. (3.157) is responsible for the particle gyrating at a lower frequency. The magnitude of the perpendicular acceleration component is then $a_\perp = v_\perp^2 / R = \omega_B v_\perp$. By inserting this into Eq. (3.152), we obtain the total emitted power:

$$\frac{dW}{dt} = \frac{2q^2}{3c^3} \gamma^4 \frac{q^2 B^2}{\gamma^2 m^2 c^2} v_\perp^2. \tag{3.159}$$

If we introduce $\beta_\perp = v_\perp / c = v \sin \theta / c$, where θ is the angle between the velocity and the magnetic field, the so-called *pitch angle*, and insert the classical electron radius defined in the context of Thomson scattering, Eq. (3.137), we have

$$\frac{dW}{dt} = \frac{2}{3} r_0^2 c \beta_\perp^2 \gamma^2 B^2. \tag{3.160}$$

---

[19] Note that this is not strictly true: the acceleration produces radiation (i.e., an electric field) that can feed back onto the motion of the particle. To conserve energy, a radiating particle obviously has to slow down.

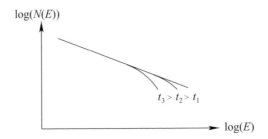

Figure 3.12   High-energy electrons will lose their energy first; therefore the electron distribution will first turn over at the high end.

Encapsulated in $\beta_\perp$, this formula still contains the pitch angle. By averaging over all pitch angles, one finds

$$\langle \beta_\perp^2 \rangle = \frac{1}{4\pi} \int \beta^2 \sin^2 \theta \, d\Omega = \frac{1}{2} \beta^2 \int_0^\pi \sin^3 \theta \, d\theta = \frac{2}{3} \beta^2. \qquad (3.161)$$

If we insert this average, using the Thomson cross section (Eq. [3.138]), and the magnetic energy density $u_B = B^2/8\pi$, the total synchrotron power reads

$$\left(\frac{dW}{dt} \right)_{\text{sync}} = \frac{4}{3} \sigma_T c \beta^2 \gamma^2 u_B, \qquad (3.162)$$

which means that the synchrotron power is $\propto \frac{1}{m^2} \cdot \gamma^2 \cdot B^2$. Because of the mass dependence, synchrotron losses are much more important for electrons than for protons. As the electron energy is $E = \gamma mc^2$, this also means that the emitted synchrotron power is proportional to E^2. One can introduce a typical timescale on which the particles lose their energy, the so-called *synchrotron cooling time*

$$\tau_{\text{sync}} \equiv \frac{E}{(dW/dt)_{\text{sync}}} = \frac{\gamma mc^2}{(dW/dt)_{\text{sync}}} = 635 \, \text{s} \left(\frac{B}{G} \right)^{-2} \left(\frac{E}{\text{erg}} \right)^{-1} \propto \frac{1}{B^2 E}. \qquad (3.163)$$

Consequently, high-energy electrons lose their energy first. If we have a power-law distribution of electrons, as shown in Fig. 3.12, the high-energy end loses its energy first and thus deviates from the original power law.

Spectrum of a single electron

In this section, we give a brief, qualitative discussion of the spectrum expected from a single electron spiraling around a magnetic field line. Because of the relativistic motion, the emission is beamed in the forward direction into a cone of opening angle $1/\gamma$, where γ is the electron Lorentz factor introduced in Chapter 1. An observer sees only emission during the short time interval in which the cone sweeps over the line of sight. Two such pulses are separated by a time interval of $2\pi/\omega_B$. Think

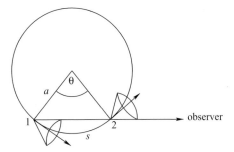

Figure 3.13   Position of the electron when the observer sees radiation from the emission cone for the first (1) and the last time (2).

of an electron moving in a circular orbit of radius a around a field line where the emission cone is always pointing in the direction of the velocity, that is, tangential to the orbit. This situation is sketched in Fig. 3.13. The point where the observer starts to see the leading edge of the emission cone is denoted by 1 in the figure; the point when the cone passes out of the line of sight by 2. The angle passed between 1 and 2 is, for geometric reasons, just twice the cone opening angle

$$\theta = \frac{2}{\gamma}.$$ (3.164)

In the frame of the electron, the time to move from 1 to 2 is

$$\Delta t_{\mathrm{e}} = \frac{s}{v} = \frac{a\theta}{v} = \frac{\theta}{\omega_B},$$ (3.165)

where ω_B is the relativistic gyration frequency introduced in Eq. (3.158) and s is the path length between 1 and 2. Similar to the Doppler effect, the time difference for the observer is shorter than this time by the time it takes the light to travel the distance s, because the electron has nearly caught up with its previously emitted light. The observed pulse duration is therefore

$$\Delta t_{\mathrm{obs}} = \Delta t_{\mathrm{e}} - \frac{s}{c} = s \left(\frac{1}{v} - \frac{1}{c} \right) = \frac{\theta}{\omega_B}(1 - \beta).$$ (3.166)

For velocities close to the speed of light, we can approximate

$$\frac{1}{\gamma^2} = 1 - \beta^2 = (1 + \beta)(1 - \beta) \approx 2(1 - \beta)$$ (3.167)

and

$$\Delta t_{\mathrm{obs}} = \frac{2}{\gamma \omega_B} \frac{1}{2\gamma^2} = \frac{1}{\gamma^3 \omega_B} = \frac{1}{\gamma^2 \omega_L}.$$ (3.168)

Fourier theory shows that if the length of a pulse is Δt, then the typical frequencies contained in it are $\omega \sim (\Delta t)^{-1}$. Therefore, we can define the *characteristic synchrotron frequency*

$$\omega_c = \gamma^2 \omega_L = \frac{\gamma^2 q B}{mc}. \tag{3.169}$$

A more careful analysis[20] of the synchrotron spectrum shows that the emission occurs over a broad range of frequencies[21] with a peak near $0.3\,\omega_c$.

Because the total energy of an electron is given $E = \gamma mc^2$, the characteristic synchrotron frequency is proportional to the square of the electron energy, $\omega_c \propto E^2$, which means that the high-energy electrons radiate at substantially higher frequencies.

Emission from a power-law distribution of electrons

The previous discussion of the spectrum referred to a single electron of a given energy. In reality, electrons have a distribution of energies often described by *power laws*. Let the number of electrons with Lorentz factors between γ and $\gamma + d\gamma$ be

$$dN = N_0 \gamma^{-p} d\gamma, \tag{3.170}$$

where N_0 is a constant and p the power law index of the electron distribution. In typical astrophysical applications, the values of p range from about 0.5 to about 2. Following our result for the total power emitted by a single electron, Eq. (3.162), we use for the power per frequency interval

$$\frac{dW_\nu}{dt}(\gamma) = \frac{4}{3}\sigma_T c \beta^2 \gamma^2 u_B \Phi_\nu(\gamma), \tag{3.171}$$

where the function $\Phi_\nu(\gamma)$ contains the spectral shape of an electron with a Lorentz factor γ. As the spectrum has a strong peak near $\nu_c = \omega_c/2\pi$, we use $\Phi_\nu(\gamma) \approx \delta(\nu - \nu_c)$, so that the integral over the electron distribution reads

$$\frac{dW_\nu}{dt} = \int \frac{dW_\nu}{dt}(\gamma)\, dN(\gamma) = \frac{4}{3} N_0 \sigma_T c u_B \int \gamma^{2-p} \delta(\nu - \nu_c) d\gamma, \tag{3.172}$$

where we have used $\beta^2 \approx 1$. Collecting the constants into C_0 and substituting $\nu' = \gamma^2 \nu_L$, where $\nu_L = \omega_L/2\pi$, we have

$$\frac{dW_\nu}{dt} = \frac{C_0}{2\nu_L} \int \left(\frac{\nu'}{\nu_L}\right)^{-(p-1/2)} \delta(\nu - \nu')d\nu' = \frac{C_0}{2\nu_L}\left(\frac{\nu}{\nu_L}\right)^{-(p-1/2)}, \tag{3.173}$$

---

[20] Ginzburg and Syrovatskii (1965), p. 297.
[21] The spectrum is obtained via a Fourier transform. Because of the narrowness of the emission cone, the signal in the time domain is close to a δ-peak. The Fourier transform of a delta distribution is a constant; therefore one expects a broad range of frequencies for synchrotron emission.

where we have used the property of the delta function $\int f(x')\delta(x - x')\,dx' = f(x)$. This equation states that *a power-law distribution of electrons with index p produces a power law of synchrotron emission with index $(p - 1)/2$.*

Polarization and self-absorption

We briefly mention further results from a more detailed treatment of synchrotron radiation. Synchrotron radiation from a distribution of particles is partially linearly polarized with the degree of *polarization* being quantified by

$$\Pi(\omega) = \frac{(dW/dt)_\perp(\omega) - (dW/dt)_{||}(\omega)}{(dW/dt)_\perp(\omega) + (dW/dt)_{||}(\omega)}, \tag{3.174}$$

where $(dW/dt)_\perp$ and $(dW/dt)_{||}$ are the powers per unit frequency perpendicular and parallel to the projection of the true magnetic field onto the plane of the sky. The polarization from electrons in a uniform magnetic field can be as high as $\sim 70\%$, but it generally depends on the power-law index of the electron distribution p[22]:

$$\Pi = \frac{p + 1}{p + 7/3}. \tag{3.175}$$

The polarization of synchrotron light can thus be used to map out the geometry of cosmic magnetic fields.

Similar to the *bremsstrahlung* case, synchrotron radiation is also subject to *self-absorption* at low frequencies. Imagine a distribution of electrons spiraling around magnetic field lines and emitting synchrotron radiation. Some fraction of emitted photons can be reabsorbed by neighboring electrons, which causes the electrons to become optically thick to their own radiation at low frequencies. A detailed analysis shows that in the optically thick regime the intensity is $\propto \nu^{5/2}$ *independent of the power-law index of the electrons*, whereas in the optically thin regime it is $\propto \nu^{-(p-1/2)}$. This is sketched in Fig. 3.14. The transition frequency between both regions can be used to determine the magnetic field strength.

Summary

The main results are as follows.

- Synchrotron radiation is produced by *relativistic electrons* spiraling around magnetic field lines.
- The emitted power goes as $(dW/dt)_{\text{sync}} \propto (1/m^2)\gamma^2 B^2$, where m is the particle rest mass, γ the Lorentz factor, and B the magnetic field strength.
- Radiation is emitted at a characteristic frequency $\omega_c = \gamma^2 \omega_L$, where ω_L is the Larmor frequency given by $\omega_L = (qB)/(mc)$.

---

[22] For more details, see Rybicki and Lightman (1979).

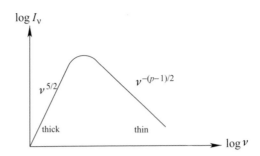

Figure 3.14   Synchrotron spectrum of a power-law distribution of electrons with index p: the optically thick, self-absorbed regime is $\propto \nu^{5/2}$, independent of p; the optically thin part of the spectrum is $\propto \nu^{-(p-1)/2}$.

- A distribution of electrons whose energies are distributed according to a power law with index p will produce a synchrotron power that is also a power law, but with an index $(p-1)/2$.
- For uniform magnetic fields, the emitted radiation is highly linearly polarized.
- At low frequencies, the electrons are optically thick to their own radiation, and so-called synchrotron self-absorption occurs.

3.8.3   Compton scattering and Comptonization

So far, we have encountered the scattering between a photon and an electron twice: once as Thomson and once as Compton scattering (see Chapter 1). The assumptions made for Thomson scattering were that the electron has little or no kinetic energy and that the incident electromagnetic wave has very little power. Under these circumstances, the electron recoil can be neglected and the photon energy is unchanged. In the general case, two modifications have to be made: first, we have to account correctly for the conservation of energy and momentum as we did in our Compton scattering example in Chapter 1. Second, the scattering cross section changes because of the quantum-electrodynamical effects.

The QED result, the so-called *Klein–Nishina cross section*,[23] is given by

$$\sigma_{\mathrm{KN}} = \frac{3}{8}\sigma_{\mathrm{T}}\frac{1}{x}\left\{\left[1 - \frac{2(x+1)}{x^2}\right]\ln(2x+1) + \frac{1}{2} + \frac{4}{x} - \frac{1}{2(2x+1)^2}\right\}, \quad (3.176)$$

where x is the energy of the incident photon measured in units of the electron rest mass energy, $x = h\nu/m_e c^2$. For very low energies, $x \ll 1$, this expression reduces, as it should, to the Thomson cross section, $\sigma_{\mathrm{KN}} \approx \sigma_{\mathrm{T}}(1 - 2x) \approx \sigma_{\mathrm{T}}$, and for very large energies, $x \gg 1$, it can be approximated by $\sigma_{\mathrm{KN}} \approx (3/8)(\sigma_{\mathrm{T}}/x)[\ln(2x + 1/2)]$. The main result from the QED corrections is that the *cross section decreases*

---

[23] The formula was derived in 1929 by the physicists Oskar Klein (1894–1977) from Sweden and Yoshio Nishina (1890–1951) from Japan.

below the Thomson value for high photon energies, roughly like $1/x$. This means that high-energy photons have a substantially longer mean free path than low-energy photons.

From the conservation of four-momentum, we had found in Chapter 1 that the change in photon wavelength for Compton scattering is given by

$$\Delta\lambda = \lambda' - \lambda = \frac{h}{m_e c}(1 - \cos\theta); \tag{3.177}$$

see Eq. (1.80). This can be translated into a relation between the photon energies before and after the collision, $E_{\gamma,\mathrm{in}}$ and $E_{\gamma,\mathrm{fin}}$:

$$E_{\gamma,\mathrm{fin}} = \frac{E_{\gamma,\mathrm{in}}}{1 + (E_{\gamma,\mathrm{in}}/m_e c^2)\cdot(1 - \cos\theta)}. \tag{3.178}$$

For very small initial photon energies, $(E_{\gamma,\mathrm{in}}/m_e c^2) \ll 1$, we recover the results that the photon energy is not changed and that the scattering is elastic. For larger photon energies, however, the electron recoil becomes important.

If $(E_{\gamma,\mathrm{in}}/m_e c^2) \ll 1$ Eq. (3.178) can be expanded in a Taylor series like $1/(1 + x) \approx 1 - x$, the fractional change in the photon energy is

$$\frac{\Delta E_\gamma}{E_{\gamma,\mathrm{in}}} = \frac{E_{\gamma,\mathrm{fin}} - E_{\gamma,\mathrm{in}}}{E_{\gamma,\mathrm{in}}} = -\frac{E_{\gamma,\mathrm{in}}}{m_e c^2}(1 - \cos\theta). \tag{3.179}$$

Averaging over the scattering angle θ, one finds that the average energy lost by the photon is

$$\langle\Delta E_\gamma\rangle = -\left(\frac{E_{\gamma,\mathrm{in}}}{m_e c^2}\right)\cdot E_{\gamma,\mathrm{in}}. \tag{3.180}$$

Now consider a plasma made of photons and electrons: in an encounter between a high-energy photon and a low-energy electron, the electron receives energy. Conversely, a high-energy electron can also up-scatter a low-energy photon to high energies. If the electron and the photon temperatures are the same, there is no net transfer of energy, but for different temperatures one component can transfer energy to the other.

As long as the photon energies are small enough that the Thomson cross section is valid, the rate of energy loss by a single electron to the photon field is given by

$$\left(\frac{dW}{dt}\right)_{\mathrm{comp}} = \frac{4}{3}\sigma_T c\beta^2\gamma^2 u_{\mathrm{rad}}. \tag{3.181}$$

Note that this is nearly exactly the same formula as for the synchrotron emission of a single electron, given by Eq. (3.162). The only difference is that the magnetic energy density u_B has been replaced by the radiation energy density u_{rad}. Thus, in a plasma embedded in both a magnetic and a radiation field, the ratio between their energy densities decides which process dominates:

$$\frac{(dW/dt)_{\text{sync}}}{(dW/dt)_{\text{comp}}} = \frac{u_B}{u_{\text{rad}}}. \tag{3.182}$$

As in the case of synchrotron radiation, one can define a *cooling time*:

$$\tau_{\text{comp}} \equiv \frac{E}{(dW/dt)_{\text{comp}}} = \frac{\gamma mc^2}{(dW/dt)_{\text{comp}}} \propto \frac{1}{\beta^2\gamma}. \tag{3.183}$$

The mean number of photons scattered per time by an electron is then given by $R_s = \sigma_T c n_{\text{rad}} = \sigma_T c u_{\text{rad}}/\bar{E}_\gamma$. Here n_{rad} is the photon number density, and $\bar{E}_\gamma$ is the average photon energy. Therefore, the mean energy gained by the photon in one collision is

$$\langle \Delta E_\gamma \rangle = \frac{(dW/dt)_{\text{comp}}}{R_s} = \frac{4}{3}\beta^2\gamma^2\bar{E}_\gamma. \tag{3.184}$$

This means that on average $\Delta E_\gamma/E_\gamma \propto \gamma^2$; that is, if relativistic electrons are involved, inverse Compton scattering can easily produce highly energetic photons. If the electrons have a Lorentz factor of 10^3, photons can be up-scattered from the radio into the ultraviolet or from the optical to the γ-ray regime.

If the electron velocities are small, $v \ll c$ and $(1/2)mv^2 \approx (3/2)k_B T_e$, where T_e is the temperature of the electrons, the typical energy gain per collision is

$$\left\langle \frac{\Delta E_\gamma}{E_\gamma} \right\rangle \sim \frac{4k_B T_e}{mc^2}, \tag{3.185}$$

or, combined with the loss term, Eq. (3.180)

$$\left\langle \frac{\Delta E_\gamma}{E_\gamma} \right\rangle \sim \frac{4k_B T_e - \bar{E}_\gamma}{mc^2}. \tag{3.186}$$

So if the average photon energy is smaller than $\sim 4k_B T_e$, net energy is transferred to the photon field (inverse Compton effect); otherwise, energy is transferred to the electrons (Compton effect).

Sunyaev–Zel'dovich effect

Inverse Compton scattering has an interesting application in the context of galaxy clusters. Galaxies are only rarely isolated in space. Most of them are members of groups ranging from a few up to several thousands of galaxies. In between the galaxies, galaxy clusters contain a very hot ($\sim 10^8$ K) and dilute gas. If photons from the cosmic microwave background, which has a spectrum very close to a perfect blackbody spectrum of a temperature 2.7 K, cross a galaxy cluster, the

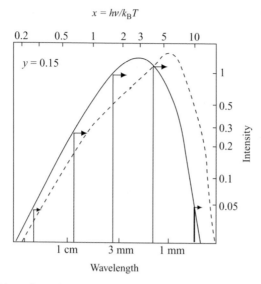

$$x = h\nu/k_B T$$

Figure 3.15   Distortion of the cosmic microwave blackbody spectrum via Compton scattering. Figure after Sunyaev and Zeldovich 1980. Reprinted with permission from the Annual Review of Astronomy and Astrophysics, Vol. 18. ©1980 by Annual Reviews. www.annualreviews.org.

photons are Compton up-scattered by the hot cluster gas electrons. This leads to a decrement of the microwave flux from the direction of the galaxy cluster. The scattering does not create or destroy photons; instead it transfers electrons from the Rayleigh–Jeans part to the Wien portion of the spectrum. This *Sunyaev–Zel'dovich effect*[24] was predicted in 1972 as a way to prove the cosmic, rather than galactic, origin of the observed microwave background.

3.9   Exercises

3.1 Photon container

A container is at a temperature of 10^6 K. What is the energy flux of photons exiting a small hole? At what frequency is the flux the greatest? What is the energy density of photons in the container? What is the approximate number density of photons in the container? What is the photon pressure on the walls of the container?

3.2 Star light

The light from a faint star has an energy flux of 10^{-7} erg cm^{-2} s^{-1}. Assuming that the light has a wavelength of approximately 5×10^{-5} cm, estimate the number of photons from this star that enter a human eye in one second.

---

[24] This effect has been predicted in 1972 by the Russian physicists Rashid Sunyaev and Yakov Zel'dovich.

3.3 Surface of uniform brightness: Relation between brightness and flux

Calculate the flux at a distance r from a sphere of uniform brightness and radius R. Show that if $r = R$, the (frequency-integrated) flux is just π times the (frequency-integrated) brightness.

3.4 Monotonicity of the Planck law with temperature

Show that Planck spectra of higher temperature always lie entirely above the lower temperature curves.

3.5 Cosmic microwave background

The cosmos is pervaded by a 3-K radiation field, which is regarded as the "echo" of the Big Bang. This radiation field is called the Cosmic Microwave Background.
(i) Calculate the wavelength of this radiation.
(ii) What is the radiation pressure due to this field?

3.6 Energy density blackbody radiation

The spatial energy density per unit frequency interval of blackbody radiation is given by

$$u_\nu = \frac{8\pi h}{c^3} \frac{\nu^3}{e^{h\nu/k_{\mathrm{B}}T} - 1},\tag{3.187}$$

where ν denotes frequency, T denotes temperature, and the constants have their usual meaning.
(i) Write down an expression for the total energy of a photon gas of volume V.
(ii) By approximating the denominator in Eq. (3.187) by $e^{h\nu/k_{\mathrm{B}}T} - 1 \approx e^{h\nu/k_{\mathrm{B}}T}$, evaluate the expression for the total energy found in part (i).
(iii) From the formula found in part (ii), derive an expression for the radiative flux emitted from a black body.

3.7 Retarded potentials I

Insert the missing steps leading to Eq. (3.89).

3.8 Retarded potentials II

Confirm that the Lorentz gauge does not change the physical fields $\vec{E}$ and $\vec{B}$.

3.9 Fourier transform

Start from Eq. (3.124), and use the definition of the Fourier transform to confirm Eq. (3.125).

3.10 Faraday rotation

Consider light that propagates parallel to the magnetic field lines through a uniformly magnetized plasma, with $\vec{B} = B_0\hat{z}$. An electron in the plasma feels the forces from the electromagnetic wave and from the external magnetic field. The equation of motion of the electron is given by

$$\vec{F} = -e\vec{E} - \frac{e}{c}\vec{v} \times \vec{B},\tag{3.188}$$

where the electric field $\vec{E}$ can be decomposed into right-circularly polarized (RCP) and left-circularly polarized (LCP) components:

$$\vec{E} = E_0(\hat{x} \mp i\hat{y})\exp[i(k \mp z - \omega t)]. \tag{3.189}$$

The upper $(-)$ sign corresponds to RCP and the lower $(+)$ sign to LCP waves. In this equation of motion, we have neglected the wave's B-field as it is smaller by v/c from the wave's E-field.

Prove that the equation of motion reads

$$\vec{v} = \frac{-ie\vec{E}}{m(\omega \pm \omega_L)}, \tag{3.190}$$

where $\omega_L = eB/m_e c$.

3.11 Synchrotron radiation

Show that the energy of a single, relativistic electron that is losing energy solely through synchrotron radiation is changing with time according to (assuming $\sin \alpha = 1$, where α is the pitch angle)

$$E(t) = \frac{E_0}{1 + t/\tau}, \tag{3.191}$$

where τ is the synchrotron loss time and E_0 is the energy at $t = 0$.

3.12 Crab Nebula

One astronomical object where synchrotron radiation plays an important role is a supernova remnant known as the Crab Nebula. Take the magnetic field strength in some of the Crab's filaments to be of order 10^{-4} G, and show that a classically moving electron would radiate at a frequency of about 300 Hz, independent of the energy. If the electron energy is 10^9 eV, show that the peak radiation will occur at about 600 MHz. For electron energies of 10^{12} eV, the radiation peaks in the visible part of the spectrum at $6 \cdot 10^{14}$ Hz.

3.10 Further reading

Brussaard, P. J. and van de Hulst, H. C. (1962). *Reviews of Modern Physics*, **34**, 507.

Blumenthal, G. R. and Gould, R. J (1970). Bremsstrahlung, synchrotron radiation, and Compton scattering of high-energy electrons traversing dilute gases. *Reviews of Modern Physics*, **42**, 237–71.

Ginzburg, V. L. and Syrovatskii, S. I. (1965). Cosmic magnetobremsstrahlung (synchrotron radiation). *Annual Review of Astronomy and Astrophysics*, **3**, 297.

Jackson, J. D. (1998). *Classical Electrodynamics*, 3rd edn. New York: Walter de Gruyter.

Karzas, W. and Latter, R. (1961). Electron radiative transitions in a Coulomb field. *Astrophysical Journal Supplement*, **6**, 167.

Longair, M. (1992). *High Energy Astrophysics*. Cambridge: Cambridge University Press.

Padmanabhan, T. (2000). *Theoretical Astrophysics I*. Cambridge: Cambridge University Press.

114

3   *Radiation processes*

Rybicki, G. B. and Lightman, A. P. (1979). *Radiative Processes in Astrophysics*. New York: Wiley.

Sunyaev, R. A. and Zel'dovich, Y. B. (1980). Microwave background radiation as a probe of the contemporary structure and history of the universe. *Annual Review of Astronomy and Astrophysics*, **18**, 537.

Sutherland, R. S. (1998). Accurate free–free Gaunt factors for astrophysical plasmas. *Monthly Notices of the Royal Astronomical Society*, **300**, 321.

Web resources

A database on atomic transition probabilities is available at http://physics.nist.gov/.

4

Supernovae

4.1 Observational overview and classification of supernovae

4.1.1 What is a supernova?

Supernovae are catastrophic explosions of stars. The name stems from the Latin *nova* for *new* because, with the explosion, an apparently new star appears on the night sky. Supernovae emit a substantial fraction of the luminosity of a typical galaxy as can be seen in the Hubble Space Telescope image shown in Fig. 4.1. Supernovae are named *SN*, followed by the year in which they are discovered and a letter that indicates alphabetically the order in which they were discovered in that year. The supernova shown in Fig. 4.1, for example, was discovered in the year 1994 and is called SN 1994D. Within a few weeks, a supernova can produce as much light as its progenitor star during its entire life. The brightest supernovae have been detected a third of the way across the observable universe and a particular class of supernova (Type Ia – see below) has become pivotal in measuring large distances in the cosmos. Observations of Type Ia supernovae have provided evidence that, at the present time, the universe appears to be accelerating, which has profound implications for cosmology and physics in general. We return to this point later.

Supernovae in our own Galaxy are rare by the standards of a human lifetime, with the last six dating back to the years 1006, 1054, 1181, 1572, 1604, and 1680 all before the invention of the telescope. At the time of writing, three supernovae have been detected where the progenitor star is known. One was observed in 1987 in our neighboring galaxy, the Large Magellanic Cloud, and is called SN 1987A; another one, SN 1993J, in the galaxy M81; and the third one, SN 2005CS, in M51.[1] Compared to cosmic timescales, however, supernovae occur rather

---

[1] Galaxies are often named after their number in the so-called Messier catalogue. The French astronomer Charles Messier (1730–1817) cataloged all kinds of astronomical objects, among them are galaxies, planetary nebulae, and globular clusters.

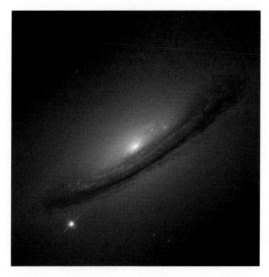

Figure 4.1   *Hubble Space Telescope* picture of the disk galaxy NGC 4526, where supernova SN 1994D, visible as the bright spot on the lower left, occurred. Photograph courtesy of High-Z Supernova Search Team, *Hubble Space Telescope*.

frequently. There is approximately one supernova exploding every second in the universe. In our Galaxy, supernovae occur every 30–50 years, although many of them are obscured by the galactic gas and dust. Type Ia supernovae are rarer and occur only about once every 300 years in the Galaxy. Supernova search teams can find dozens of supernovae, mostly of the brighter Ia type, per night.

Supernovae play a fundamental role in the history of the Universe because they produce the bulk of the heavy elements[2] and disperse them throughout the interstellar medium. These elements are essential for the formation of planets and the evolution of life. During the existence of the Milky Way, about 10^8 supernovae have exploded. They have produced most of the oxygen we breathe, the calcium contained in our bones, and the silicon that forms the rocks under our feet.

Moreover, supernovae are closely related to many other astrophysical topics. Each supernova pumps an energy of about 10^{51} erg into its host galaxy, each time blowing a hole of several hundred parsec radius into the gas of the galaxy. This feedback is important for the evolution of galaxies and is thought to trigger the collapse of molecular clouds, which, in turn, form new stars. Supernovae also produce a good fraction of the highly relativistic particles, so-called cosmic rays, that continuously bombard the Earth's atmosphere. Finally, a very rare type of core-collapse supernova is probably related to gamma-ray bursts (see Chapter 7).

---

[2] For astronomers, all elements heavier than hydrogen and helium are "heavy." These elements are also called "metals" in astrophysics (unlike in chemistry).

4.1.2 Classification of supernovae

Traditionally, the fundamental classification scheme for supernovae is based on their spectra near maximum light and was devised in 1940. This purely phenomenological scheme persists until today and distinguishes, on the coarsest level, between two types of supernovae: Type I and Type II.

The essential distinguishing feature between the two major types of supernovae is the occurrence of hydrogen lines in their spectra. Although Type II supernovae show evidence of hydrogen in their spectra, Type I supernovae do not. The Type II supernovae can be classified further by the shape of their light curves. The luminosities of supernovae range from 10^{42} erg s^{-1}, which corresponds to $3 \cdot 10^8$ $L_\odot$, where $L_\odot = 4 \cdot 10^{33}$ erg s^{-1} is the luminosity of the Sun, for the faint Type II supernovae, to $2 \cdot 10^{43}$ erg s^{-1}, about $6 \cdot 10^9$ $L_\odot$, for the Type Ia supernovae.

The Type I supernovae are categorized according to whether their spectra show certain silicon lines, in which case they are classified as Type Ia supernovae, or not. Among those that do not, some show helium lines and are classified as Type Ib. Those that show no helium lines are called Type Ic. For a sample of supernova spectra, see Fig. 4.2.

Moreover, it has become apparent over the past years, as hundreds of supernovae have been detected, that all supernovae display some peculiarity if observed closely enough. As a result, the classification scheme has become increasingly complex. Nonetheless, we describe the main features of the most important types of supernovae.

Supernovae Type Ia

The spectral properties, absolute magnitudes, and light curve shapes of the majority of supernovae of Type Ia are remarkably homogeneous and show only subtle spectroscopic and photometric differences. In addition to silicon lines, their optical spectra at maximum light contain lines of calcium, magnesium, sulfur, and oxygen. These elements are called intermediate-mass elements. Their most stable isotopes are all particularly strongly bound nuclei, which are multiples of ^{4}He nuclei. The prevailing theory is that Type Ia supernovae are caused by thermonuclear explosions of carbon-oxygen white dwarfs that accrete matter from a binary companion. We explain the reasoning for this in Section 4.3.

Supernovae Type Ib/c

Type Ib supernovae have no evident Balmer hydrogen lines,[3] weak or absent Si II lines, and strong helium lines. At times longer than 100 days, a strong emission line of O I is seen at a wavelength of 630 nm. All Type Ib SNe have been discovered in

---

[3] After Johann Jakob Balmer (1825–1898), Swiss mathematician and physicist.

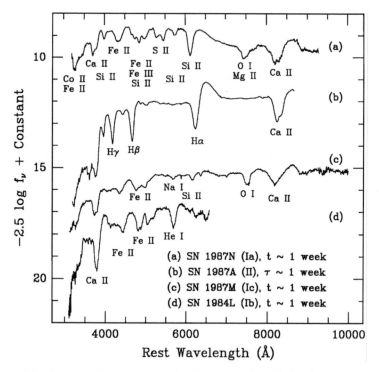

Figure 4.2   Spectra of supernovae, showing early-time distinctions between the four major types. The parent galaxies and their redshifts (kilometers per second) are as follows: SN 1987N (NGC 7606; 2171), SN 1987A (LMC; 291), SN 1987M (NGC 2715; 1339), and SN 1984L (NGC 991; 1532). The variables t and τ represent time after observed B-band maximum and time after core collapse, respectively. Courtesy of Alex Filippenko. Reprinted, with permission, from the *Annual Review of Astronomy and Astrophysics*, Vol. 35. ©1997 by Annual Reviews. www.annualreviews.org.

spiral galaxies, in or near H II regions, and are therefore related to the death of a massive star. Such stellar deaths are not likely to be observed in elliptical galaxies because in elliptical galaxies no new stars are formed and the short-lived massive stars have already died. Type Ic supernovae show no or weak hydrogen, helium, and Si II lines.

Type Ib/c supernovae are believed to be caused by the explosion of evolved helium cores of massive stars that have lost their hydrogen envelope via stellar winds or mass exchange in a binary system (to explain the lack of hydrogen lines). The amount of ^{56}Ni synthesized in the explosion of the helium core of a massive star is much smaller and the light curve is much fainter than in Type Ia supernovae. It remains unclear why Si II features are suppressed at peak light. A particular subset of Type Ib/c supernovae seems to be associated with a gamma-ray burst (see Chapter 7).

Supernovae Type II

The light curves of Type II supernovae are quite diverse, and their spectra are dominated by Balmer lines. A Type II supernova is thought to be the result of the collapse of a massive star ($M > 8\ M_\odot$). In such a collapse, protons capture electrons forming neutrons and emitting neutrinos. These neutrinos carry away the largest share of the energy, giving rise to a neutrino burst, which is absent in the Type Ia supernovae. Type II supernovae often leave a remnant behind, such as a neutron star or a black hole of a few solar masses, whereas the Type Ia supernovae are thought to lead to a complete disruption of the star.

We point out that this phenomenological classification of supernovae does not reflect their different physical mechanisms. The classification into Type I and Type II supernovae does *not* imply that the former are formed by thermonuclear explosions and the latter by core collapses. Types Ib and Ic are also core-collapse supernovae.

4.1.3 Light curves

Light curves show the temporal evolution of the light emitted by supernovae and contain information about the supernova, its explosion mechanism, and its progenitor.

We begin by discussing the light curves of Type Ia supernovae because their uniformity has made them extremely valuable for modern astrophysics. Supernovae of Type Ia rise to maximum light (absolute magnitude $M \sim -19.30$) in approximately 20 days, which is followed by a rapid decline at about 0.065 mag/day (for more on magnitudes, see Appendix D). After about 50 days, the decline of the light curves slows to ~ 0.01 mag/day, as illustrated in Fig. 4.3. The source of this prolonged high luminosity of the supernova has long been a mystery. The thermal radiation from the ejected matter cannot sustain this luminosity because the adiabatic expansion of the ejecta would render it invisible within about an hour. Today we know that the sustained luminosity is powered by the radioactive decay of nickel that is copiously produced in supernova explosions. The first phase of the light curve is powered by the inverse-β decay of ^{56}Ni:

$$^{56}\text{Ni} + e^- \rightarrow\ ^{56}\text{Co}^* + \nu_e$$

with a half-life of $\tau = 6.1$ days. The product ^{56}Co* is an excited state of Cobalt that decays down to its ground state by emitting γ-rays that eventually transfer their energy to the stellar plasma. This reaction releases an energy of $3.0 \cdot 10^{16}$ erg g^{-1}. The produced isotope ^{56}Co again is unstable and decays via another electron capture to iron

$$^{56}\text{Co} + e^- \rightarrow\ ^{56}\text{Fe} + \nu_e$$

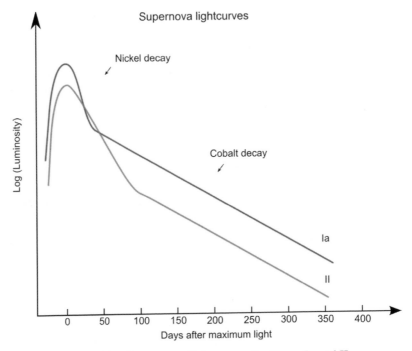

Figure 4.3　Supernovae light curves for Types Ia and II.

with a half-life of $\tau = 77.7$ days. This decay releases an energy of $6.4 \cdot 10^{16}$ erg g^{-1} and is primarily responsible for the slow decline after the peak of the light curve (see Fig. 4.3). According to the laws of radioactive decay, the number of cobalt atoms varies with time as

$$N(t) = N_0 \exp(-\lambda t), \tag{4.1}$$

where the decay constant is related to the half-life τ via $\lambda = \ln 2/\tau$. The rate of energy deposition by the decaying nuclei is proportional to the rate of their decay (i.e., dN/dt). This leads to a variation in luminosity according to

$$\frac{d \log_{10} L}{dt} = \frac{d \log_{10} N}{dt} = -0.434\lambda, \tag{4.2}$$

where the factor 0.434 comes from taking the log to basis 10 of Eq. (4.1). This can be converted into a rate of change of the bolometric magnitude, that is, $dM_{bol}/dt = 1.086\lambda$ (for more on magnitudes, see Appendix D). For a half-life of 77.7 days, this yields a slope of 0.01 mag/day, in good agreement with the observations. From the total bolometric luminosity, one can also work out the original mass of ^{56}Ni that has been produced in the supernova explosion. A total radiated energy of $1.1 \cdot 10^{50}$ erg (about 20 times the rest-mass energy of the Earth!) requires a mass of $\sim 0.6\,M_\odot$ of radioactive nickel.

Apart from providing a good explanation for the shape of the Type Ia light curves, this model also explains the relatively high abundance of ^{56}Fe in the universe. In passing, we should also mention that the shape of the light curve varies with wavelength in a fashion such that at later times Type Ia supernovae become bluer.

Type Ia supernovae can be used as standardized light sources because an empirical law relates their intrinsic luminosities and their dimming rate. The dimming rate is characterized by the decrement in magnitude in a certain frequency band (of the the the so-called B-band) 15 days after its maximum, and this decrement is denoted by $\Delta m_{15}(B)$. This so-called *Phillips relation* shows that Type Ia supernovae with smaller values of $\Delta m_{15}(B)$ are intrinsically more luminous than those with larger values of $\Delta m_{15}(B)$. The empirical relation states that the maximum B-band magnitude is given by

$$M_{\max}(B) = -21.726 + 2.698\Delta m_{15}(B). \tag{4.3}$$

This can be used to infer the absolute magnitude of supernovae. Once $\Delta m_{15}(B)$ is measured, one can infer the absolute magnitude, $M_{\max}(B)$, from the Phillips relation. Comparing this inferred absolute magnitude with the measured apparent magnitude at maximum light, we can infer the distance (for more on magnitudes, see Appendix D). Thus, in combination with the apparent magnitude, the distance to the supernova can be determined. Because supernovae are so bright, they can serve as distance indicators out to very large distances where other distance indicators cannot be used.

While the light curves of Type Ia supernovae display a staggering homogeneity, the light curves of Type II supernovae are much more varied (see Fig. 4.3). They can be coarsely divided into Type II-L ("linear"), which show a behavior similar to the Type Ia supernovae, and the Type II-P ("plateau"), which stay within about one magnitude of maximum brightness for a period of many days before they fade more or less as the Type II-L. This nearly constant luminosity is thought to arise from supernovae that expand at such a rate that the increase in radius (the amount of light emitted is proportional to the surface area and thus to the square of the radius) compensates for the decreasing surface temperature (the amount of light emitted is proportional to the fourth power of the surface temperature).

As always, there are some observed Type II supernovae that display a strange behavior. For example, SN 1993J started to dim after it reached its peak brightness but then brightened again before it faded away.

The initial phase of light curves of Type II supernovae can be explained by the instantaneous release of a vast amount of energy into the envelope of their

Table 4.1   *Features of Type Ia and Type II supernovae*

Type Ia	Type II
no hydrogen in spectrum	hydrogen in spectrum
thermonuclear explosion of WD	core collapse of $M > 8\,M_\odot$ star
probably no bound remnant	iron core collapses to NS or BH
kinetic energy: $\sim 10^{51}$ erg	kinetic energy: $\sim 10^{51}$ erg
$v_{ejecta} \sim 5\,000\text{--}30\,000\,\mathrm{km\ s^{-1}}$	$v_{ejecta} \sim 2\,000\text{--}30\,000\,\mathrm{km\ s^{-1}}$
probably no neutrino burst	neutrino burst $\sim 3 \cdot 10^{53}$ erg
$E_{optical} \sim 10^{49}$ erg	$E_{optical} \sim 10^{49}$ erg
$L_{peak} \sim 10^{43}\,\mathrm{erg\ s^{-1}}$ for ~ 2 weeks	$L_{peak} \sim 3 \cdot 10^{42}\,\mathrm{erg\ s^{-1}}$ for ~ 3 months
radioactive tail from decay of ^{56}Co	radioactive tail from decay of ^{56}Co
event rate $\sim 1/300$ yr in Galaxy	event rate $\sim 2/100$ yr in Galaxy
produces $\sim 2/3$ of iron in Galaxy	produces $\sim 1/3$ of iron and all the oxygen
occurs in spiral and ellipticals	occurs mainly in spiral galaxies

WD stands for white dwarf, NS for neutron star, and BH for black hole.

massive progenitor star, and this phase typically lasts for some hours to a couple of days. The peak magnitude of supernovae II-P show substantial variation, whereas most SNe II-L have a nearly uniform peak absolute magnitude that is about 2.5 magnitudes less than that of Type Ia supernovae. After some initial cooling, the light curves of supernovae with large progenitors exhibit a plateau, whereas those with smaller progenitors continue to decline until the supernova brightens again to reach a plateau. After about 150 days, the light curves of most Type II supernovae resemble each other: their luminosity declines and is now again solely powered by the radioactive decay of ^{56}Ni. Typically, Type II supernovae only produce about a tenth of the amount of nickel of a Type Ia supernova.

Finally, the main distinguishing features of Type Ia and Type II supernovae are summarized in Table 4.1.

4.2   Type II supernovae

As early as 1934, just after the discovery of the neutron, Baade and Zwicky suggested that a star made entirely of densely packed neutrons might be formed in a supernova explosion. According to their estimates, the collapse of a massive star to form such a compact object would release more than enough energy to power a supernova. This basic picture has been confirmed over the years, most convincingly by several observations of neutron stars located in the interiors of supernova remnants.

A typical star of $M \sim 10\,M_\odot$ that will end its life in a core-collapse supernova has a life span of less than 10^7 years. This is very short compared to the evolutionary timescale of a galaxy; therefore core-collapse supernovae occur close to the regions in which the stars were born (i.e., near the star-forming regions in galactic spiral arms).

4.2.1   *The star prior to collapse*

In the initial burning stage of a star, the hydrogen in its core is burnt into helium. This longest-lasting burning stage is called the *main sequence phase*. Our Sun is a typical example of a star in this phase. Once the central hydrogen is exhausted after a few billion years, the star begins to contract and heats up until the hydrogen in a shell around the central helium core ignites. This is the point when the star moves off the main sequence and develops into a *red giant* star. Once the hydrogen in the surrounding shell is used up, the star contracts and heats up again. For stars with masses of more than $0.5\ M_\odot$, the temperature rises to $\sim 10^8$ K to ignite the helium in its core. Helium burning proceeds via the *triple α reaction* in which three α particles fuse, mainly producing carbon and oxygen. The main products of this helium burning are carbon and oxygen. Rarer products are neon, magnesium, and silicon.

Low-mass stars do not proceed beyond this point; their self-gravity is not strong enough to lead to a large enough contraction that could ignite the carbon-oxygen core. Instead, such stars get rid of their envelopes via thermal pulses that are caused by the enormous temperature dependence of the helium burning reactions. The ejected envelopes are observable as (often beautifully colored) *planetary nebulae*. The remains of such stars, their dense inner cores, are called *white dwarfs* (see Box: White dwarfs).

White dwarfs

White dwarfs are the remains of stars with masses of less than about $8\ M_\odot$ (see Fig. 4.5). As in the even more compact neutron stars, the pressure in their interior is provided by a quantum-mechanical effect, the so-called *degeneracy pressure*. It is closely related to the fact that the pressure-providing particles, the electrons in the white dwarf material, are fermions and therefore obey the Pauli exclusion principle. The physics of such Fermi gases is discussed more rigorously in Section 5.6.1 in the context of neutron stars. Here we will give a simple order-of-magnitude estimate that already yields some important relations.

The quantum-mechanical uncertainty principle states that the momentum and position of a particle cannot be measured exactly at the same time. Instead the product of the uncertainties in momentum and position is

$$\Delta p \Delta x \geq \hbar. \qquad (4.4)$$

When the density becomes very high, the electrons are confined to a smaller space, and therefore the position uncertainty, Δx, decreases. To fulfill Eq. (4.4), the

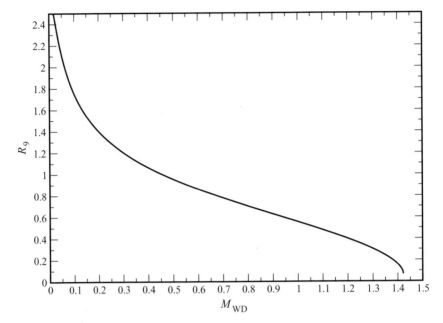

Figure 4.4   Mass–radius relationship for a white dwarf. Masses are given in solar units; R_9 is the radius in units of 10^9 cm.

momentum uncertainty has to increase. The corresponding high momenta are responsible for extra pressure. This pressure is not related to temperature and therefore is even present in very cold stars.

Consider a box of electrons with number density n_e. Half of them are moving with speed v_x in the positive x-direction, half of them are moving in the negative x-direction. The number of electrons hitting the wall per second per unit surface area is $\frac{1}{2} n v_x$. The momentum transferred per second per unit area to the wall is $2 n v_x p_x$, where p_x is the momentum of an electron in the x-direction. The momentum per second per unit area is the pressure, that is,

$$P \sim n_e v_x p_x = \frac{n_e}{m_e} p_x^2, \tag{4.5}$$

where we have used the relation $p_x = m v_x$ for nonrelativistic particles. If the number density of electrons is n_e, then each electron occupies on average a volume $\sim 1/n_e$, and the average distance between electrons is $\Delta x \sim n_e^{-1/3}$. Assuming that the average momentum is of the order of the uncertainty in momentum, then

$$p_x \sim \frac{\hbar}{\Delta x} \sim \hbar n_e^{1/3}. \tag{4.6}$$

Inserting this into Eq. (4.5) yields

$$P \sim \frac{\hbar^2 n_e^{5/3}}{m_e}, \tag{4.7}$$

which differs from only the more careful calculation, in Section 5.6.1, by a factor of about two. As each ion carries Z charges and the overall plasma is assumed to be neutral, the ion number density is

$$n_i = \frac{1}{Z} n_e. \tag{4.8}$$

We denote the ion mass by $A m_p$,[4] and the mass density is given by

$$\rho = \rho_i + \rho_e = A m_p n_i + n_e m_e = \left(\frac{A}{Z} m_p + m_e \right) n_e \approx \frac{A}{Z} m_p n_e \tag{4.9}$$

or

$$n_e \approx \frac{Z}{A} \frac{\rho}{m_p}. \tag{4.10}$$

By inserting this back into Eq. (4.7), the pressure can be written as

$$P \sim \frac{\hbar^2}{m_e} \left(\frac{Z}{A} \right)^{5/3} \left(\frac{\rho}{m_p} \right)^{5/3}. \tag{4.11}$$

As remarked previously, the pressure depends on only density and not on temperature.
By dimensional analysis, the central pressure in a star can be estimated as

$$P \sim \frac{G M^2}{R^4}. \tag{4.12}$$

This follows, for example, from the stellar structure equation, Eq. (5.31), by replacing dP/dr by P/R, $m(r)$ by M, r by R, and $\rho(r)$ by $\bar{\rho} = M/(4/3\pi R^3)$. Here r is the radial coordinate, R is the stellar radius, $m(r)$ is the mass enclosed at r, and M is the mass of the star; for more details see Section 5.5. When we equate Eqs. (4.11) and (4.12), we obtain

$$G \frac{M^2}{R^4} = \frac{\hbar^2}{m_e} \left(\frac{Z}{A} \right)^{5/3} \left(\frac{\rho}{m_p} \right)^{5/3}. \tag{4.13}$$

Now, replacing ρ by $\bar{\rho}$, we find a relationship between mass and radius

$$R = \frac{\hbar^2}{G m_e} \left(\frac{3Z}{4\pi A m_p} \right)^{5/3} M^{-1/3}, \tag{4.14}$$

which implies

$$R \propto M^{-1/3}. \tag{4.15}$$

This relation is surprising in that, contrary to normal stars, a white dwarf shrinks when its mass grows.

Degenerate fluids have a low opacity and are good thermal conductors. As a result, the fluid cannot sustain large temperature differences for long times, and the

---

[4] This means that we are neglecting the effect of the nuclear binding energy, which is of the order of a few MeV per nucleon, see Fig. 4.6, as compared to the rest mass energy $m_p c^2 \approx 938$ MeV.

temperature inside the white dwarf is fairly homogeneous at about 10^7 K. The outermost layer of the white dwarf is not degenerate, and in this layer the temperature falls from $\sim 10^7$ K to $\sim 10^4$ K.

After a star has become a white dwarf, it radiates away its thermal energy. Effective temperatures for white dwarfs range from around 4 000 K to well more than 100 000 K. Most observed white dwarfs have temperatures much higher than the surface of the Sun, which is why they are called *white*. Because the degeneracy pressure does not depend on temperature, the star maintains its size as it cools.

For high densities, it is no longer correct to neglect relativistic effects. In particular, we can no longer assume that $v_x = p_x/m_e$; we have to use the correct relativistic expression. A relativistic calculation of the degenerate pressure yields (Section 5.6.1)

$$P \approx 0.8 \, \hbar c \left(\frac{Z}{A} \right)^{4/3} \left(\frac{\rho}{m_{\rm p}} \right)^{4/3} . \tag{4.16}$$

When we use this pressure in Eq. (4.13) to determine the mass–radius relationship, we find that the radius drops out, and we are simply left with an expression of the mass. This mass is called *Chandrasekhar mass* and has a value of about 1.4 $M_\odot$. It is the maximum mass that a degenerate electron gas can support.

The Chandrasekhar mass is named after Subrahmanyan Chandrasekhar,[5] who worked out the equations governing the structure of white dwarfs during his trip from India to England to take up graduate study at Cambridge University. Using Eddington's theory of stellar polytropes in combination with a recently published equation of state for a nonrelativistic degenerate electron gas, Chandrasekhar demonstrated that the density and radius of a white dwarf are very simple functions of its mass.

After calculating the central density for the white dwarf Sirius B, however, Chandrasekhar realized that white dwarfs of mass 1 $M_\odot$ reach high enough-densities in their cores for the electron gas to becomes fully relativistic. By allowing for the relativistic increase in the electron's momentum, Chandrasekhar was able to deduce that the dependence of pressure on electron density changes from $n_e^{5/3}$ to $n_e^{4/3}$. This seemingly small change has drastic consequences. A fully relativistic polytropic gas can be in equilibrium only up to a maximum mass. In an article published in 1931, Chandrasekhar interpreted this value as the maximum mass attainable by a white dwarf as it approaches fully relativistic, degenerate conditions. It is now named after him. For this conclusion, Chandrasekhar was publicly ridiculed by the most influential physicist of the time, Sir Arthur Eddington. Only much later, Eddington admitted that he had been wrong.

In 1932, the Russian physicist Lev Davidovich Landau – who had completed a similar calculation for degenerate neutron stars – described what would happen if the mass was increased above the maximum. The increase would lead to a collapse of the star: the star would shrink to a point. This remarkable result has a simple physical explanation. For a star of small enough mass, the degeneracy pressure can always be

---

[5] Indian-American astrophysicist (1910–1995).

brought into balance with gravity by increasing the density via contraction. As the star becomes more massive, however, further contractions eventually lead to high-enough core densities to bring the kinetic energy at the top of the Fermi sea to levels comparable to the rest energy of the particles; the gas then becomes relativistic. Because of the softening in the equation of state that follows the transition into the relativistic regime, the increase in quantum pressure that derives from further contraction always falls short of balancing gravity: the contraction cannot be halted.

At the end of this box, let us summarize the main properties of white dwarfs:

- Their size is comparable to the Earth.
- Their masses are roughly the same as the Sun with a peak in the distribution at about 0.6 $M_\odot$.
- Larger masses yield smaller radii.
- They cannot become arbitrarily massive; the upper mass limit is the Chandrasekhar mass of 1.4 $M_\odot$.
- As their size is small, they are not very luminous and therefore difficult to detect. They were first discovered in binary star systems.
- As they have no source of energy, they slowly cool and fade away.

Stars that are more massive than $\sim 8\ M_\odot$ can proceed to burn carbon (at $T > 6 \cdot 10^8$ K) to produce mostly oxygen, neon, and magnesium. If a star is more massive than ~ 9–$10\ M_\odot$, it can achieve sufficiently high temperatures to burn these ashes into silicon, sulfur, calcium, and argon and finally into the strongest bound nuclei of the iron group (see Fig. 4.6). The various burning stages become successively shorter. The oxygen-burning phase takes about two weeks, silicon burning about a day. The curve of the nuclear binding energy as a function of the atomic mass number reaches its maximum near these nuclei (see Fig. 4.6). The light nuclei are weakly bound because each nucleon is not surrounded by a full complement of neighbors. For the same reason, small droplets of water or mercury coalesce into larger drops, thus reducing the fraction of the constituent molecules in the surface. Atomic nuclei, similar to these liquids, have a surface tension that makes it favorable to minimize the surface. The weaker binding of the heavy nuclei, and the fact that they contain more neutrons than protons, both are a result from the electrostatic repulsion of the protons.

As a consequence, nuclear burning is exothermic only up to these elements: the fusion of lighter into heavier nuclei releases energy if the resulting nuclei are not beyond the maximum of the curve, but the production of a very heavy nucleus beyond the iron peak out of two lighter nuclei would *cost* energy. Hence, nuclear burning will not proceed beyond the stage where iron group nuclei have been formed. At the end the thermonuclear life of a star, it possesses an onion-like structure with shells of different compositions. The central core of the star is

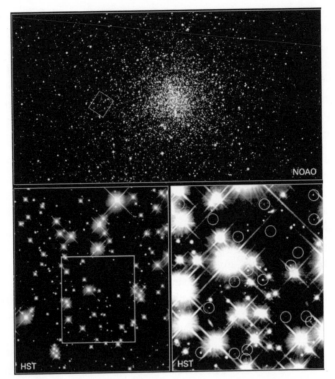

Figure 4.5   Top: Ground-based image of the globular star cluster M4 (closest globular cluster to Earth). Bottom: *Hubble Space Telescope* image of a field of 12- to 13-billion-year-old white dwarfs in the same globular cluster (M4). These stars are so feeble (at 30th magnitude) that they are less than one-billionth the apparent brightness of the faintest stars that can be seen by the naked eye. Hubble telescope photos courtesy of NASA and H. Richer (University of British Columbia); Ground-based photo courtesy of NOAO/AURA/NSF.

composed mainly of iron. The shell above this iron core is composed mainly of silicon, followed by shells of oxygen, carbon, helium, and, finally, hydrogen.

In the late stages of the evolution of massive stars, the temperatures are high enough for the very temperature-sensitive neutrino reactions to become important. In fact, the luminosity in neutrinos at the end of the stellar life is orders of magnitudes higher than that in photons. Because the neutrino reaction cross sections are vanishingly small, their mean free path in the stellar material is huge (of the order of light-years). Thus, while the energy transformed into photons is efficiently trapped inside the star, the neutrinos carry away a substantial fraction of the produced energy.

Interestingly, the specific entropy in the stellar center *decreases* from its birth to its death: although the central temperatures are high, the high densities and the ordered arrangement of the neutrons and protons into nuclei leads to a low specific

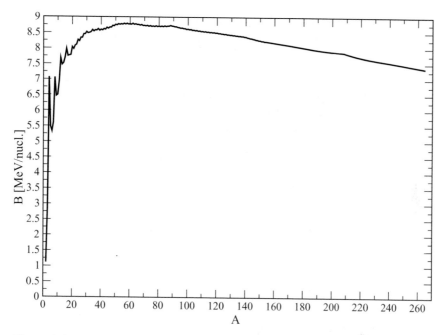

Figure 4.6   Average nuclear binding energy per nucleon in MeV as a function of
the nucleon number.

entropy. Under these conditions, the central pressure is provided by the *electron degeneracy pressure* that is explained in more detail in the box on white dwarfs and in Section 5.6.1.

4.2.2   *The onset of collapse*

Once the central iron core has reached the Chandrasekhar mass of $M_{ch} = 1.44\,(Y_e/0.5)^2\,M_\odot$, its electrons are ultrarelativistic and their adiabatic exponent has become equal to 4/3; see Eq. (5.53). A stability analysis shows that a gaseous sphere with an adiabatic exponent less than or equal to 4/3 is unstable to gravitational collapse. Therefore, at M_{ch} the iron core can no longer be stabilized by the electron degeneracy pressure and starts to contract.

During the oxygen and silicon burning phases, the electron fraction in the center has been lowered by electron captures to $Y_e \approx 0.44$. This means that there are fewer electrons available to support the star against gravity via electron degeneracy pressure. In previous evolutionary stages, a further contraction caused a temperature increase and a subsequent stabilization, but now electron captures in the iron core reduce the pressure. Moreover, temperatures in the core become so high that they destabilize the star through *photodisintegration*.

A typical photodissociation process is $\gamma + {}^{56}_{26}\mathrm{Fe} \leftrightarrow 13\alpha + 4n$. This process costs an energy of (mass difference times c^2)

$$Q = c^2(13m_\alpha + 4m_n - m_{\mathrm{Fe}}) = 124.4 \text{ MeV}, \qquad (4.17)$$

which is taken from the thermal energy of the plasma. The lower thermal energy leads to a lower pressure, which speeds up the contraction. The increasing densities lead to higher electron Fermi energies, which, in turn, speed up electron captures on free protons (compare with Eq. [2.71]),

$$e^- + p \rightarrow n + \nu_e \qquad (4.18)$$

and on nuclei

$$e^- + (Z, A) \rightarrow (Z - 1, A) + \nu_e, \qquad (4.19)$$

which, again, further reduce the electron pressure component.

4.2.3   Infall, bounce, and explosion

Infall

The collapsing core consists of two parts: the inner core of $\sim 0.8\ M_\odot$ that collapses homologously and an outer, supersonically infalling core.[6] Homologously means that the infall velocity increases linearly with the radius, $|\vec{v}| \propto r$. The collapse happens so fast that the outer mantle of the star has no chance to realize what is going on in its interior, so that the collapsing core is completely decoupled from the rest of the star.

At this stage of the collapse, temperatures of several times 10^9 K have been reached and the destruction of nuclei by energetic photons has become as fast as the formation of new nuclei. As a result, a reaction equilibrium, the so-called *nuclear statistical equilibrium*, is established (see Box: Nuclear statistical equilibrium). Under these conditions, the composition of the in-falling stellar material is completely determined by the density ρ, the temperature T, and the electron fraction Y_e. As can be seen from Eq. (4.22), the high temperatures favor smaller nuclei, corresponding to, for example, the photodisintegration of iron nuclei into alpha particles. This effect competes with the increasing density, which tends to establish high-mass nuclei (see Eq. [4.22]). The last quantity that influences the composition, the electron fraction Y_e, can be changed only by weak interactions, such as the transformation of a proton into a neutron through an electron capture. In the early stages of the infall, neutrinos can escape the star more or less unhindered.

---

[6] The mass of the homologously collapsing core corresponds approximately to the Chandrasekhar mass at the local Y_e; see Goldreich, P. and Weber, S. V. (1980). *Astrophysical Journal*, **238**, 991.

This changes once the neutrino escape time becomes comparable to the infall time. This happens at densities of $\rho_{\text{trap}} \sim 10^{12}$ g cm^{-3}, where the neutrinos are dragged along with the infalling material: they are essentially trapped. Once trapped, electron and neutrino captures are in reaction equilibrium, $e + p \leftrightarrow \nu_e + n$, and the lepton fraction $Y_L = Y_e + Y_\nu$, where $Y_e = Y_{e^-} - Y_{e^+}$ and $Y_\nu = Y_{\nu_e} - Y_{\bar{\nu}e}$, no longer changes (on a timescale relevant for the collapse).[7]

Bounce

During the collapse, the adiabatic exponent of the collapsing material is close to the 4/3 of an ultrarelativistic electron gas. At $\rho > 4 \cdot 10^{12}$ g cm^{-3}, the pressure from the free neutrons becomes dominant. This, however, has only a minor influence on the infall. At even higher densities, the competition between the attractive nuclear and the repulsive Coulomb forces leads to a bizarre matter state often referred to as *nuclear pasta* (Section 5.7.2). Once the densities are higher than the density necessary for nucleons to "touch each other," at about nuclear density $\rho_{\text{nuc}} = 2.6 \cdot 10^{14}$ g cm^{-3}, the short-range, repulsive part of the nuclear force becomes noticeable, and the adiabatic exponent suddenly increases very steeply. It is this very incompressible nuclear matter that stops the collapse and makes the infalling material "bounce back." This bounce causes an outward-moving shock wave that initiates the supernova explosion. Interestingly, it is the nature of microphysical forces, the nuclear forces at subfermion distances, that is essential for one of the most spectacular explosions in the universe.

The outgoing shock forms at a radius of about 20–30 km. Once on its way out, it has to plow through several tenths of solar masses of infalling material. The passage of the shock through the still infalling iron breaks up the iron nuclei into neutrons and protons. This costs about $\epsilon_b = 8.7$ MeV per nucleon (see Fig. 4.6). If the shock possesses an energy E_{sh}, its energy is used up once it has plowed through $\Delta m = E_{\text{sh}}/\epsilon_b \cdot m_u = 0.35 \, M_\odot (E_{\text{sh}}/6 \cdot 10^{51}$ erg), where $m_u = 1.6 \cdot 10^{-24}$ g is the atomic mass unit. In reality, Δm is smaller than this estimate, because photodissociations make protons available, on which further electron captures, $e + p \rightarrow n + \nu_e$, can occur. The produced neutrinos further rob the shock of its energy. Thus, whether such a prompt shock can be successful depends on the initial size of the iron core and of the fraction of it that undergoes the homologous collapse. Most numerical calculations indicate that the shock stalls 10–20 ms after bounce. It forms a quasistationary *accretion shock* at a distance between 100 and 200 km; this situation is sketched in Fig. 4.7. The mass

---

[7] Neutrinos still can diffuse out of the newly formed *protoneutron star*, but only on time scales of seconds. During the cooling phase, the protoneutron settles into a reaction equilibrium, the so-called *β-equilibrium*, in which no more weak interactions occur. The final neutron star then has Y_e values of ~ 0.1.

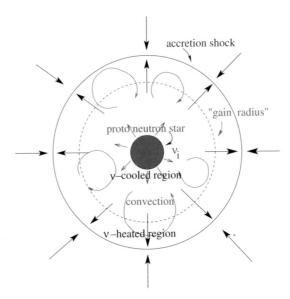

Figure 4.7   Sketch of the standard picture of the Type II supernova mechanism: after the shock has stalled by plowing through the outer iron core, it is re-energized by the neutrinos emitted from the newborn protoneutron star.

enclosed by the shock keeps increasing as matter is falling in at velocities of $\sim 50\,000$ km s^{-1}, but the shock is stalled at a constant radius.

Consequently, it is generally agreed that this prompt explosion scenario is unlikely to work. However, supernova explosions are frequently observed, so what could re-energize the stalled explosion?

Explosion

The bulk of the released gravitational energy has been channelled into neutrinos. They diffuse out of the freshly formed, hot protoneutron star on a diffusion timescale. Assuming neutrinos perform a random walk out of the star, this diffusion time is given by $\tau_{\text{diff}} \approx 3R^2/l_\nu c \sim 5$ s, where R is the stellar radius and l_ν the neutrino mean free path (also see Exercise 4.3 at the end of this chapter). The typical neutrino luminosity of a Type II supernova can be estimated as $L_\nu \sim \Delta E_{\text{grav}}/\tau_{\text{diff}} \approx 3 \cdot 10^{53} \, \text{erg}/5\,\text{s} = 6 \cdot 10^{52} \, \text{erg s}^{-1}$, where the gained gravitational binding energy, ΔE_{grav}, is approximately the binding energy of the newly formed neutron star, and we have assumed that the bulk of the energy goes into neutrinos.

Although most neutrinos leave the star and carry away energy, some neutrinos interact with stellar layers further out and deposit energy there. Neutrino heating processes such as $\nu_e + n \rightarrow p + e$ fall off more slowly with radius than the cooling processes. As a result, there must be a radius, the so-called gain radius, beyond which neutrino heating dominates over cooling. The deposition of only a small fraction of their energy in this region behind the stalled shock could suffice to re-accelerate the shock and drive the explosion. This gain region is just inside the stalled accretion shock. What keeps the shock bottled up is the matter that falls in. Heating by neutrinos that are emitted by the protoneutron star could drive the shock out of the star. Interestingly, the success of a supernova depends on the competition between the neutrino luminosity and the mass accretion rate, both of which are decreasing with time. Between several hundred milliseconds and a few seconds, the critical condition for a successful explosion could be reached. This mechanism is called the *delayed explosion mechanism*.

Once under way, it takes hours to days – depending on the stellar radius – until the shock wave breaks through the surface of the star. Until then, the outer layers of the star are unaffected by the collapse of the stellar interior. The bulk of the composition of the debris ejected into space is the result of the quiescent burning stages during the life span of the star. However, the inner regions of the stellar mantle are heated by the shock to high enough temperatures such that further nuclei can be synthesized through nuclear fusion. For example, parts of the silicon shell are burnt into iron-peak nuclei, parts of the oxygen into intermediate-mass nuclei, and some of the carbon into oxygen. Most important, radioactive isotopes such as ^{56}Ni, ^{57}Ni, and ^{44}Ti are produced that are important for powering the late-time light curve.

The general picture of Type II supernova explosions as we have described it has turned out to be fairly robust. But still, even now, many decades after Baade and Zwicky's suggestion, several questions about the explosion mechanism are still unsettled. The explosion energy of a Type II supernova, $E_{kin} \approx 10^{51}$ erg, is only a small fraction of the total gravitational binding energy, $E_{bind} \approx 3 \cdot 10^{53}$ erg, that is released during the collapse. Whether the launched material has just about enough energy for a successful explosion or whether it gets stuck in the mantle of the star may be a question of the details of the explosion. The described delayed explosion mechanism via neutrino heating is generally considered to be plausible but still far from being a certainty. To date, the most sophisticated computer models using the best available input physics and the most advanced numerical methods still do not yield robust explosions. Notoriously, the supernova fails in computer simulations.

What is missing that could help the explosion? For a long time, computer models of supernova explosions assumed spherical symmetry and were one dimensional. Only recently have multidimensional simulations of supernovae become feasible. Multidimensional effects can help the explosion process as we discuss in the context of Type Ia supernovae. In fact, there are several hints from observations that nature's supernovae deviate from spherical symmetry. It is easy to show that even slight asymmetries in the explosion process impart large "kicks" on the resulting neutron stars (also see Exercise 5.3). This is confirmed by observations. Typically, neutron stars have relative velocities of several hundred kilometers per second.

In the course of the supernova explosion, convective motions are likely to develop, both in the protoneutron stars and behind the stalled shock. This could increase the neutrino luminosity by dredging up hotter material from deeper regions and thus provide the struggling shock with the extra portion of energy to boost the explosion. But also moderate changes in the interactions between neutrinos and dense matter (due to so-called many-body effects) could increase the neutrino luminosities by the slight amount needed to make the explosion a success. Further help could come from magnetic fields that may be strongly amplified in the stellar collapse. These issues are currently subject to intense research.

Nuclear statistical equilibrium (NSE)

Under conditions close to core collapse (i.e., at very high temperatures and densities), nuclear and electromagnetic interactions can become sufficiently fast for forward and backward reactions to come into equilibrium. This state is called *nuclear statistical equilibrium* (NSE) and usually is reached at temperatures of $T \approx 4 \cdot 10^9$ K. From thermodynamics, we know that in reaction equilibrium, $A + B \leftrightarrow C + D$, the chemical potentials of the constituents are related by

$$\mu_A + \mu_B = \mu_C + \mu_D. \tag{4.20}$$

When all reactions of the form $(Z, A) + n \leftrightarrow (Z, A + 1) + \gamma$ or $(Z, A) + p \leftrightarrow (Z + 1, A + 1) + \gamma$ are in equilibrium, the condition for the chemical potentials, Eq. (4.20), reads

$$\mu_{Z,A} = Z\mu_p + N\mu_n, \tag{4.21}$$

where $N = A - Z$, and we have treated the chemical potential of photons as zero. Under the conditions of interest, all the constituents can be treated as a nonrelativistic Maxwell–Boltzmann gas (see Section 2.5). When we solve Eq. (2.67) for the chemical potential and then insert the chemical potentials of the constituents into Eq. (4.21), we find a relation between the abundances of a nucleus (Z, A), $Y_{Z,A}$, and the abundances of the nucleons, Y_p and Y_n. The abundances of species i are defined as $Y_i = n_i/n_{baryon} = n_i/\rho N_A$, where N_A is Avogadro's constant. If we express all masses in terms of the atomic mass unit m_u, use the approximations $m_n \approx m_p \approx m_u$, $M_{Z,A} \approx A \cdot m_u$, and introduce the abbreviation $\theta \equiv (m_u k_B T / 2\pi \hbar)^{3/2}$, then the abundance of a given nucleus is

$$Y_{Z,A} = \frac{G_{Z,A} A^{3/2}}{2^A \theta^{A-1}} (\rho N_A)^{A-1} Y_p^Z Y_n^N \exp\left(\frac{Q}{k_B T}\right). \tag{4.22}$$

Here we have used the so-called Q-value, $Q = c^2[Zm_p + (A - Z)m_n - M(Z, A)]$, which is the energy difference between the constituents of the nucleus and the nucleus itself. Moreover, we have replaced the statistical factor, g from Eq. (2.67), by the nuclear partition function, $G_{Z,A}$, which accounts for the possibility of excited states within the nucleus. The partition function gives an average spin factor where the spin factor of each individual state of spin I_k has been weighted by a Boltzmann factor, $\exp(-E/k_B T)$:

$$G_{Z,A} = \sum_k (2I_k + 1) \exp\left(-\frac{E_k}{k_B T}\right). \tag{4.23}$$

In addition to Eq. (4.22), we can invoke baryon conservation, which requires that the individual mass fractions $X_i = A_i Y_i$ add up to unity, $\sum_i A_i Y_i = 1$. Finally, if we require charge neutrality, $\sum_i Z_i Y_i = Y_e$, the abundances are completely determined.

Equation (4.22) is very useful as it enables us to determine the preferred equilibrium composition as a function of the physical conditions:

- high temperatures favor small nuclei ("$Y_{Z,A} \propto (k_B T)^{-3/2(A-1)}$" hidden in the θ-term),
- high densities favor large nuclei ("$Y_{Z,A} \propto \rho^{A-1}$"), and
- intermediate conditions favor tightly bound, large Q-value nuclei ("$Y_{Z,A} \propto \exp[Q/k_B T]$").

4.3 Type Ia supernovae

Type Ia supernovae are extremely powerful explosions in which a large amount of ^{56}Ni and intermediate-mass elements are produced. Their defining characteristics are their lack of hydrogen lines and the presence of strong, red Si II absorption features. This implies that their progenitor must contain no or only little ($<0.1\ M_\odot$) hydrogen and that some nuclear processing must have taken place to produce intermediate-mass elements such as silicon. Type Ia supernovae form a fairly homogeneous class of explosions. Most of them, about 85%, have nearly identical light curves, spectra, and peak absolute magnitudes of[8]

$$M_B \approx M_V \approx -19.30 + 5\log\left(\frac{H_0}{60\ \text{kms}^{-1}\text{Mpc}^{-1}}\right), \tag{4.24}$$

where H_0 is the current value of the Hubble constant as defined in Eq. (4.41).

Because stars come with a wide range of masses and compositions, this homogeneity implies that only under very specific circumstances can a Type Ia supernova

---

[8] Hamuy, M., et al. (1996). *Astronomical Journal*, **112**, 2391.

occur. Near the peak of their light curves, their spectra are characterized by high velocity (8 000–30 000 km) and intermediate mass elements from oxygen to calcium. At later times, prominent iron lines appear.

Unlike for Type II supernovae, by the time of this writing no progenitor system has been identified for any of the observed Type Ia supernovae. This is not too astonishing given the fact that the progenitor stars are most likely faint white dwarfs and not giant stars as in the Type II case. Therefore, we have only indirect hints on what the progenitors might be. Type Ia supernovae are produced most efficiently by relatively young populations near the star-forming regions in the spiral arms of galaxies. However, they also occur in old, elliptical galaxies in which star formation has ceased a long time ago and that contain only very old stellar populations. This immediately implies that Type Ia supernovae cannot be the result of the deaths of very massive stars because these are no longer present in old elliptical galaxies. (Remember that the lifetimes of stars decrease quickly with increasing mass of the star.)

Instead, it is generally agreed that a Type Ia supernova is caused by a thermonuclear explosion of a white dwarf because

- the specific kinetic energy of the ejecta $e_{kin} = (1/2)v^2 \sim (1/2)(10\,000 \text{ km/s})^2 = 5 \cdot 10^{17}$ erg g^{-1} is very similar to the energy per mass that is gained by transforming carbon and oxygen into iron, $\epsilon_{burn} \approx 1 \text{ MeV/nucleon} = 1.603 \cdot 10^{-3}/1.66 \cdot 10^{-24}$ g $\approx 9.7 \cdot 10^{17}$ erg g^{-1} (see the nuclear binding energy curve in Fig. 4.6);
- the explosion may suggest that degeneracy is involved; and
- they can occur long after star formation has expired.

To ignite, the supernova progenitor star must have accreted mass from a companion star. It is generally agreed that Type Ia supernovae are caused by thermonuclear explosions of white dwarfs in a binary system. Unfortunately, this is where the consensus ends. The mass of the exploding white dwarf, its composition, the explosion mechanism, and the nature of the companion star are all uncertain.

4.3.1   The exploding white dwarf

The distribution of white dwarf masses starts at around 0.2 $M_\odot$ and goes up to nearly the Chandrasekhar mass, with a peak in the distribution at about 0.6 $M_\odot$ (see the box on white dwarfs). The most massive known white dwarf to date is an object called RE J0317−853[9] with a mass of 1.35 $M_\odot$. The composition of white dwarfs is closely correlated with their mass: low-mass white dwarfs,

---

[9] Barstow, M. A., et al. (1995). *Monthly Notices of the Royal Astronomical Society*, **277**, 971. It has been speculated that this white dwarf may be the result of a coalescence of two lower-mass white dwarfs.

$M_{\mathrm{WD}} \lesssim 0.45\,M_{\odot}$, are made of helium; the bulk of the white dwarfs consists of carbon and oxygen; and some smaller fraction may be made of oxygen and neon.

Composition

When low-mass, helium white dwarfs accrete mass, the helium tends to ignite when the mass has reached $\sim 0.7\,M_{\odot}$. The resulting explosion energetics and the spectra do not seem to match the observations of most Type Ia supernovae. Consequently, such an explosion can be ruled out, at least for the bulk of Type Ia supernovae.

White dwarfs made of oxygen and neon are produced by stars that have main-sequence masses close to $10\,M_{\odot}$. Such stars are probably not common enough to form the bulk of the Type Ia supernovae. Moreover, it is doubtful whether white dwarfs with this composition produce an energetic explosion. Instead, they may form a neutron star in a *accretion-induced collapse*.

There may be some variety in Type Ia supernovae, but the bulk is probably produced by white dwarfs made of carbon and oxygen.

Mass

We discuss two classes of explosion models: the ignition of carbon in the center of a white dwarf close to the Chandrasekhar mass (*Chandrasekhar mass carbon igniters*) and the off-center ignition of a helium shell on top of a carbon-oxygen white dwarf, where the total mass is less than the Chandrasekhar mass (*sub-Chandrasekhar mass helium igniters*). Both cases have their strengths and weaknesses.

In Chandrasekhar mass central carbon ignition, the release of some 10^{51} erg of nuclear binding energy can account for the observed kinetic energy in the ejecta. This scenario produces enough $^{56}\mathrm{Ni}$ to power the supernova light curve and the resulting spectra are consistent with observations. Probably, the most attractive feature of this model is that the explosion always occurs at the Chandrasekhar mass, which would naturally explain the observed homogeneity of Type Ia supernovae. However, it seems, difficult to accrete enough mass to reach the Chandrasekhar mass frequently enough to explain the observed supernova rate. As the star accretes mass, it is quite likely to go through episodes where it loses mass again. For example, some of the accreted material may explode and result in a nova.[10] Moreover, there seem to be problems to accommodate the calculated late-time-light curves with the observations.

In the second class of models, helium piles up a carbon-oxygen white dwarf. Once a layer of $\sim 0.15\,M_{\odot}$ has been reached, but the total mass is still less than

---

[10] Nuclear explosion caused by the accretion of hydrogen onto the surface of a white dwarf star. Originally the name comes from *stella nova* new star.

the Chandrasekhar mass, the helium ignites at the inner edge of the accreted layer and produces what is sometimes called an edge-lit detonation. The detonation propagates outward through the helium shell, but at the same time compresses the carbon-oxygen core and triggers an off-center explosion. These models have the advantage that they need to accrete less material, and therefore they can more easily explain the observed Type Ia rate. The late-time spectra predicted by these models seem in better agreement with observations than for the Chandrasekhar mass models. Recent calculations,[11] however, indicate that a spin-up of the white dwarf due to the accreted angular momentum favors an earlier ignition of the helium layer. Such a premature explosion would produce a phenomenon more similar to a nova than to a Type Ia supernova.

At the time of writing, the Chandrasekhar mass explosion of a carbon-oxygen white dwarf is the most popular model. However, there is increasing evidence for a larger diversity of Type Ia supernovae than previously thought, which may indicate that there is more than one mechanism behind this phenomenon.

4.3.2 *The explosion mechanism*

In an ordinary, nondegenerate star, the energy produced by nuclear reactions in its center is offset by an expansion of the star against its gravity. This expansion lowers the central temperature of the star and thus prevents a runaway of the nuclear reactions in the center. This interplay between gravity and nuclear reactions acts as a sort of safety valve so that a normal star regulates itself. Crucial for this self-regulation is the temperature dependence of the pressure in a nondegenerate gas. In a white dwarf, however, the pressure is provided by degenerate electrons and is practically independent of the temperature (see the box on white dwarfs and Section 5.6.1). When a white dwarf approaches the Chandrasekhar mass, its electrons become more and more relativistic and the equation of state becomes increasingly softer. Thus, a small increase in mass leads to a substantial contraction of the star, which, in turn, raises its central temperature. Once the temperature is larger than about 10^7 K, hydrostatic burning of carbon in the core of the white dwarf sets in. Initially, excess energy is carried away by neutrinos that are produced in various processes. However, at some stage the white dwarf can no longer rid itself of its energy via neutrino emission, and a convective core develops in its center. This stage of the presupernova evolution is poorly understood, but it is believed that this last phase before the ignition of the supernova lasts for about 10^3 years. When the core temperature reaches about $6-7 \cdot 10^8$ K, the convective zone encompasses most of the white dwarf mass. At a temperature $T \sim 10^9$ K, the

---

[11] Yoon, S. C. and Langer, N. (2004). *Astronomy and Astrophysics*, **419**, 645.

critical temperature for explosive carbon burning is reached, which means that the energy generation rate starts to exceed the rate at which heat can be carried away. Now a thermonuclear runaway starts, and the white dwarf ignites. The nuclear reactions begin with carbon and oxygen nuclei and end in nickel and other iron-group elements. The conditions of this ignition are still unclear and subject of current research.

The computational modeling of the explosion is very difficult because the nuclear reactions take place in a very thin zone, called a thermonuclear flame, that cannot be spatially resolved in global computer models because the width of the flame is less than a centimeter but the radius of a white dwarf in this stage is about $\sim 10^8$ cm (see also Fig. 4.4). The reaction zone is so thin because the nuclear reactions are very sensitive to the temperature: at a temperature of $T \sim 10^{10}$ K, the energy production rate is proportional to the twelfth power of the temperature. One of the key questions in this field is how the flame propagates through the white dwarf. It is believed that the flame is laminar at the beginning. As the flame propagates through the white dwarf, hydrodynamic instabilities may wrinkle the flame. If the flame is wrinkled, it has a bigger surface area and can generate more energy. Consequently, the flame will accelerate. How this exactly works and whether the flame accelerates enough to make a transition from a slow, subsonic regime, called *deflagration*, into a supersonic regime, called *detonation*, is still unsolved. It is also unclear how the flame accelerates and how hydrodynamic instabilities change the morphology of the flame. Currently a lot of effort is invested in three-dimensional computer simulations with the necessary nuclear and flame physics. An example of such a computer simulation is shown in Fig. 4.8.

4.3.3   The nature of the companion star

Although it has emerged that the progenitor for Type Ia supernovae is a white dwarf, the nature of the companion star remains unclear. One possibility is that the companion star is also a white dwarf; this is called the *double-degenerate scenario*. The companion could also be a nondegenerate star, for example, an evolved giant or a main-sequence star. This case is called the *single-degenerate scenario*.

The double-degenerate scenario

White dwarfs are the most common end products of stellar evolution, and many of them form binary systems with other white dwarfs. As is shown in Section 6.2.3, such compact binaries emit gravitational waves that extract energy and angular momentum from the binary orbit. As a result, the white dwarfs slowly spiral toward each other. The time it takes until the stars come into contact is very sensitive to the separation of the two stars (see Eq. [6.50]). However, a good fraction of

Figure 4.8   Simulation of a Type Ia supernova explosion. The upper-left snapshot shows the ignition of the thermonuclear flame (isosurface) in many sparks around the center of the white dwarf star (indicated by the volume rendering of the logarithm of the density). From the center, it propagates outward as it is accelerated by turbulence (upper-right snapshot). At $t \sim 3$ s, the flame has consumed large parts of the star and burning terminates (lower-left snapshot). The lower-right snapshot displays the density structure of the remnant, which shows traces of turbulent combustion. Courtesy of Friedrich Röpke, Max-Planck-Institut für Astrophysik 2005.

systems should be close enough to merge within the age of the universe. Once the stars are close enough such that the less massive (and therefore larger; see Fig. 4.4) white dwarf fills its Roche lobe,[12] it is disrupted by tidal forces and, after a few orbital periods, forms a thick accretion disk around the more massive white dwarf.

[12] For an explanation of a Roche lobe, see Section 6.2.4.

This scenario naturally explains the nonobservation of hydrogen because none of the stars involved contains any hydrogen. Also, according to stellar population synthesis calculations, such encounters should occur frequently enough to be in accord with observations. The explosion is expected to occur at the Chandrasekhar mass and would therefore plausibly explain the homogeneity of Type Ia supernova explosions. However, it is not obvious that such a merger between white dwarfs leads to an explosion. Computer simulations indicate that ignition may start far from the center of the remnant. This could lead to a gradual transformation of the carbon and oxygen into a oxygen-neon-magnesium composition, and, triggered by electron captures on ^{24}Mg, it could end with a collapse to a neutron star in a so-called *accretion-induced collapse*.

The single-degenerate scenario

In the single-degenerate scenario, hydrogen is transferred from a nondegenerate companion star onto a carbon-oxygen white dwarf. Systems in which the accreted hydrogen is steadily burnt on the surface of the white dwarf are observed as supersoft X-ray sources. As mentioned before, most white dwarfs have masses around $\sim0.6\ M_\odot$. Consequently, a lot of hydrogen has to be accreted to reach the Chandrasekhar mass. However, the details of this accretion process are complicated and not entirely resolved. For example, some accretion rates can lead to explosive outbursts that expel more material than has been accreted. Or, if the accretion rate is too high, the white dwarf expands into a red giant-like configuration, and this can go along with strong mass loss either in a common envelope phase or via winds (see Section 6.2.7). Despite these difficulties, this scenario is regarded today as the most promising one for Type Ia supernovae.

From the single-degenerate scenario, one would expect to detect at least some hydrogen in the supernova spectrum, but to date no convincing detection exists. Such a detection would give the most conclusive hints to distinguish between both scenarios.

Type Ia supernovae as tools

So far, we have stressed the point that Type Ia supernovae are a remarkably homogenous class of objects. Largely for this reason, the thermonuclear explosion of a Chandrasekhar-mass white dwarf in a binary system has emerged as the most likely scenario. If this is true, this raises the opposite problem. Now the remaining diversity among the observed Type Ia events is difficult to explain. The masses of ^{56}Ni vary from explosion to explosion from about 0.07 $M_\odot$ (SN 1991bg) to at least 0.92 $M_\odot$ (SN 1991T). The remaining diversity among Type Ia supernovae may be attributed to variations among the progenitor white dwarfs, such as its composition, its rotation rate, or the environment. There is some evidence that bright and slowly declining supernovae are rare in elliptical galaxies, which would indicate that there is some dependence on the environment. It also cannot be excluded that at least some Type Ia supernovae are produced via an entirely different mechanism, maybe both the single- and the double-degenerate scenarios produce thermonuclear explosions. Because Type Ia supernovae are so crucial for distance measurements on a cosmological scale, it has turned out to be essential for our understanding of the universe to understand the cause for the diversity among Type Ia supernovae. As a result of distance measurements with supernovae in combination with other observations, it has been claimed that the universe is accelerating at the current epoch because of something that the cosmologists call *dark energy*.[13]

4.4   Supernova remnants

4.4.1   *Dynamics of supernova remnants*

In a supernova explosion, stellar material is ejected into the dilute plasma that fills the space between stars, the so-called interstellar medium (ISM). The enormous explosion of a supernova creates a cavity in the ISM that gradually expands until it reaches up to several hundred parsecs in diameter. The ISM is strongly disturbed by supernova explosions, which are believed to have a strong effect on how gas is distributed in our Galaxy. It is widely acknowledged that supernova remnants (SNRs) play crucial roles in the heating and chemical evolution of galaxies. The shock waves from these explosions may also be responsible for collapsing interstellar clouds to form new stars, thus closing the cycle of stellar evolution.

Traveling at high speeds, the remains of the supernova interact with the ambient medium and often give a colorful display. Although the supernova itself only lasts for a short time, an SNR emits strong electromagnetic radiation for about 100 000 years. The spectra from the different phases of the supernova remnant yield valuable information about the conditions in the ISM. Moreover, SNRs are important because here we can observe how heavy elements that have been synthesized in stars are dispersed through the ISM and how the galaxies are thus enriched with metals. Here, metal means what astronomers call metals, which are all elements except hydrogen and helium. These elements form the material out of which the next generation of stars and their accompanying planets are going to form. Making some simple assumptions about the physics of SNRs, we try to make some rough estimates about their evolution.

---

[13] For further reading on the accelerating universe, we refer the reader to the book by Liddle, A. (2003). *An Introduction to Modern Cosmology*. Chichester: Wiley.

A supernova with a typical kinetic energy of $E_{SN} \sim 10^{51}$ erg ejects matter with a velocity

$$E_{SN} \sim \frac{1}{2} M_{ej} v_{ej}^2, \tag{4.25}$$

which gives

$$v_{ej} \sim 10^4 \text{ km/s} \left(\frac{E_{SN}}{10^{51} \text{ erg}}\right)^{1/2} \left(\frac{M_{ej}}{M_\odot}\right)^{-1/2}. \tag{4.26}$$

The expansion of the supernova ejecta can be divided roughly into four distinct phases, each of which is dominated by different physical principles: In the first phase of the explosion, the stellar ejecta retain their initial velocity and expand such that the radius of the blast wave is $v_{ej}t$, where t is time. As the blast wave sweeps up mass, its mass increases, and momentum conservation demands that it slows down. Generally, the point when the mass swept up by the blast wave equals the ejected mass is reached as the end of this first phase. The swept-up mass is given by $M_{swept-up} \sim (4\pi/3) r^3 \rho_{ISM}$. Hence, in this phase the remnant reaches a radius

$$r_1 \sim 2 \text{ pc} \left(\frac{M_{ej}}{M_\odot}\right)^{1/3} \left(\frac{\rho_{ISM}}{10^{-24} \text{ g cm}^{-3}}\right)^{-1/3}. \tag{4.27}$$

Dividing this by the velocity gives us an estimate for the time when this occurs, that is,

$$t_1 \sim \frac{r_1}{v_{ej}} \sim 200 \text{ yr} \left(\frac{E_{SN}}{10^{51} \text{ erg}}\right)^{-1/2} \left(\frac{M_{ej}}{M_\odot}\right)^{5/6} \left(\frac{\rho_{ISM}}{10^{-24} \text{ g cm}^{-3}}\right)^{-1/3}. \tag{4.28}$$

In the second phase, the energy losses through radiation are still negligible, and we can assume that $E_{SN} = $ constant.

Because energy, E_{SN}, is proportional to $\rho_{ISM} r^3 v^2$ and $v = \dot{r}$, this can be rewritten as

$$E_{SN}/\rho_{ISM} \propto r^3 \dot{r}^2, \tag{4.29}$$

which can be integrated to yield

$$r \propto (E_{SN}/\rho_{ISM})^{1/5} t^{2/5}. \tag{4.30}$$

The spherical blast wave that we have just sketched is called a *Sedov explosion*. Unlike in the first phase, in the second phase the expanding wave decelerates as it expands. Clearly, material further out decelerates first, so that material further in starts to run into the outer shells. This heats up the outer shell and can produce complex flow patterns.

The temperature in the shocked gas is still very high. In the limit of strong shocks, the ratio of the temperatures behind and in front of the shock is given by Eqs. (2.106) and (2.107):

$$\frac{T_2}{T_1} = \frac{2\Gamma(\Gamma - 1)\mathcal{M}_\infty^2}{(\Gamma + 1)^2} = \frac{5}{16}\mathcal{M}_\infty^2, \tag{4.31}$$

where $\mathcal{M}_\infty$ is the Mach number of the shock wave with respect to the sound speed in the unshocked gas, and the last equality is based on the assumption that the adiabatic exponent is $\Gamma = 5/3$.

For example, the supernova remnant *Cygnus Loop* has reached the end of this second phase about now – 5 000 to 8 000 years after the supernova explosion. It has a radius of about 20 pc and velocities of 115 km s^{-1}. Temperatures in this phase are of the order of 10^6 K. At these temperatures, the SNR strongly emits X-rays, mainly through thermal *bremsstrahlung* (see Section 3.8.1).

When temperatures fall to less than a million K, some ions start to recombine to form atoms. These atoms show emission lines and lead to strong cooling of the gas. When radiative losses start to affect the kinetic energy, the *Sedov* phase ends. This occurs at times of about 10^4 yr.

At some point, the outer shell radiates so much and cools so fast that eventually it forms a cold, dense shell that is driven by a hot interior. The shell propagates with constant radial momentum and piles up the ambient material like a snowplow (why this third phase is sometimes called *snowplow phase*). Now, the radial momentum is constant, which can be described by the equation

$$\frac{d}{dt}[Mv] = \frac{d}{dt}\left[\left(\frac{4\pi}{3}\right)\rho r^3 \dot{r}\right] = 0. \tag{4.32}$$

With the initial condition that a thin shell formed at time t_0 with radius r_0 and velocity v_0, we can write

$$\left(\frac{4\pi}{3}\right)\rho r^3 \dot{r} = \left(\frac{4\pi}{3}\right)\rho r_0^3 v_0. \tag{4.33}$$

Integrating this with respect to time yields

$$r = r_0\left[1 + \frac{4v_0(t - t_0)}{r_0}\right]^{1/4}, \tag{4.34}$$

which, at late times, has the solution $r \propto t^{1/4}$. At this stage, the gas has temperatures of around 10^4 K, emits strong line emission, and radiates strongly in the optical waveband.

In the final stage of the SNR, the stellar debris merges with the ISM. This happens when the speed of the ejecta become comparable to the sound speed of the

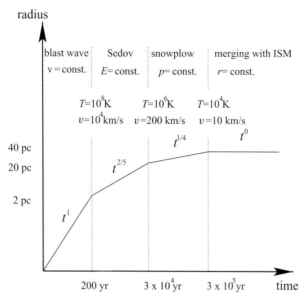

Figure 4.9   An illustration of the four different phases in the dynamics of a supernova remnant (SNR). As described in the text, we distinguish between the blast-wave stage, the Sedov phase, the snowplow phase, and the final phase where the SNR merges with the ISM. Rough estimates for the temperatures and velocities at the end of each phase are given.

ISM, which usually is at a time of about 10^5 yrs after the supernova explosion. The evolution of a SNR in the four phases described previously is depicted in Fig. 4.9.

SNRs are typically divided into two categories: (i) the shell-like remnants, in which the radio, optical, and X-ray luminosity comes almost entirely from a distinct shell, and (ii) the Crab-like or filled-center remnants, which have a central radio source and whose brightness is maximal at the center and decreases radially. The observed features of the first category can be reasonably described by the four stages mentioned previously. In contrast, the dynamics of the Crab-like nebulae is somewhat different because the central source forms a steady supply of electromagnetic radiation and high-energy particles. It is not always easy to identify which type of supernova has produced a given SNR. Clearly, those SNRs that have a pulsar at their centers are the result of Type II supernovae. In contrast, remnants of Type Ia supernovae are characterized by an increase of the iron abundance toward the center. This is because Type Ia supernovae produce $\sim 0.5\, M_\odot$ of ^{56}Ni, which then decays into ^{56}Fe, and this is a factor of 10 more iron than what one would expect from a Type II supernova. The X-ray satellite CHANDRA (Section 9.5) has identified this central iron peak in a number of young supernova remnants in our neighbor galaxy, the Large Magellanic Cloud. The historical supernova SN 1006, dating from the year 1006 AD, is a peculiar case. The dynamics, in particular the pronounced shell-like appearance, points to a Type Ia supernova. However, the spectra in the optical, ultraviolet, and X-ray ranges show a deficiency of iron. The reason for this is unclear. It is possible that the plasma is so far out of ionization equilibrium that the emission lines are not a good indicator for the abundances. This case shows that the plasma physics of SNRs is very complex and the interpretation of observations a tricky business.

Redshift

Observations of the spectral lines of extragalactic objects show that the observed wavelength, λ_0, is longer than the emitted wavelength, λ_e. This shift in the wavelength is quantified by a dimensionless quantity called redshift, which is defined as

$$z = \frac{\lambda_{\text{rec}} - \lambda_{\text{em}}}{\lambda_{\text{em}}}. \tag{4.35}$$

If z is positive, the spectral lines are shifted toward the red (long-wavelength) part of the spectrum. Negative z corresponds to a blueshift, which indicates that the wavelengths are shortened.

There are three causes of astronomical redshift:

- *Doppler redshift* is a result of relative radial motion between emitter and observer caused by the Doppler effect.
- *Gravitational redshift* occurs when photons lose energy when they have to overcome a gravitational field on their way from source to observer.
- *Cosmological redshift* is caused by the expansion of the universe, as quantified by the Hubble constant, H_0. According to Hubble's law, the farther a galaxy is from Earth, the faster it is receding from us.

Although we assume that the reader is familiar with the Doppler redshift, the other sources of redshift are less straightforward. Let us begin with the gravitational redshift. A simple derivation of the formula for the gravitational redshift of photons is based on an energy-conservation argument. A photon has an energy of $E = h\nu$; therefore it can be assigned an effective mass, $m = h\nu/c^2$. Hence, its total energy in a gravitational field is

$$E_{\text{tot}} = h\nu + m\Phi = h\nu \left(1 + \frac{\Phi}{c^2} \right). \tag{4.36}$$

If we label quantities at the point where the photon is measured by "rec" and those belonging the point of emission by "em," equating $E_{\text{tot,rec}}$ and $E_{\text{tot,em}}$ yields

$$\frac{\nu_{\text{rec}}}{\nu_{\text{em}}} = \frac{(1 + \Phi/c^2)_{\text{em}}}{(1 + \Phi/c^2)_{\text{rec}}} \tag{4.37}$$

or

$$\frac{\Delta\nu}{\nu_{\text{em}}} = \frac{\nu_{\text{rec}} - \nu_{\text{em}}}{\nu_{\text{em}}} = \frac{(1 + \Phi/c^2)_{\text{em}}}{(1 + \Phi/c^2)_{\text{rec}}} - 1. \tag{4.38}$$

In the limit of weak gravitational fields, $\Phi/c^2 \ll 1$, we can use

$$\frac{1}{1 + \Phi/c^2} = 1 - \frac{\Phi}{c^2} + O\left[\left(\frac{\Phi}{c^2}\right)^2\right]. \tag{4.39}$$

If we insert this into Eq. (4.38) and keep only terms that are linear in Φ/c^2, we find

$$\frac{\Delta\nu}{\nu_{em}} = -\frac{\Phi_{rec} - \Phi_{em}}{c^2} = -\frac{\Delta\Phi}{c^2}. \tag{4.40}$$

Cosmological redshifts become dominant over Doppler redshifts caused by peculiar motions for objects outside our Local Group of Galaxies. These redshifts are a result of the stretching of space-time, as the universe expands. Light moving in this space-time is also stretched. Its wavelength increases, and it becomes redshifted.

Edwin Hubble made the very important observation that there is a simple relationship between the amount of redshift of a galaxy and its distance from Earth. This discovery, published in 1929 and now known as *Hubble's law*, can be expressed by the relationship

$$v = H_0 \cdot d, \tag{4.41}$$

where v is the recession velocity, d the distance from us, and H_0 the Hubble constant. The Hubble constant specifies the rate of expansion of the universe. This rate changes with time; its current rate is denoted by H_0. The value of the Hubble constant is important because it allows us to convert spectral features into true distances. The precise value of the Hubble constant has been the subject of much debate over several decades. Today, the value is known with good accuracy. The best determination comes from observations of the Cosmic Microwave Background, which puts the value at $H_0 = 73$ km s^{-1}Mpc^{-1}.[14]

For relatively nearby galaxies, the velocity v can be estimated from the galaxy's redshift z using the formula $v = zc$, where c is the speed of light. On large scales where z is more than 0.1, General Relativity predicts departures from a strictly linear Hubble law, and the relationship between redshift and distance becomes more complicated and depends on various cosmological parameters, such as the average density of the universe. For more information, we recommend *An Introduction to Modern Cosmology* (Liddle, 2003).

4.4.2   *Particle acceleration in supernova remnants*

The shock wave of an SNR is also a place where there is clear evidence that particles are accelerated to relativistic energies via Fermi-I acceleration (see Section 2.8). High-resolution observations of SNRs show strong emission in the radio as well as

----

[14] Spergel, D. (2006). arXiv:astro-ph/0603449.

in X-ray bands. The SNR Cassiopeia A is the brightest radio source in the northern sky and the youngest known Galactic SNR. Another famous example is the shell-like SNR, SN 1006, shown in Fig. 4.10. The radio emission is synchrotron radiation from relativistic electrons with energies of several GeV that gyrate around magnetic field lines. This is supported by spectral and polarization measurements of the radio emission. For the spectrum and polarization properties of synchrotron radiation, see Section 3.8.2. The radio spectrum follows a power law, and the emission is moderately polarized. The power-law slopes vary significantly from remnant to remnant. The average polarized fractions are generally less than 20%, implying that magnetic fields are tangled on scales that are smaller than those resolvable in observations (i.e., on scales below 1 parsec).

Detailed comparisons between the radio and X-ray maps show that the sites of radio emission and the sites of X-ray emission coincide. The X-ray emission is partly thermal emission from very hot gas ($T \sim 10^6$ K) that has been heated by the outgoing shock wave, but there is also a nonthermal, power-law component in the X-ray emission. This nonthermal X-ray emission is synchrotron emission from electrons. If this X-ray emission is synchrotron radiation from electrons, what energies must these electrons have? In Eq. (3.169), we found that the characteristic synchrotron frequency of electrons is given by

$$\omega_c = \gamma^2 \omega_L = \frac{\gamma^2 e B}{m_e c} = \left(\frac{E}{m_e c^2}\right)^2 \frac{e B}{m_e c}, \qquad (4.42)$$

where γ is the Lorentz factor of the electrons, E is the electron energy, B is the magnetic field strength, and the other constants have their usual meaning. If we plug in the constants e, m_e, and c and write $\nu_c = \omega_c / 2\pi$, we find

$$\nu_c \approx 4.2 \cdot 10^{19} \left(\frac{E}{\text{erg}}\right)^2 \left(\frac{B}{G}\right) \text{Hz}. \qquad (4.43)$$

In SNRs, nonthermal X-rays are detected with photon energies of 10 keV. This photon energy corresponds to a frequency of $2.4 \cdot 10^{19}$ Hz (see Appendix B). Assuming a magnetic field strength of $B \sim 10^{-5}$ G (as we justify later in this section), we find an electron energy of $E \sim 760$ erg, which corresponds to nearly 5×10^{14} eV or 500 TeV. Apparently, in SNRs, particles are accelerated to energies of up to hundreds of TeV. This nonthermal X-ray emission has been seen in about 10 young SNRs. There is also evidence that ions are accelerated in shocks in SNRs, but this evidence is still circumstantial. However, in other places in the universe, for example, in our Solar System, there is clear evidence that protons are accelerated at least as efficiently as electrons at shock waves.

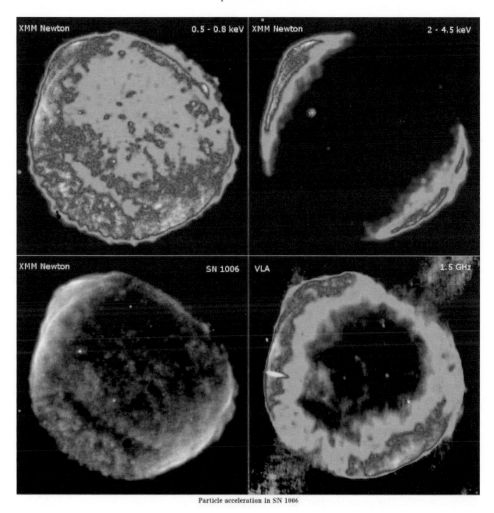

Figure 4.10   The X-ray images by the satellite XMM-NEWTON of supernova remnant SN 1006 reveal particle acceleration by supernova remnants. By comparing the top-left and top-right images, it is clear that the gas is more or less homogeneous, and the accelerated particles are concentrated in two bright limbs. Those crescents are thought to be polar caps (the supernova remnant is a sphere), corresponding to the direction of the presupernova magnetic field. More particles are accelerated where the shock propagates along the magnetic field lines. By comparing the top-right and bottom-right images, it is apparent that the high-energy particles (emitting in X-rays) are more concentrated in the limbs than the lower energy particles (emitting in the radio). This means that not only are more particles accelerated there, but they are also accelerated faster. Image courtesy of CEA/DSM/DAPNIA/SAp and ESA; Rothenflug, R. et al. (2004). *Astronomy and Astrophysics*, **425**, 121.

Let us revisit the theory of Fermi-I acceleration introduced in Section 2.8 to work out the maximum energies to which particles can be accelerated. The typical distance a particle diffuses in time t is $(D_2 t)^{1/2}$, where D_2 is the diffusion coefficient in the downstream region.[15] The distance advected in this time is $v_2 t$. One can set $(D_2 t)^{1/2} = v_2 t$ to define a distance D_2/v_2 downstream from the shock, which marks a boundary between the region closer to the shock where the particles can diffuse back through the shock and a region farther from the shock in which the particles are advected downstream and never return to the shock.

There are $n_{\text{rel}} D_2/v_2$ relativistic particles per unit area of shock between the shock and this boundary, and the rate at which they cross the shock is $n_{\text{rel}} c/4$, which we know from Eq. (2.118). The average time spent downstream before diffusing back through the shock can be estimated as

$$t_2 \sim \frac{n_{\text{rel}} D_2}{v_2} \frac{4}{n_{\text{rel}} c} = \frac{4 D_2}{c v_2}. \tag{4.44}$$

Next we look at the second half of the acceleration cycle where the particle has to diffuse from the upstream region through the shock to the downstream region. Just like what we did for the downstream region, we can define a boundary at a distance D_1/v_1 upstream of the shock such that nearly all particles upstream of this boundary have never encountered the shock, and nearly all the particles between this boundary and the shock have diffused there from the shock. Again, dividing the number of particles per unit shock area, $n_{\text{rel}} D_1/v_1$ by $n_{\text{rel}} c/4$, we find the average time spent upstream before returning to the shock,

$$t_1 \sim \frac{n_{\text{rel}} D_1}{v_1} \frac{4}{n_{\text{rel}} c} = \frac{4 D_1}{c v_1}. \tag{4.45}$$

Consequently, the time for the full cycle is

$$t_{\text{cycle}} \sim t_1 + t_2 \sim \frac{4}{c} \left(\frac{D_1}{v_1} + \frac{D_2}{v_2} \right). \tag{4.46}$$

The acceleration time at energy E, defined by $E/(dE/dt)$, is then given by

$$t_{\text{acc}} \sim \frac{4}{v_1} \left(\frac{D_1}{v_1} + \frac{D_2}{v_2} \right). \tag{4.47}$$

The diffusion coefficient for dilute plasmas as found in an SNR is approximated by

$$D = \frac{1}{3} r_g c, \tag{4.48}$$

---

[15] Protheroe, R. J. (2000). *Physics Reports*, **327**, 109–247.

where $r_g = qB/mc$ is the gyroradius and q the charge. This diffusion coefficient is called the *Bohm diffusion coefficient*. In the case where $B_1 = B_2$, one can approximate $D_1 = D_2$ and obtain

$$t_{acc} \propto \frac{E}{qBv_1^2}. \tag{4.49}$$

The rate of gain of energy can be written as

$$\left(\frac{dE}{dt}\right)_{acc} = \xi qc^2 B, \tag{4.50}$$

where $\xi < 1$ is an efficiency that depends on the acceleration mechanism. For first-order Fermi acceleration at a shock with a speed of 10% of the speed of light, ξ can be of the order of 0.05. Note that the rate at which particles are accelerated is inversely proportional to the diffusion coefficient because faster diffusion means that the particle can get away from the shock region more easily. The relativistic particles in SNRs can lose energy through synchrotron radiation or inverse Compton scattering with photons from the Cosmic Microwave Background. For electrons, the rate of energy loss via synchrotron radiation is given by Eq. (3.162),

$$\left(\frac{dW}{dt}\right)_{sync} = \frac{4}{3}\sigma_T c u_B \left(\frac{E}{m_e c^2}\right)^2, \tag{4.51}$$

where β^2 has been approximated as 1. The equivalent rate for inverse Compton scattering is

$$\left(\frac{dW}{dt}\right)_{IC} = \frac{u_{CMB}}{u_B}\left(\frac{dW}{dt}\right)_{sync}, \tag{4.52}$$

where $u_B = B^2/8\pi$ is the energy density in the magnetic field, $\sigma_T = 6.65 \cdot 10^{-25}$ cm^2 is the Thomson cross section, and u_{CMB} is the energy density in the Cosmic Microwave Background. At the present epoch, $u_{CMB} = aT_{CMB}^4 \sim 4.2 \cdot 10^{-13}$ erg cm$^{-3} \sim 0.26$ eV cm^{-3}. Here a is the radiation constant given in the context of the Planck law (see Section 3.5.3), and $T_{CMB} = 2.725$ K is the temperature of the Cosmic Microwave Background. Therefore, synchrotron cooling is faster, provided $u_B > u_{CMB}$, which translates into $B > 3\mu$G.

Any mechanism to accelerate relativistic particles yields some maximum particle energy. The energy can be limited by a variety of factors including the time available for acceleration, the escape of the highest-energy particles from the finite-sized acceleration region, or the balance between energy gains and losses. For example, by equating the rate of energy gain with the rate of energy loss by synchrotron radiation, we can obtain the maximum energy achievable by electrons and protons, respectively:

$$E_{e,max} \sim 6 \cdot 10^4 \xi^{1/2} \left(\frac{B}{1\,G} \right)^{-1/2} \text{GeV}, \tag{4.53}$$

$$E_{p,max} \sim 2 \cdot 10^{11} \xi^{1/2} \left(\frac{B}{1\,G} \right)^{-1/2} \text{GeV}, \tag{4.54}$$

where ξ is the efficiency defined in Eq. (4.50). One can note that protons (and nuclei) can be accelerated to much higher energies than the lighter electrons, as they lose less energy to synchrotron radiation; see Eq. (3.162).

Shock waves from SNRs remain strong enough to accelerate particles to relativistic velocities for about 1000 years, and this also limits the energy to which particles can be accelerated. Every time a particle crosses the shock front, its energy increases approximately by 1/100. A careful analysis of various limitations of particle acceleration in SNRs yields a maximum electron energy of about 10^{15} eV. SNRs can therefore account for a good part of the particles observed in cosmic rays. Some cosmic rays, however, have energies of up to 10^{21} eV; they must come from different sources.

The SNR SN 1006 was also detected at energies of up to 100 TeV, which corresponds to $1.6 \cdot 10^2$ erg. At this energy, the gyroradius of the particles is of the same scale as the size of the SNR. It is not entirely clear what causes such high energies. The γ-ray detections of some supernova remnants have been interpreted as the inverse Compton emission from these high-energy electrons against the Cosmic Microwave Background. However, the source of such high energies remains the subject of much controversy.

As we showed in Section 2.8, Fermi-I acceleration gives rise to a power law of index 2. For the sake of simplicity, we assume that the accelerated particles are mainly electrons, but obviously, this can readily be generalized to protons and nuclei. Then, for the distribution of the relativistic electrons we can write

$$dn_e = AE^{-2}\,dE, \tag{4.55}$$

where n_e is the number density of electrons and A is a normalization constant. This constant can be fixed by assuming that a fraction ξ_e of the postshock thermal energy density is transferred to the relativistic electrons. Then for the energy density of the relativistic electrons we can write

$$u_e \approx \xi_e \frac{3}{2} k_B (T_2 - T_1) \frac{\rho_2}{\mu m_H}, \tag{4.56}$$

where k_B is Boltzmann's constant and the subscripts 1 and 2 refer to the up- and downstream variables, respectively. In Eq. (4.56), we assumed that $\Gamma = 5/3$. We can also express the energy density in terms of the distribution function, that is,

$$u_e = \int_0^{E_{\max}} E \frac{dn_e}{dE} \, dE = m_e c^2 \int_1^{\gamma_{\max}} \gamma \frac{dn_e}{d\gamma} \, d\gamma, \qquad (4.57)$$

where $\gamma = (1 - v^2/c^2)^{-1/2}$ is the Lorentz factor, $E_{\max}$ the maximal energy of the electrons, and $\gamma_{\max}$ the corresponding Lorentz factor. Inserting Eq. (4.55) into Eq. (4.57) yields

$$u_e = \int_1^{\gamma_{\max}} \gamma^{-1} \, d\gamma = A \ln \gamma_{\max}. \qquad (4.58)$$

Equating Eqs. (4.56) and (4.58) and assuming $T_2 \gg T_1$ yields an equation for A:

$$A = \frac{3\xi_e \rho_2 k_B T_2}{2 \ln \gamma_{\max} \mu m_H}. \qquad (4.59)$$

Multifrequency observations of SNRs have helped to shed light on the microphysical parameters of shock acceleration. For example, in SN 1006 shock efficiencies of $\xi_e \sim 5\%$ have been found. Similar values have been inferred from other SNRs. One of the results of this work is that the total energy in relativistic electrons constitutes about 1–2% of the total supernova explosion energy.

In diffusive shock acceleration, electrons must diffuse ahead of the shock to gain energy. In this upstream region, they also emit synchrotron radiation. That radiation is weaker than downstream because the magnetic field is not compressed, and the ambient field may also lie nearly along the line of sight to the observer, suppressing synchrotron emission. However, the very sharp rims observed in several SNRs can be used[16] to infer an upper limit on the diffusion length of electrons. For the sharpest rims, the mean free path of electrons is typically $\sim 10^{-3}$ pc, which is less than 1% of that derived for the mean free path in the ISM in the Galaxy. This result is interpreted in terms of an enhanced hydromagnetic wave intensity generated by diffusive shock acceleration.

With the X-ray satellite CHANDRA, thin filaments of synchrotron-emitting gas near the shock fronts of many young SNRs have been resolved for the first time. The measured width of the filaments can be used to constrain the magnitude of the magnetic field inside the plasma. The width of the filaments is determined by two factors, the velocity of the emitting particles and the timescale, after which the radiation from these particles fades away. First, the relativistic electrons are advected away from the shock with a downstream velocity v_2. The velocity of

---

[16] Achterberg, A., et al. (1994). *Astronomy and Astrophysics*, **281**, 220.

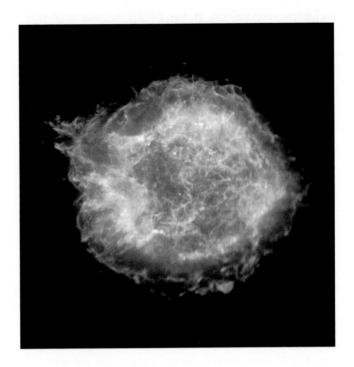

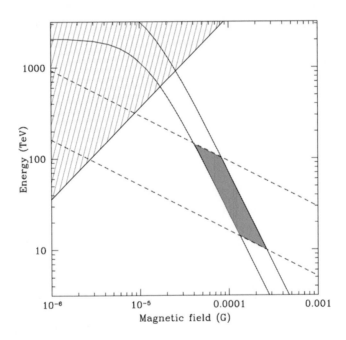

plasma downstream from the shock front, for the shock compression ratio of 4 valid for strong shocks and an adiabatic exponent of 5/3, is $v_2 = v_s/4$ (see Eq. [2.95]), where v_s is the velocity of the shock front. As the relativistic particles move away from the shock, they cool via synchrotron radiation until a time t_{sync}, after which they have lost a significant fraction of their energy and are no longer visible in X-rays. This loss time is given by Eq. (3.163):

$$t_{sync} = \frac{E}{(dW/dt)_{sync}} \approx 635 \left(\frac{B}{G}\right)^{-2} \left(\frac{E}{erg}\right)^{-1} \text{s}. \qquad (4.60)$$

Hence the width of the filaments is given by

$$l_{fila} \sim v_2 t_{sync}(E, B), \qquad (4.61)$$

and thus is a function of the downstream velocity v_2, the particle energy E, and the average magnetic field B. Now, we do not know the energy of the particles. However, the energy of the particles in the expression for t_{sync} can be expressed in terms of the magnetic field via the relation between the characteristic synchrotron frequency and electron energy (Eq. [3.169]):

$$\nu_{sync} \sim 10^{18} \left(\frac{B}{G}\right) \left(\frac{E}{erg}\right)^2 \text{Hz}. \qquad (4.62)$$

For the case of Cassiopeia A, the filaments are best visible in the X-ray continuum–dominated band of 4–6 keV. By taking 5 keV as the typical photon energy, which corresponds to a frequency of $\sim 10^{18}$ Hz, one obtains

$$\left(\frac{B}{G}\right) \left(\frac{E}{erg}\right)^2 \sim 1. \qquad (4.63)$$

Figure 4.11 Top: A deep CHANDRA image of Cassiopeia A in the 4–6 keV continuum band (Hwang, U. et al. 2004). Bottom: Most likely values of E_{max} and B just behind the shock front (shaded region). This area is formed by the two dashed lines, which mark the electron energies that contribute to the continuum emission around 5 keV, allowing for emission at 1/3 and 10 times the peak frequency for synchrotron emission, and the solid lines, which are formed by assuming that the width of the rims is determined by radiative losses (decay times of 1850 yr and a shock velocity of 5 200 km s^{-1} are assumed). The width of the rims should be at least larger than twice the gyroradius. This means that for a typical width of 3, the hatched region is excluded. From Vink, J. and Laming, J. M. (2003). *Astrophysical Journal*, **584**, 758–769.

The shock velocity in Cassiopeia A has been estimated[17] from proper motion studies using EINSTEIN and ROSAT HRI X-ray images and is thought to be $v_s = 5\,200$ km s^{-1}. This yields a downstream velocity of $v_2 \sim v_s/4 \sim 1\,300$ km s^{-1}.

For a distance of 3.4 kpc, the measured angular widths of the rims of $\sim 1.5''$ corresponds to a linear width of roughly $l_{\text{fila}} \sim 0.025$ pc. Substituting the values for v_2 and l_{fila} into Eq. (4.61) yields $B^2 E \sim 10^{-6}$, where B is in Gauss and E in erg. Combining this with Eq. (4.63) yields magnetic fields of the order of $100\ \mu G$. A careful analysis of the data on the thin filaments that also takes into account inverse Compton losses is shown in Fig. 4.11. Generally, the estimates of the magnetic field near the shock front give values of roughly one fifth of the average magnetic field strength and indicate a preshock magnetic field of $B \sim (2 - 4) \cdot 10^{-5}$ G, a factor of 10 higher than the typical interstellar medium value.

The high average magnetic field in Cassiopeia A is inconsistent with simple shock compression of the interstellar magnetic field and indicates that the magnetic field inside the remnant must have been enhanced by turbulence.

A likely place for turbulence is the contact discontinuity between the ejecta and the interstellar medium. This may explain why the radio emission shows a bright ring because this is the region in the remnant where magnetic field amplification is important.

It was first suggested[18] that Rayleigh–Taylor instabilities (see Section 2.9.2) in the decelerating bright ring of Cassiopeia A and other young SNRs would lead to radial magnetic fields. Detailed numerical simulations since that time confirm the existence of these filamentary structures and their role in the amplification and alignment of the magnetic fields.[19] Approximately radial filamentary structures have been seen in the Crab SNR and more recently in Hubble Space Telescope images of Cassiopeia A.[20] They show large velocity shears in the line profiles of several emission lines from ground-based, optical spectra of knots near the remnant's center. The nonradial orientation of several well-defined systems of linear filaments suggests the presence of large-scale shear flows behind the reverse shock.

In any case, the close correspondence between the Rayleigh–Taylor finger orientations and the local magnetic field directions strongly supports a connection between the two. It has also been shown[21] that the Rayleigh–Taylor instabilities

[17] Vink, J., Bloemen, H., Kaastra, J. S., and Bleeker, J. A. M. (1998). *Astronomy and Astrophysics*, **339**, 201.
[18] Gull, S. F. (1975). *Monthly Notices of the Royal Astronomical Society*, **171**, 263.
[19] Jun, B. and Norman, M. L. (1996). *Astrophysical Journal*, **472**, 245; Jun, B. and Jones, T. W. (1999). *Astrophysical Journal*, **511**, 774.
[20] Fesen, R. A. et al. (2001). *Astronomical Journal*, **122**, 2644.
[21] Blondin, J. M. and Ellison, D. C. (2001). *Astrophysical Journal*, **560**, 244.

generated at the reverse shock can extend all the way out to the outer shock front, providing radial magnetic fields instead of the tangential ones otherwise expected at shocks.

4.4.3   Supernovae and Galactic cosmic rays

The interstellar medium in the Galaxy is filled with relativistic particles that are also called cosmic rays. The energy density of this high-energy population is $u_{CR} \sim 1\,\text{eV cm}^{-3}$. One could now ask whether particle acceleration by SNRs can account for the entire energy density of cosmic rays in the Galaxy as already proposed by Baade and Zwicky in 1934. Let us denote the average time between supernovae by t_{SN}, the volume within which the cosmic rays are confined by V, the escape time from this volume by t_{esc}, and the mechanical energy release per supernova by E_{SN}. A typical value for the kinetic energy of a supernova is E_{SN} is 10^{51} erg and for the time between two successive supernovae $t_{SN} \sim 100$ years. Hence, the mechanical power by supernovae is E_{SN}/t_{SN}. Now, the fraction of this power that is needed to maintain the cosmic-ray population against escape equals

$$\eta \sim 0.05 \left(\frac{t_{SN}}{100\,\text{yr}} \right) \left(\frac{E_{SN}}{10^{51}\text{erg}} \right)^{-1} \left(\frac{t_{esc}}{10^7 \text{yr}} \right)^{-1}, \tag{4.64}$$

where we assumed for the volume that the Galactic cosmic rays occupy $V \sim 200$ kpc^3. The bottom line of this crude calculation is that it is entirely conceivable that SNRs are able to supply the bulk of the cosmic rays within our Galaxy.[22]

4.5   Exercises

4.1 **Radioactive power**

Assuming that 0.075 $M_\odot$ of ^{56}Co was produced by the decay of ^{56}Ni following the explosion of SN 1987A, estimate the amount of energy released per second through the radioactive decay of ^{56}Co just after its formation and 1 yr after the explosion.

4.2 **Standard candles**

SN 1996E, discovered by the high-redshift supernova search team, attained a maximum B-band apparent brightness at maximum of $m_B = 22.72$ mag. It is estimated that it suffered from a B-band extinction of $A_B = 0.10$ mag and had a $\Delta m_{15}(B) = 1.18$ mag. (Riess et al. [1988] AJ, 116, 1009).

(i) Estimate the distance modulus to SN 1996E, corrected for extinction.

(ii) Compute the linear distance to the supernova.

---

[22] However, supernovae are unlikely to explain the existence of the so-called ultra-high-energy cosmic rays.

4.3 Neutrinos

For neutrino energies much less than 1 GeV, the cross section for the nucleon scattering is

$$\sigma_\nu = \sigma_0 \left(\frac{E_\nu}{m_e c^2} \right)^2 , \tag{4.65}$$

where

$$\sigma_0 = \frac{4 G_F^2 m_e^2}{\hbar^4} = 1.76 \cdot 10^{-44} \text{ cm}^2, \tag{4.66}$$

where G_F is Fermi's constant.

(i) Show that the mean free path for scattering is

$$l_\nu = 2 \cdot 10^5 \left(\frac{E_\nu}{10 \text{ MeV}} \right)^{-2} \rho_{12}^{-1} \text{ cm}, \tag{4.67}$$

where ρ_{12} is the density in units of 10^{12} g cm^{-3}. The typical neutrino energy is 20 MeV, so the mean free path is only $0.5\rho_{12}^{-1}$ km.

(ii) If the neutrinos perform a random walk out of the neutron star, show that the time to diffuse a radial distance, R, is given by

$$t_{\text{diff}} = \frac{3 R^2}{l_\nu c}. \tag{4.68}$$

(iii) Show that for a uniform density sphere of mass $1.4 M_\odot$ we get

$$t_{\text{diff}} = 3.9 \cdot 10^{-3} \rho_{12}^{1/3} \left(\frac{E_\nu}{10 \text{ MeV}} \right)^2 \text{ s.} \tag{4.69}$$

4.4 Supernova remnants

For Cassiopeia A, the interpretation of both the hard X-ray and the TeV emission is critically dependent on the magnetic field strength. Equipartition arguments to explain the radio luminosity imply magnetic fields of B 0.42 mG,[23] far in excess of the canonical value for the ISM of 3 μG. Show that the typical loss time of an electron with energy E due to synchrotron radiation is

$$t_{\text{sync}} = \frac{E}{(dW/dt)_{\text{sync}}} \approx 635 \left(\frac{B}{G} \right)^{-2} \left(\frac{E}{\text{erg}} \right)^{-1} \text{ s.} \tag{4.70}$$

Show that for a magnetic field of 1 mG, a particle with $E = 8$ TeV loses its energy in 2 yr. The 2-yr lifetime is much shorter than the lifetime of Cassiopeia A, which is probably the remnant of a supernova that exploded in year 1680. Discuss possible origins of such high energies.

4.5 White dwarfs I

Calculate the escape velocity from the surface of a solar-mass white dwarf with the same radius as the Earth. If the surface temperature is 30 000 K, calculate the typical

---

[23] Wright, M. et al. (1999). *Astrophysical Journal*, **518**, 284.

velocity of a hydrogen atom near the surface. Comparing these two values, will much mass be lost from the surface of the white dwarf?

4.6 **White dwarfs II**

For a white dwarf of density 10^6 g cm^{-3} consisting only of ^{16}C, estimate the degeneracy pressure and compare it with the thermal pressure of a gas of temperature 10^7 K.

4.7 **White dwarfs III**

(i) What is the thermal energy stored in a white dwarf of temperature $T = 10^7$ K and a mass of 1 $M_\odot$?

(ii) Assuming that the white dwarf radiates like a black body of temperature $T = 10^4$ K, estimate its lifetime as a luminous object.

4.6 Further reading

Arnett, D. (1996). Supernovae and nucleosynthesis: An investigation of the history of matter, from the Big Bang to the present. Princeton Series in Astrophysics. Princeton, NJ: Princeton University Press.

Bethe, H. A. (1990). Supernova mechanisms. *Reviews of Modern Physics*, **62**, 801–66.

Bruenn, S. (1985). Stellar core collapse—Numerical model and infall epoch. *Astrophysical Journal Supplement Series*, **58**, 771–841.

Burrows, A. (2000). Supernova explosions in the universe. *Nature*, **403**(6771), 727–33.

Goldreich, P. and Weber, S. V. (1980). Homologously collapsing stellar cores. *Astrophysical Journal*, **238**, 991.

Hansen, C. J., Kawaler, S. D., and Trimble, V. (2004). *Stellar Interiors : Physical Principles, Structure, and Evolution*, 2nd edn. New York: Springer-Verlag.

Hillebrandt, W. and Niemeyer, J. C. (2000). Type IA supernova explosion models. *Annual Review of Astronomy and Astrophysics*, **38**, 191–230.

Janka, T. (2000). Core-collapse supernovae—successes, problems, and perspectives. *Nucleur Physics A*, **663**, 119.

Kotake, K., Sato, K., and Takahashi, K. (2006). Explosion mechanism, neutrino burst, and gravitational wave in core-collapse supernovae. *Reports on Progress in Physics*, **69**, 971.

Liddle, A. (2003). *An Introduction to Cosmology*, Chichester, UK: Wiley.

Livio, M. (2000). The progenitors of Type Ia supernovae. In *Type Ia Supernovae: Theory and Cosmology*, eds. J. C. Niemeyer and J. W. Truran.

Longair, M. (1992). *High Energy Astrophysics*. Cambridge: Cambridge University Press.

Padmanabhan, T. (2001). *Theoretical Astrophysics*. Cambridge: Cambridge University Press.

Prialnik, D. (2000). *An Introduction to the Theory of Stellar Structure and Evolution*. Cambridge: Cambridge University Press.

Rothenflug, R. et al. (2004). *Astronomy and Astrophysics*, **425**, 121.

Stephenson, F. R. and Clark, D. A. (2002). *The Historical Supernovae and Their Remnants*. Oxford: Oxford University Press.

Thielemann, F.-K., Nomoto, K., and Yokoi, K. (1986). Explosive nucleosynthesis in carbon deflagration models of Type I supernovae. *Astronomy and Astrophysics*, **158**, 17–33.

Thielemann, F.-K., Nomoto, K., and Hashimoto, M. (1996). Core-collapse supernovae and their ejecta. *Astrophysical Journal*, **460**, 408.

Vink, J., Bloemen, H., Kaastra, J. S., and Bleeker, J. A. M. (1998). *Astronomy and Astrophysics*, **339**, 201.

Weiler, K. (2003). *Supernovae and Gamma-Ray Bursters,* Vol. 598 of *Lecture Notes in Physics*. Berlin: Springer-Verlag.

Woosley, S. E. and Weaver, T. A. (1986). The physics of supernova explosions. In *Annual Review of Astronomy and Astrophysics*, Vol. 24. Palo Alto, CA: Annual Reviews, 205–53.

5

Neutron stars, pulsars, and magnetars

5.1 History

Already in 1934, Baade and Zwicky[1] had speculated that core-collapse supernovae could produce compact stars made of neutrons. In 1939, Oppenheimer and Volkoff[2] solved the general relativistic stellar structure equations for a pure neutron gas to obtain the interior structure of a neutron star. The observational discovery of these exotic stars, however, had to wait until 1967. By that time, Anthony Hewish from the Cavendish Laboratory in Cambridge had built a radio telescope. His idea was to use the so-called interplanetary scintillation to search for compact, quasistellar radio sources or quasars (Chapter 8). Like stars, whose apparent position on the sky changes erratically because of the air motion in the Earth's atmosphere (scintillation), the position of radio sources can exhibit a similar effect as a result of the solar wind in the interplanetary space. This interplanetary scintillation is stronger for compact than for extended sources, and thus Hewish hoped to identify quasars by looking for strongly scintillating sources. One of his PhD students, Jocelyn Bell, who was analyzing the output of the radio telescope, discovered a series of regular pulses of 1.3-s period in the data. After they had excluded terrestrial sources and were sure that they had picked up a regularly pulsed radio signal from outside the Solar System, they realized that the period was too short to come from the oscillation of a star. Although it seemed extremely unlikely, one of the first discussed explanations was that the signal they had picked up had been sent by another civilization in our Galaxy. Therefore, the source was initially dubbed "LGM" for little green men. After further data analysis, they found three more of these pulsing radio sources, or *pulsars*, and this made the explanation of other civilizations even more unlikely. Soon after they had published their discovery,

---

[1] Walter Baade (1893–1960), German astronomer and astrophysicist. Fritz Zwicky (1898–1974), Swiss-American physicist and astronomer.
[2] Julius Robert Oppenheimer (1904–1967), American theoretical physicist. George Michael Volkoff (1914–2000), Russian-Canadian theoretical physicist.

Thomas Gold and Franco Pacini made a strong case that pulsars are highly magnetized, rotating neutron stars. In 1974, Anthony Hewish was awarded the Nobel Prize for physics for his decisive role in the discovery of pulsars.

The first theoretical models of Oppenheimer and Volkoff just had to explain the structure of very compact stars. Today's theorists are challenged to explain a large zoo of exotic astrophysical objects that are now thought to be related to neutron stars: magnetically braking pulsars, accreting neutron stars in binary systems, soft gamma repeaters, anomalous X-ray pulsars, isolated neutron stars that cool as lonely black bodies, sources of astrophysical jets and so on.

5.2   Observational properties of pulsars

5.2.1   Location in the Galaxy

By the time of writing, more than 1700 pulsars have been discovered. An example of a newborn star inside a supernova remnant is shown in Fig. 5.1. They are usually found by identifying a dispersed, periodic signal hidden in the noisy time series that are collected by large radio telescopes. Pulsars populate preferentially the disk of our Galaxy, but some are found at very high latitudes, as apparent from

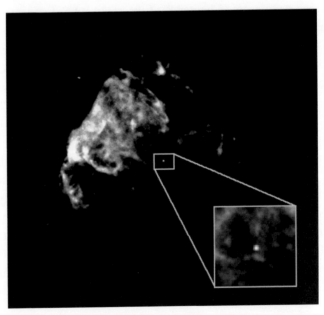

Figure 5.1   A young neutron star in the center of SNR Puppis A. Courtesy of S. Snowden, R. Petre (LHEA/GSFC), C. Becker (MIT), et al., ROSAT Project, NASA.

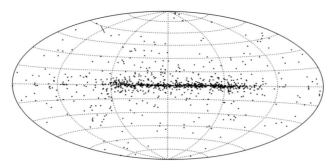

Figure 5.2   Sky distribution of 1026 pulsars in Galactic coordinates. The central horizontal line refers to the Galactic plane. Figure from D. R. Lorimer (2005). http://relativity.livingreviews.org/Articles/lrr-2005-7/.

Fig. 5.2. The distances to pulsars can be estimated using an effect known as *pulse dispersion.*

Assume the emitted pulse is spread over a finite frequency range. The dispersion has the effect that high frequencies travel with a larger group velocity through the interstellar medium than lower frequencies. Therefore, they arrive somewhat earlier at the radio telescope. It can be shown (see Exercise 5.1) that the resulting time delay between two frequencies, ν_{low} and ν_{high}, is

$$\Delta t = 4150 \text{ s} \left(\frac{1}{\nu_{low}^2} - \frac{1}{\nu_{high}^2} \right) \text{DM}, \tag{5.1}$$

where DM is the *dispersion measure*, which is defined as

$$\text{DM} = \int_0^d n_e(x) \, dx. \tag{5.2}$$

Here, d is the distance to the pulsar and n_e is the number density of free electrons along the line of sight to the pulsar. From the observations and Eq. (5.2), the dispersion measure can be determined. To infer the distance, d, one needs a model of the free electron distribution in the Galaxy. Such models can be calibrated with pulsars whose distance is known via other methods, such as the trigonometric parallax or the association with objects of known distances (e.g., supernova remnants). Distances to individual pulsars may be off by up to a factor of 2, but in a statistical sense the dispersion measure yields reliable distance estimates.

5.2.2 Spins

As all stars rotate, one expects that angular momentum conservation in a collapsing star leads to a dramatic increase in spin of the collapse product; see Exercise 5.2.

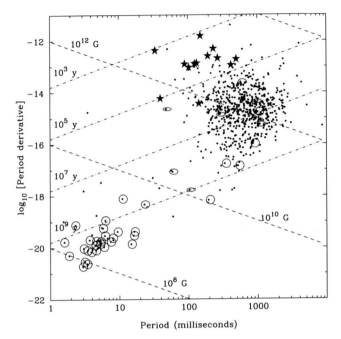

Figure 5.3 P–$\dot{P}$ diagram for a sample of radio pulsars. Circles indicate pulsars
that are known to be members of binary systems; ellipses refer to eccentric binary
systems. Stars indicate the association with an SNR. Figure from D. R. Lorimer
(2005). http://relativity.livingreviews.org/Articles/lrr-2005-7/.

Therefore, neutron stars are expected to rotate rapidly. They do indeed; observed
rotation periods are between 1.5 ms and several seconds. Generally, pulsars are
good clocks; that is, they emit their pulses very regularly (apart from abrupt small
changes in their rotational periods, called *glitches*, that occur from time to time).
They are, however, found to spin down very slowly. It is instructive to plot the
observed pulsars in a P–$\dot{P}$ diagram as shown in Fig. 5.3, where P refers to the
period and $\dot{P}$ to the absolute value of its time derivative. Obviously, pulsars group
in two broad classes: normal and millisecond pulsars. Normal pulsars have rotation
periods from ~ 0.1 to ~ 1 s and spin down at a rate of $\sim 10^{-15}$ s s^{-1}, whereas
millisecond pulsars spin with periods between 1.5 and 30 ms and become slower
at a rate of $\dot{P} \sim 10^{-19}$ s s^{-1}.

Pulsar observations are often interpreted in the framework of the magnetic dipole
model, which will be presented in detail in Section 5.4. Its main assumption is that
the observed slowdown of pulsars is caused entirely by the emission of magnetic
dipole radiation. Although we know from the measurement of the braking index
(see Section 5.4) that this model does not tell the full story, it is relatively simple
and provides a straightforward interpretation of many of the observed phenomena.
Let us quote two important predictions of the dipole model already here: (i) the

spin-down is related to the magnetic field (more precisely to the magnetic field at the pole), $\dot{P} \propto B^2$, and (ii) the magnetic field strength can be deduced from measuring P and $\dot{P}$: $B \propto \sqrt{P\dot{P}}$. Lines of constant field strength are shown in the diagram as dashed lines. The following tendencies can be read off this diagram:

- young pulsars (those with a clear association with a supernova remnant, indicated in the diagram with a star) have large magnetic fields of the order of 10^{12} G and therefore spin down rapidly (slow rotation, high B-field, rapid spin-down), and
- millisecond pulsars are preferentially found in binary systems and spin down very slowly (fast rotation, low B-field, slow spin-down).

Although the details of the magnetic field evolution are still not fully understood, the field is generally assumed to decay with the age of the pulsar. Looking at the class of normal pulsars, there seems to be an evolution from high B-fields and spin frequencies to lower ones. So how come the millisecond pulsars seem to be old (as inferred from their magnetic fields) and at the same time young (as inferred from their spin periods)? The explanation lies in their binary nature. They are thought to be neutron stars that have aged (i.e., their rotation has slowed down and their field has decayed) while orbiting a companion star. During its evolution, the companion star has transferred mass and angular momentum to the neutron star, which is spun up to fast rotation – or "recycled" – in this way. For further details of this evolution, see Chapter 6.

5.2.3   *Velocities*

Many pulsars have very large velocities, some as high as $\sim$1000 km s^{-1}, that are much higher than the velocities of their progenitor stars, which typically have velocities of 10 to 50 km s^{-1}. These velocities are often referred to as *kick velocities*. Given their violent birth process (see Chapter 4), this does not come as much of a surprise. Even a very small asymmetry in the core-collapse/explosion process imparts a kick velocity of several 100 km s^{-1} to the neutron star (see Exercise 5.3). If the pulsar progenitor was part of a binary system that was disrupted during the explosion (see Section 6.2.6), the orbital velocity also contributes to the pulsar kick.

As pulsars are the remnants of short-lived massive stars, they are born close to the place where their progenitors formed, usually in star-forming regions in the plane of the Galaxy. The high birth-kick velocities allow the pulsars to reach very high latitudes above the Galactic plane, as can be seen in Fig. 5.2. Given the broad distribution of initial neutron star kicks, as much as one half of all pulsars may be able to escape the gravitational potential of their parent galaxy and may end up in intergalactic space.

5.2.4 *Masses*

Generally, it is not trivial to measure the masses of isolated stars. Fortunately, about one half of the stars are members of binary (or multiple) stellar systems, which allows us to measure at least the total system mass (e.g., via Kepler's third law; see Eq. [6.32]). Among known pulsars, only about 4% are members of a binary system. This rarity is probably a result of the violent neutron star-formation process in a supernova explosion, which disrupts most binary systems (see Section 6.2.6). In the framework of Newtonian gravity, a binary orbit is completely determined by five Keplerian parameters. If general relativistic effects are important, up to five more post-Keplerian parameters may be needed, as we describe in Section 6.4.4. If we know all five Keplerian parameters and two more post-Keplerian parameters, then we can determine the individual binary component masses and the inclination angle of the orbit with respect to the line of sight. For systems whose orbital parameters are well known, this yields astonishingly accurate mass measurements. The mass of the famous binary pulsar PSR 1913+16, for example, is $1.4414 \pm 0.0002\ M_\odot$.[3]

The mass measurements of several pulsars are displayed in Fig. 5.4. Generally, pulsar masses are very close to $1.35\ M_\odot$. The maximum neutron star mass that can be stabilized against gravitational collapse by the matter pressure may be substantially higher, but there is no general agreement about the numerical value. To date, it is constrained to in the range given by the largest, well-known pulsar mass, the one of PSR $1913 + 16$, and an upper limit derived from causality arguments of about $3.2\ M_\odot$.

The standard formation mechanism in a Type II supernova (collapse of the iron core at the Chandrasekhar mass, radiating away parts of the binding energy in neutrinos and subsequent accretion of $\sim 0.1\ M_\odot$) may produce neutron stars with masses close to the observed $1.35\ M_\odot$ and possibly well less than the maximum value allowed by fundamental baryonic interactions. Other astrophysical processes such as accretion from a companion star may lead over time to more massive neutron stars. For example, neutron stars that accrete from a white dwarf companion seem to have masses that are consistently higher than the canonical value of $1.35\ M_\odot$ (see also Section 6.3.1). At the time of writing (summer 2006), the highest reported neutron star mass is $2.1 \pm 0.2\ M_\odot$ (1σ error).[4] If this mass stands the test of time, it would pose a severe challenge to our microscopic understanding of nuclear matter.

---

[3] Weisberg, J. M. and Taylor, J. H. (2005). *Proc. Aspen Conference*, ASP Conf. Series, ed. Rasio, F. A. and Stairs, I. H. San Francisco: Astronomical Society of the Pacific.
[4] Nice, D. J. et al. (2005). *The Astronomical Journal*, **634**, 1242.

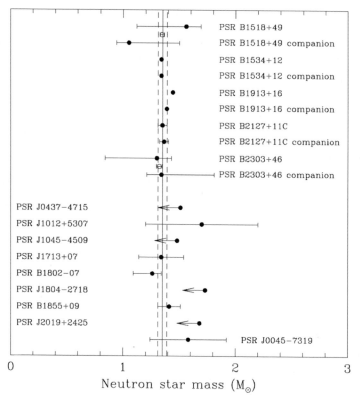

Figure 5.4   Distribution of pulsar masses in binary systems. The mass distribution in double neutron star systems is consistent with a Gaussian mass distribution centred on $m = 1.35 \pm 0.04\,M_\odot$, indicated by the vertical lines. Figure from Thorsett and Chakrabarty (1999). *Astrophysical Journal*, **7512**, 288.

5.3   Why rotating neutron stars?

So why do we believe that pulsars are rotating neutron stars? The following argumentation is based on three of the described pulsar properties:

- short periods down to 1.5 ms,
- periods do *increase* slowly, with a rate of $\dot{P} \sim 10^{-15}$ s s^{-1}, and
- they are good clocks, in the sense that they emit a very regular signal.

Simple causality arguments suggest that the diameter, D, of a source that emits a signal with a period, τ, should be smaller than the light-travel distance, $D < c\tau$[5] (Section 8.2.3). By inserting a period of 1.5 ms here, we see that the source must

---

[5] This argument is often used, for example, in the context of active galactic nuclei (AGN). The inequality is in principle only valid for nonrelativistic bulk motion. In the case of ultrarelativistic motion, the right-hand side has to be multiplied by $2\gamma^2$, where γ is the bulk Lorentz factor. This correction has decisive consequences in the context of gamma-ray bursts and quasars.

be smaller than about 450 km. Such a small size rules out white dwarfs as possible sources, as a white dwarf of 1.35 $M_\odot$ (i.e., just below the Chandrasekhar mass) has a radius of about 3000 km; see Fig. 4.4. Therefore, we are left with either neutron stars or black holes.

It is obviously difficult to attach any radiating entity to a black hole to produce the observed periodicity. It might still be argued that accretion onto the hole could be responsible for the periodicity. However, calculations show that black-hole accretion produces a rather irregular signal. Hence, we can rule out also black holes as the object behind the pulsar phenomenon.

Now that we are left with neutron stars, the question remains whether they pulsate or rotate. The oscillation period τ_{osc} is of the order of the dynamical time scale, which is given by

$$\tau_{osc} \approx \tau_{dyn} = \frac{1}{\sqrt{G\bar{\rho}}}, \qquad (5.3)$$

where $\bar{\rho}$ is the average density of the star. If a star were to pulsate at a period $\tau_p = 1.5$ ms, its average density would have to be

$$\bar{\rho} \approx \frac{1}{G\tau_p^2} \approx 7 \cdot 10^{12} \text{ g cm}^{-3}. \qquad (5.4)$$

As is shown later in this chapter, this density is about two orders of magnitude less than the average density of neutron stars.

Let us now consider rotation. The maximum possible rotation frequency of a star is determined by its mass-shedding limit. This occurs when the centrifugal forces on a mass element at the stellar surface start to overwhelm the gravitational attraction from the star.[6] Equating the centrifugal to the gravitational acceleration at the surface yields (ignoring a possible stellar deformation and general relativistic effects) a limiting rotation frequency of

$$\omega^2 = G\frac{M}{R^3} = G\frac{4}{3}\pi\bar{\rho} \qquad (5.5)$$

or

$$\bar{\rho} = \frac{3\pi}{GP^2}. \qquad (5.6)$$

Because it is unlikely that the observed neutron star spins at exactly the mass-shedding limit, the average densities derived from this equation represent a lower limit to the real average density. Again inserting a period $P = 1.5$ ms yields an

---

[6] Note that this refers to *rigid* rotation. If the star is rotating *differentially*, it can, for example, rotate very rapidly in the interior without shedding mass at the (slower rotating) equator. As the differential rotation is dissipated quickly, rigid rotation is still the relevant case for our discussion here.

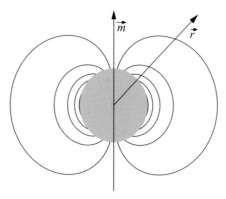

Figure 5.5   Sketch of dipole field.

average density of about $6 \cdot 10^{13}$ g cm^{-3}, which is still lower than but close to the real neutron star density.[7] The decisive argument against pulsation and in favor of rotation comes from the fact that the observed pulsars show a slow period *increase*. Pulsars emit large amounts of energy via radio emission and via relativistic winds. The Crab pulsar, for example, emits $\sim 10^{38}$ erg s^{-1} into the surrounding nebula. Neutron stars do not have a source of internal heating such as nuclear burning. They can, if at all, only contract, and their density should *increase*. As we have seen before, $\tau_{\mathrm{osc}} \propto \bar{\rho}^{-1/2}$, so a possible increase in density would lead to an increase in speed of the signal. In contrast, pulsars are observed to slow down. Therefore, it is concluded that pulsars are rotating neutron stars.

5.4   The magnetic dipole model

The magnetic dipole model is a very simple model of a pulsar. Nevertheless, it is able to explain many of the observed properties of pulsars. It assumes that a pulsar is a rigid, magnetized sphere that rotates in vacuum. The magnetic field is assumed to be dipolar with a magnetic moment, $\vec{m}$, that forms an angle α with respect to the rotation axis (see the sketch in Fig. 5.5).

5.4.1   *Emitted power*

Consider a dipolar magnetic field

$$\vec{B}(\vec{r}) = \frac{3\vec{n}(\vec{m} \cdot \vec{n}) - \vec{m}}{r^3}, \tag{5.7}$$

---

[7] Note that both these density estimates are well more than the average density of white dwarfs. The average density of a white dwarf of 1.35 $M_\odot$, for example, is about $6.6 \cdot 10^8$ g cm^{-3}.

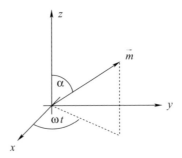

Figure 5.6   Sketch of the magnetic moment, $\vec{m}$, rotating around the z-axis.

where $r = |\vec{r}|$, $\vec{n} = \vec{r}/r$, and $\vec{m}$ is the magnetic dipole moment. This can be used to express the magnetic field at the pole, using $\vec{m} \parallel \vec{n}$, $r = R$, where R is the stellar radius,

$$\vec{B}_p = \frac{2\vec{m}}{R^3},$$
(5.8)

or, vice versa,

$$\vec{m} = \frac{\vec{B}_p R^3}{2}.$$
(5.9)

Completely analogously to the power emitted by an electric dipole, $(dW/dt)_{\text{dip}} = -(2/3c^3)|\ddot{\vec{d}}|^2$ (Larmor's formula; $\vec{d}$ is the electric dipole moment; see Section 3.7.6), a time-varying magnetic dipole radiates a power

$$\left(\frac{dW}{dt}\right)_{\text{md}} = -\frac{2}{3c^3}|\ddot{\vec{m}}|^2.$$
(5.10)

Now we need the second time derivative of a dipole moment, $\vec{m}$, that forms an angle α with the rotation axis and rotates with an angular frequency ω. If we assume that the rotation axis coincides with the z-axis of our coordinate system, we can express at any given time the magnetic moment, $\vec{m}$, in polar coordinates as $\vec{m} = m_0 \hat{e}_m$, where $m_0 = B_p R^3/2$ and $\hat{e}_m = (\sin \alpha \cos \omega t, \sin \alpha \sin \omega t, \cos \alpha)$; see Fig. 5.6.

Taking two derivatives with respect to time gives

$$|\ddot{\vec{m}}|^2 = m_0^2 \omega^4 \sin^2 \alpha,$$
(5.11)

and the total emitted power is

$$\left(\frac{dW}{dt}\right)_{\text{md}} = -\frac{B_p^2 R^6}{6c^3} \omega^4 \sin^2 \alpha$$
(5.12)

$$= -6.2 \cdot 10^{27} \text{erg s}^{-1} \left(\frac{B_p}{10^{12}G}\right)^2 \left(\frac{R}{10^6 \text{cm}}\right)^6 \left(\frac{\omega}{1 \text{s}^{-1}}\right)^4 \sin^2 \alpha.$$
(5.13)

The dipole model assumes that the emitted energy is tapped from the rotational energy of the pulsar

$$E_{\rm rot} = \frac{1}{2} I \omega^2, \tag{5.14}$$

where I is the moment of inertia. Assuming the pulsar to be a rigid sphere ($dI/dt = 0$), its change of energy is given by

$$\frac{dE_{\rm rot}}{dt} = I \omega \dot{\omega}. \tag{5.15}$$

By solving for $\dot{\omega}$ and equating $dE_{\rm rot}/dt$ and $(dW/dt)_{\rm md}$ given by (5.12), one finds

$$\dot{\omega} = -\frac{B_{\rm p}^2 R^6 \omega^3 \sin^2 \alpha}{6c^3 I} < 0; \tag{5.16}$$

that is, the pulsar slows down as expected.

5.4.2 Pulsar ages

It is convenient to define a characteristic timescale for this slowdown, the so-called dipole age of a pulsar, which is defined as

$$\tau_{\rm dipole} \equiv -\left(\frac{\omega}{2\dot{\omega}}\right) = \frac{3c^3 I}{B_{\rm p}^2 R^6 \omega^2 \sin^2 \alpha}, \tag{5.17}$$

where we have used Eq. (5.16). The values of ω and $\dot{\omega}$ are the measured values of the angular frequency and its time derivative. Let us apply this estimate to the Crab Pulsar, which is shown in Fig. 5.7. The Crab pulsar has an observed period $P = 0.033$ s and a period derivative $\dot{P} = 4.2 \cdot 10^{-13}$ s s^{-1}. If we insert the measured P and $\dot{P}$ for a neutron star into Eq. (5.17), we find $\tau_{\rm dipole} \approx 1247$ years; the supernova in which the Crab pulsar was born has been observed by Chinese astronomers in AD 1054. This gives us some confidence that the dipole age is a reasonable order-of-magnitude estimate.

This reasoning can be generalized without specifying the details of the energy-loss process. Let us assume that the pulsar spins down (e.g., via emission of magnetic dipole radiation) according to

$$\dot{\omega} = -K \omega^n. \tag{5.18}$$

The exponent n is called the *braking index*, and K is a proportionality constant. Integrating this equation from t_0 to t and from ω_0 to ω, respectively, where the

Figure 5.7   A giant Hubble mosaic of the Crab Nebula. Figure courtesy of NASA and STScI.

subscript 0 refers to the value at birth, yields

$$t = \frac{1}{K(n-1)} \left(\frac{1}{\omega^{n-1}} - \frac{1}{\omega_0^{n-1}} \right),$$
(5.19)

where we have set $t_0 = 0$. If we now insert $K = -\dot{\omega}/\omega^n$ from Eq. (5.18), we have

$$t = -\frac{1}{n-1} \left(\frac{\omega}{\dot{\omega}} \right) \left[1 - \left(\frac{\omega}{\omega_0} \right)^{n-1} \right].$$
(5.20)

If we assume that the observed pulsar is "old" and has spun down considerably since birth, $\omega_0 \gg \omega$, we can neglect the last term in the curly brackets. The remaining expression is called the *characteristic pulsar age*:

$$\tau_{pa} = -\frac{1}{n-1} \left(\frac{\omega}{\dot{\omega}} \right).$$
(5.21)

By comparing Eq. (5.16) with Eq. (5.18), one finds that the braking index in the magnetic dipole model is 3. When we insert this into Eq. (5.21), we recover the dipole age.

Braking indices that are determined observationally lie in the range between 2 and 3, so magnetic dipole emission definitely cannot be the full story.

5.4.3   *Estimating the magnetic field strength*

Assuming that the dipole model is a fair description of a real pulsar, we can use observable quantities, such as the period, P, and its time derivative, $\dot{P}$, to estimate the strength of the magnetic field of a pulsar. By using $\omega = 2\pi/P$ and $dE_{rot}/dt = I\omega\dot{\omega} = (dW/dt)_{md}$ and Eq. (5.12), we can solve for $B_p \sin\alpha$:

$$B_p \sin\alpha = \sqrt{\frac{3c^3 I}{2\pi^2 R^6} \dot{P} P}, \tag{5.22}$$

or, if we only need a rough estimate, we can insert typical numbers and $\sin\alpha \approx 1$ to find

$$B \approx 3.2 \cdot 10^{19} G \left(\frac{P\dot{P}}{1s}\right)^{1/2}. \tag{5.23}$$

Using the dipole model, we can now estimate the magnetic field strength of the Crab pulsar. By inserting $P = 0.033$ s and $\dot{P} = 4.2 \cdot 10^{-13}$ s s^{-1}, we estimate the magnetic field as $B_{crab} \approx 10^{12}$ G. As we can see from Fig. 5.3, this field strength is typical for a young pulsar.

Are such high field strengths plausible? If a typical star with a magnetic field of ~ 100 G collapses from 10^6 km to 10 km, its magnetic field strength is expected be amplified to just about 10^{12} G (see Exercise 5.4).

5.5   The stellar structure equations

The structure of a star is determined by *hydrostatic equilibrium*; see Eq. (2.57). This equation describes the balance between gravity and the pressure forces that are a result of the microscopic interactions in the stellar material. Neutron stars are the only stars whose structure is noticeably influenced by General Relativity. Therefore, to calculate a neutron star model, one has to solve the relativistic generalization of the stellar structure equations, the *Tolman–Oppenheimer–Volkoff equations*.

Neutron star matter cannot be described realistically as an ideal gas. As we have seen in the Chapter 4, the typical separations between nucleons in a neutron star are comparable to their radius. In this sense, one can think of the nucleons as touching each other. In fact, they are even so close that the short-range repulsive core of the strong interaction becomes important. Therefore, the structure of neutron stars is determined by all fundamental physical interactions: gravity controls the attraction, the strong interaction provides the most important pressure component, the weak interaction sets the neutron-to-proton ratio via the so-called β-equilibrium

condition, and the electromagnetic interaction determines the lattice structure of
the atomic nuclei that form the neutron star crust.

5.5.1   *The Newtonian case*

In general, the equations of stellar structure contain an equation that relates the
density to the mass (mass equation), an equation that expresses the hydrostatic
equilibrium or the balance of momentum (momentum equation), an equation for
the gravitational potential (potential equation), an equation that relates nuclear
energy generation to the luminosity (energy equation), and finally, if radiation is
assumed to be transported via diffusion, a radiative diffusion equation.

Here, we restrict ourselves to the case of cold neutron stars, so that we are
only concerned with the first three of these equations. To derive the Newtonian
stellar-structure equations, think of the star as being built up of infinitesimally thin,
homogeneous shells, at radius r with a thickness of dr, as illustrated in Fig. 5.8.
The mass equation can be found trivially by relating the mass of a shell, dm, to its
density ρ:

$$dm = 4\pi r^2 \rho \, dr, \tag{5.24}$$

or

$$\frac{dm(r)}{dr} = 4\pi r^2 \rho(r). \tag{5.25}$$

For a more natural transition to the Tolman–Oppenheimer–Volkoff equations, it
is favorable to replace the mass density ρ by the mass energy density $\epsilon = \rho c^2$

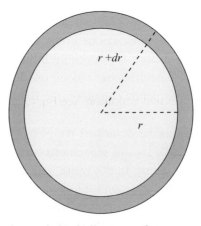

Figure 5.8   Consider the star being built up out of a sequence of concentric shells.
In hydrostatic equilibrium, the gravity on the shell has to be balanced its pressure
gradients; see the text.

(completely analogous to $E = mc^2$):

$$\frac{dm(r)}{dr} = \frac{4\pi r^2 \epsilon(r)}{c^2}, \qquad (5.26)$$

from which the total mass of the star can be found by integration from the center to the stellar surface at R, which is defined as the location where the pressure vanishes:

$$M = \int_0^R \frac{dm}{dr} dr = \frac{4\pi}{c^2} \int_0^R \epsilon(r) r^2 dr. \qquad (5.27)$$

The momentum equation expresses the hydrostatic equilibrium condition: gravity has to balance pressure. Consider a shell bounded by the radii r and $r + dr$. Its net force per area (i.e., its pressure change) is given by

$$P(r) - P(r + dr) = -\frac{dP}{dr} dr, \qquad (5.28)$$

where the minus sign on the right-hand side appears as the pressure decreases outward. The gravitational force on the shell is determined by its enclosed mass:

$$f_{\text{grav}} = \frac{Gm(r)}{r^2} dm = \frac{Gm(r)}{r^2} \rho 4\pi r^2 dr. \qquad (5.29)$$

For local equilibrium, the gravitational force *per area* has to balance the pressure change; that is,

$$-\frac{dP}{dr} dr = \frac{Gm(r)}{r^2} \rho \, dr, \qquad (5.30)$$

or

$$\frac{dP}{dr} = -\frac{Gm(r)}{r^2} \rho. \qquad (5.31)$$

This is the hydrostatic equilibrium condition; see Eq. (2.57). Again, one may want to replace ρ by ϵ/c^2. The equation for the radial change of the gravitational potential Φ,

$$\frac{d\Phi}{dr} = \frac{Gm(r)}{r^2}, \qquad (5.32)$$

can be rewritten using Eq. (5.31):

$$\frac{d\Phi}{dr} = -\frac{1}{\rho}\frac{dP}{dr}. \qquad (5.33)$$

5.5.2   *The Tolman–Oppenheimer–Volkoff equations*

As mentioned previously, the relativistic generalization of the previous equations goes back to Tolman, Oppenheimer, and Volkoff. The derivation of the Tolman–Oppenheimer–Volkoff equations is beyond the scope of this book, but it can be found in most textbooks on General Relativity (GR); see, for example, *Gravitation* by Misner, Thorne, and Wheeler or *Gravitation and Cosmology* by Weinberg.

The Tolman–Oppenheimer–Volkoff equations can be written in a form that corresponds to the Newtonian stellar-structure equations supplemented by correction factors:

$$\frac{dm}{dr} = \frac{4\pi r^2 \epsilon}{c^2} \tag{5.34}$$

$$\frac{dP}{dr} = -\frac{G\epsilon m(r)}{c^2 r^2} \left(1 + \frac{P}{\epsilon}\right) \left[1 + \frac{4\pi r^3 P}{m(r)c^2}\right] \left[1 - \frac{2Gm(r)}{c^2 r}\right]^{-1} \tag{5.35}$$

$$\frac{d\Phi}{dr} = -\frac{c^2}{\epsilon}\frac{dP}{dr}\left(1 + \frac{P}{\epsilon}\right)^{-1}. \tag{5.36}$$

The mass equation looks exactly like the Newtonian version. The momentum Eq. (5.35) is the Newtonian one extended by three correction factors, of which the first two arise because of special relativistic effects, and the last one is of general relativistic nature.

This last correction factor provides us with a criterion of whether General Relativity is important (and has to be accounted for). To be unimportant, the condition

$$\zeta \equiv \frac{2GM}{c^2 R} \ll 1 \tag{5.37}$$

has to be fulfilled. To be dimensionally correct, the quantity $2GM/c^2$ has to have the dimension of a length. This length often appears in the context of black holes and is generally referred to as *Schwarzschild radius*; see the box on black holes in the next chapter. The ratio of an object's Schwarzschild radius and its actual physical radius indicates the importance of general relativistic effects. One often speaks about the *compactness parameter* GM/Rc^2, or M/R if $G = c = 1$ is used, a convention that is frequently adopted in the context of General Relativity. In this convention, velocities are measured in units of the speed of light, and masses are measured in units of lengths: 1 $M_\odot$ corresponds to a length of $1.99 \cdot 10^{33}$g $G/c^2 \approx 1.5$ km. Thus, the Schwarzschild radius of our Sun is 2 $M_\odot = 2 \cdot 1.5$ km $= 3$ km. Using the radius of our Sun, $\approx 7 \cdot 10^5$ km, we find $\zeta = 3$ km/$7 \cdot 10^5$ km $= 4 \cdot 10^{-6}$; that is, our Sun is a completely nonrelativistic object.

For a typical white dwarf of 0.6 $M_\odot$ and a radius of about 10^9 cm, ζ is approximately $0.6 \cdot 2 \cdot 1.5$ km/10^4 km $\approx 1.8 \cdot 10^{-4}$, so General Relativity corrections can be safely ignored. But for a neutron star with 1.4 $M_\odot$ and 10-km radius, $\zeta \approx 1.4 \cdot 2 \cdot 1.5$ km/10 km ≈ 0.4; therefore general relativistic effects are important for its structure.

In a general relativistic calculation, the integration of Eq. (5.34) yields the *gravitational mass* M_{grav}. This quantity is different from the baryonic mass defined as the sum of all the N baryons in the star, $M_{bar} = N m_{bar}$, where $m_{bar} \approx m_u = 1.66 \cdot 10^{-24}$ g. The difference comes from the negatively counted gravitational binding energy, B, that is, the energy that is needed to disperse all the baryons of the star to infinity. Therefore, $M_{bar} = M_{grav} + |B|/c^2$. The value of B depends on the structure of a neutron star: a compressible, "soft" equation of state leads to a very compact star and therefore a large gravitational binding energy. Typically, the binding energy of a neutron star corresponds to about 0.2 $M_\odot c^2$.

5.6 The equation of state of neutron stars

Now that we have established the set of *macroscopic* equations, we have to address the question of how the pressure is related to energy density (and in principle also to the composition and the temperature). This means we have to find the *microscopic* equation of state of neutron star matter. This is one of the most challenging questions in astrophysics, and to date there are still many unsettled issues.

We follow the historical evolution in the sense that we start by treating in some detail the equation of state of a noninteracting, ideal Fermi gas, just as in the seminal work of Tolman, Oppenheimer, and Volkoff in 1939. They found that neutron stars have a radius of about 10 km, not too far away from modern results, but a maximum mass of only 0.7 $M_\odot$. We subsequently discuss effects that are not included in this simplified treatment. This, however, remains on a rather qualitative level, as quantitative calculations require techniques that are beyond the scope of this book.

5.6.1 Toy equation of state: Noninteracting fermions

We start by assuming that the neutron star is made entirely out of neutrons. This is not entirely true. We know from laboratory experiments that neutrons decay with a half-life of about 10.5 minutes into protons. In a real neutron star, some fraction of the neutrons will decay via $n \rightarrow p + e + \bar{\nu}_e$, where the neutrino escapes, until a sufficiently large "sea" of electrons has been built up. Because both electrons and protons are fermions themselves, they obey the Pauli exclusion principle; that is, they cannot occupy the same energy state. Therefore, if more and more of them are produced, they have to occupy higher and higher energy levels. At some stage, the fermions produced in a neutron decay would not be energetic enough to populate the free high-energy states. That means that this decay process is not possible anymore, because it would violate energy conservation. This suppression of reactions due to the fermionic nature of the final state products is referred to as *Pauli blocking*. Thus, rather than just pure neutrons, a neutron star contains (at least) a mixture of neutrons, protons, and electrons but probably also more exotic ingredients, as we describe later.

Here, we assume that the neutrons constitute a noninteracting Fermi gas, very much as in the white dwarf case, but now the pressure is provided by the neutrons rather than the electrons. Again, this is not entirely true. The interaction of the nucleons plays a very important role, and neglecting them leads to the small maximum mass that has been found by Tolman, Oppenheimer, and Volkoff.

Let us start with the most simple case where we only have one species. As we had seen in Chapter 2, once we have the statistical weight for the involved particles and know their distribution function (here Fermi–Dirac distribution), we can derive the number density n, the pressure P, and the energy density ϵ by integration over the distribution; see Eqs. (2.62), (2.63), and (2.64). We treat only cold ($T = 0$) neutron stars. Apart from the very early stages after the neutron star birth, this is a very good approximation, because typical Fermi energies in a neutron star are of the order MeV. This corresponds to a temperature of $T = 1$ MeV$/k_B \approx 10^{10}$ K. Therefore, the temperature only plays a role for very high temperatures; otherwise $T = 0$ is an excellent approximation.

In the $T = 0$ case, the distribution function becomes a step function that vanishes beyond the Fermi momentum p_F (see Chapter 2). The more general case with $T \neq 0$ is technically more complicated but conceptually very similar. In the $T = 0$ limit, the number density is

$$n = \int dn = \frac{8\pi}{h^3} \int_0^{p_F} p^2 dp = \frac{8\pi}{3h^3} p_F^3, \tag{5.38}$$

or, if the number density is known, the Fermi momentum reads

$$p_F = \left(\frac{3h^3 n}{8\pi} \right)^{1/3}. \tag{5.39}$$

The energy that corresponds to this momentum is called *Fermi energy*:

$$E_F = \sqrt{p_F^2 c^2 + m^2 c^4}. \tag{5.40}$$

The energy density is given by

$$u(p_F) = \int E \, dn = \frac{8\pi}{h^3} \int_0^{p_F} \sqrt{p^2 c^2 + m^2 c^4} \, p^2 dp \tag{5.41}$$

$$= \frac{8\pi}{h^3} m^4 c^5 \int_0^{p_F/mc} \sqrt{1 + z^2}\, z^2 dz \tag{5.42}$$

$$= u_0[\sqrt{1 + x^2}(x + 2x^3) - \sinh^{-1} x], \tag{5.43}$$

where we have substituted $z = pc/mc^2$ and used the dimensionless momentum variable $x = p_F/mc$ and defined the quantity $u_0 = \pi m^4 c^5/h^3$, which has the dimensions of energy per volume.

A very useful concept in this context is the *chemical potential*. Let us first introduce the abundance of species i, Y_i, as the ratio of the number density of this particular species to the nucleon number density:

$$Y_i = \frac{n_i}{n} = \frac{n_i}{\rho/m_{\mathrm{u}}}, \tag{5.44}$$

where m_{u} is the atomic mass unit and $m_{\mathrm{u}} = 1.66 \cdot 10^{-24}\mathrm{g} = 931.49\ \mathrm{MeV}/c^2$. Generally, the energy density depends on the number density n, the specific entropy (i.e., entropy *per mass*) s, and the abundances Y_i:

$$u = u(n, s, Y_i). \tag{5.45}$$

With this notation, we can write the first law of thermodynamics as

$$d\left(\frac{u}{n}\right) = -P d\left(\frac{1}{n}\right) + T ds + \sum_i \bar{\mu}_i dY_i, \tag{5.46}$$

where pressure is

$$P \equiv -\left.\frac{\partial(u/n)}{\partial(1/n)}\right|_{s,Y_i}, \tag{5.47}$$

the temperature is

$$T \equiv \left.\frac{\partial(u/n)}{\partial s}\right|_{n,Y_i}, \tag{5.48}$$

and the chemical potential of species i is

$$\bar{\mu}_i \equiv \left.\frac{\partial(u/n)}{\partial Y_i}\right|_{n,s} = \left.\frac{\partial u}{\partial n_i}\right|_{n,s}. \tag{5.49}$$

The rest mass is included in the chemical potential as it is included in u. If we refer to the chemical potential including rest mass, we use the bar notation; to denote the chemical potential without rest mass, we use $\mu_i = \bar{\mu}_i - m_i c^2$. From Eq. (5.49) we see that the chemical potential is the energy required to add a new particle to the system. For the case of vanishing temperature, $\bar{\mu}_i$ just corresponds to the Fermi energy of Eq. (5.40).

Similar to the calculation of the energy density, we can integrate Eq. (2.62) to find the pressure. To this end, we need to express the velocity in terms of energy and momentum. By using $p = \gamma m v$ and $E = \gamma m c^2$, we find $v = pc^2/E$ and therefore

$$
\begin{aligned}
P(p_{\mathrm{F}}) &= \frac{1}{3} \int_0^{p_{\mathrm{F}}} p v \frac{8\pi p^2 dp}{h^3} \\
&= \frac{8\pi c^2}{3h^3} \int_0^{p_{\mathrm{F}}} \frac{p^4 dp}{(p^2 c^2 + m^2 c^4)^{1/2}} \\
&= \frac{u_0}{3} \int_0^{p_{\mathrm{F}}/mc} \frac{z^4 dz}{\sqrt{1 + z^2}} \\
&= \frac{u_0}{24} [(2x^3 - 3x)\sqrt{1 + x^2} + 3 \sinh^{-1}(x)],
\end{aligned}
\tag{5.50}
$$

where we have again used the abbreviations u, u_0, and x from before. Although the last expressions for ϵ and p are not particularly enlightening, we keep them here for general reference.

Let us consider the limits of the general expression for the pressure, Eq. (5.50). If the considered particles are non-relativistic, we can use $v = p/m$, in Eq. (2.62), and therefore the pressure is given by

$$
P = \frac{8\pi}{3h^3 m} \int_0^{p_{\mathrm{F}}} p^4 dp = \frac{8\pi}{15 h^3 m} p_{\mathrm{F}}^5
\tag{5.51}
$$

If we now insert Eq. (5.39), we have an equation of state with $P \propto \rho^{5/3}$. Such an equation of state of the form $P = K\rho^{\Gamma}$ is called a *polytrope*, where Γ is called the polytropic exponent.

In the ultrarelativistic limit, where $v \approx c$, Eq. (5.50) reduces to

$$
P = \frac{8\pi c}{3h^3} \int_0^{p_{\mathrm{F}}} p^3 dp = \frac{2\pi c}{3h^3} p_{\mathrm{F}}^4,
\tag{5.52}
$$

and by using $p_{\mathrm{F}} \propto n^{1/3}$, we have $P \propto \rho^{4/3}$. These limits are particularly useful in the context of white dwarfs (see Box in Chapter 4).

If we specify this ideal fermion equation of state for neutrons, we can construct neutron star models by solving Eq. (5.35) together with Eq. (5.50).

We had mentioned before that some fraction of electrons and protons is present as neutrons have the tendency to decay. But how large will that fraction be, and does it change with the radius within the neutron star? As the equilibrium is determined by β-decays and their inverse reaction, one speaks of *β-equilibrium*; the decay reaction $n \rightarrow p + e + \bar{\nu}_e$ and the electron capture reaction $e + p \rightarrow n + \nu_e$ have to be equally fast. You may remember from your thermodynamics lectures that such a reaction equilibrium, $A + B \leftrightarrow C + D$, sets a condition for the involved chemical potentials

$$\bar{\mu}_A + \bar{\mu}_B = \bar{\mu}_C + \bar{\mu}_D. \tag{5.53}$$

As the neutrino does escape, the β-equilibrium condition for the chemical potentials reads

$$\bar{\mu}_n = \bar{\mu}_p + \bar{\mu}_e, \tag{5.54}$$

where the mass energies are included, $\bar{\mu}_i = \mu_i + m_i c^2$. These chemical potentials are related to the number densities; therefore at each point inside the neutron star, Eq. (5.54) allows us to determine the fraction of available electrons. As we see in the section about the cooling of neutron stars, a high electron fraction is necessary for the fast neutrino cooling processes.

If we have such a mixture of Fermi gases, say neutrons, protons, and electrons, the total energy density and pressure are given as the sum of the constituents:

$$u = \sum_{i=e,p,n} u_i \quad \text{and} \quad P = \sum_{i=e,p,n} P_i. \tag{5.55}$$

The neutron stars resulting from such an equation of state are not too different from the one with a pure neutron Fermi gas, which we discussed previously.

5.6.2   *The maximum neutron star mass*

Let us look again at the Tolman–Oppenheimer–Volkoff version of the hydrostatic pressure equation (5.35). As in its Newtonian version, the pressure gradients on the left-hand side have to balance the gravitational terms on the right-hand side. Contrary to the Newtonian case, in the Tolman–Oppenheimer–Volkoff equations the pressure appears both on the left- and the right-hand side of the equation. It provides support against gravity, but it also appears in the special relativistic extra terms on the right-hand side of Eq. (5.35), and being a form of energy, the pressure is a source of gravity itself. Pressure therefore has the pivotal role for the general relativistic hydrostatic equilibrium: if the mass becomes too large, the required high pressure itself enhances gravity, which ultimately leads to the collapse of the neutron star to a black hole. This occurs even for the (academic) case of a completely incompressible equation of state.

What is the numerical value of this maximum or *limiting neutron star mass* beyond which no stable neutron stars are possible? As we show in the next chapter, there are extremely accurate neutron star mass measurements from binary systems. The highest well-known neutron star mass is the one of the pulsar in the binary system PSR 1913+16, $1.4414 \pm 0.0002\, M_\odot$[8]; this is a strict lower limit on the limiting mass. As mentioned before, there are indications that the true limiting mass may be substantially higher than this value. The best theoretical neutron star models cannot be used to severely constrain the maximum mass because their predictions show a rather large spread. Rhoads and Ruffini[9] showed from very general and robust physical principles such as causality and the requirement that the pressure should increase with increasing density (*Le Chatelier's principle*) that the maximum mass cannot be larger than $3.2\, M_\odot$. Thus, the limiting mass must lie somewhere between 1.44 and $3.2\, M_\odot$.

5.6.3   More realistic equations of state

In all of the previous discussions of the neutron star equation of state, we have neglected the interactions between the nucleons. The mass of a typical neutron star, $M = 1.4\, M_\odot$, corresponds to about $1.7 \cdot 10^{57}$ nucleons. If we assume that the nucleons are distributed approximately homogeneously in a volume of radius $R = 12$ km, the typical distance between two nucleons is about $1\,\mathrm{fm} = 10^{-13}$ cm. From scattering experiments, we know that a nucleon is of comparable radius, about 0.8 fm (i.e., the nucleons are practically touching each other). Therefore, their interaction in a neutron star cannot be ignored. If the interactions between the nucleons are taken into account, the resulting neutron stars can be much more massive than the $0.7\, M_\odot$ found by Tolmann, Oppenheimer, and Volkoff in 1939.

We now briefly introduce the most important possible matter constituents of neutron stars. The lightest particles are spin-1/2 fermions that do not experience the strong nuclear force, so-called *leptons*. The lepton family consists of six particles of three flavors: electron (rest mass energy 0.511 MeV) and electron-neutrino, muon (105.6 MeV) and μ-neutrino, and tau (1777 MeV) and τ-neutrino. The masses of the neutrinos are not known exactly, but very small.[10] Of course, for each of these particles, there exists an antiparticle.

Baryons are fermions that are strongly interacting; that is, they experience the nuclear force. Their name comes from Greek and means "heavy particles." The most important baryons are those that form the *baryon octet* shown in Fig. 5.9. The lightest and most familiar baryons are the proton and the neutron. The remaining $\Lambda-$, $\Sigma-$, and $\Xi-$ particles also contain strange quarks and

---

[8] Weisberg, J. M. and Taylor, J. H. (2005). *Proc. Aspen Conference*, ASP Conf. Series, ed. Rasio, F. A. and Stairs, I. H., San Francisco: Astronomical Society of the Pacific.

[9] Rhoades, C. E. and Ruffini, R. (1974). *Physical Review Letters*, **32**(6), 324.

[10] Cosmological arguments suggest that the sum of the neutrino masses is less than 2 eV; see Fukugita et al. (2006). *Physical Review D*, **74**, 027302.

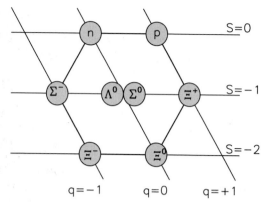

Figure 5.9 The most important baryons form the so-called baryon octet, q refers to the electric charge and S to the strangeness quantum number.

therefore carry a nonzero strangeness quantum number S. Such strange baryons are called *hyperons*. The properties of the most important baryons are summarized in Table 5.1.

The baryons interact with each other via the *strong interaction* that is mediated by the exchange of so-called *mesons*. Mesons are "messenger" particles made of an even number of quarks and antiquarks, and being made of particles and antiparticles, they are short-lived. Obviously, as their spins add up to integer numbers, they are bosons and therefore follow the Bose–Einstein statistics. The mesons that are most important in the neutron star context are shown in Table 5.2. The most important messengers between the baryons are the σ-, ω-, and ρ-mesons. Pions (π) and kaons (K^+, K^0, $\bar{K}^0$, K^-) can possibly occur as condensates in the inner cores of neutron stars. Baryons and mesons are collectively referred to as *hadrons*.

The most conservative matter constitution would be a mixture of neutrons, protons, electrons, and, at higher densities, muons, particles very similar to electrons but with higher masses. At matter densities of a few times the matter density inside an atomic nucleus, the Fermi energies of the nucleons can become so large that it is energetically favorable to transform some of the nucleons via electroweak interaction into heavier baryons such as $\Lambda-$ and $\Sigma-$hyperons. This leads to a larger number of species with lower Fermi momenta. As the highest momentum nucleons are removed, this process leads to a pressure reduction and thus a "softening" of the equation of state. Generally, a softer equation of state leads to more compact stars as it takes higher densities to reach the pressure needed to balance gravity. A soft equation of state leads to neutron stars with small radii and a relatively small maximum mass.

Table 5.1   *Important baryon states for neutron stars*

Particle	m (MeV)	J	I	b	q	S
N	939	1/2	1/2	1	0, 1	0
Λ	1115	1/2	0	1	0	−1
Σ	1190	1/2	1	1	−1, 0, 1	−1
Ξ	1315	1/2	1/2	1	−1, 0	−2
Δ	1232	3/2	3/2	1	−1, 0, 1, 2	0
Ω	1673	3/2	0	1	−1	−3

Letter m refers to the particle mass, J its spin, I its isospin, b the baryon number, q the electric charge (in units of elementary charges), and S the strangeness. Further details concerning the quantum numbers can be found in introductory texts on particle physics, for example, in Perkins, D. H. (2000). *Introduction to High Energy Physics*. Oxford: University of Oxford.

Table 5.2   *Mesons important in the context of neutron stars*

Particle	m (MeV)	J	I	b	q	S
σ	ca. 800	0	0	0	0	0
ω	782	1	0	0	0	0
ρ	770	1	1	0	−1, 0, 1	0
π	139	0	1	0	−1, 0, 1	0
K^+	494	0	1/2	0	1	1
K^0	494	0	1/2	0	0	1
$\bar{K}^0$	494	0	1/2	0	0	−1
K^-	494	0	1/2	0	−1	−1

5.7   Realistic neutron stars

We now give an overview of more realistic neutron star models. Because the technical details of the various approaches are beyond the scope of this text, the following discussion remains mostly qualitative.

The structure of a neutron star can be subdivided into different regions: the surface, the crust, and the outer and inner core. We briefly review each of these regions.

5.7.1   Surface

The surface of a neutron star, sometimes subdivided into atmosphere and envelope, has densities less than 10^6 g cm^{-3} and contributes only a negligible amount to the total mass of the neutron star. It can, however, be important for some observable quantities. For example, the only 1-cm-thick neutron star atmosphere shapes the emergent photon spectrum, and it can be crucial for the interpretation of observations to know whether a blackbody spectrum is a good approximation. To interpret observations properly, one has to understand the atomic physics in the atmosphere. Here it is important to keep in mind that neutron stars have strong magnetic surface fields, $\sim 10^{12}$ Gauss (see Section 5.4). It is easy to imagine that such strong fields will influence the electron motion inside atoms. The magnetic fields compress atoms into thin, needle-like shapes. Such atoms have very different energy levels than at zero magnetic field strength. This has to be taken into account when interpreting spectra from neutron stars. The layer directly below the atmosphere, the so-called envelope, is crucial for transport properties and the release of thermal energy at the neutron star surface, for example.

5.7.2 *Crust*

The crust is usually separated into *outer* and *inner crust*. It has a thickness of 1–2 km and contains mainly nuclei. In the competition between attractive nuclear and repulsive Coulomb force, these nuclei arrange themselves into a configuration of lowest energy. At low density, $\rho < 10^6$ g cm^{-3}, the dominant nucleus is ^{56}Fe with $Y_e = Z/A = 26/56 = 0.46$. As the density increases, matter becomes ionized, and the atomic nuclei are embedded in a nearly uniform medium of background electrons. The preferred nuclei become more and more neutron rich with $Y_e \sim 0.1 - 0.2$. Such nuclei are very different from nuclei under terrestrial conditions: they contain nucleon numbers of up to several hundred and, because of the interplay of nuclear and Coulomb force, they arrange in a crystalline form. At a particluar density, the *neutron drip density* $\rho_{\text{drip}} = 4 \cdot 10^{11}$ g cm^{-3}, the chemical potential of the neutrons inside the nuclei, becomes zero, and the neutrons start to "drip" out of the nuclei, immersing the nuclei in a background sea of neutrons. If the temperatures are low enough, these neutrons are probably in a superfluid state, that is, they form a fluid that is characterized by the complete absence of viscosity. At even higher density, the nuclear lattice can turn inside out so that it forms a lattice of voids. The geometry of nuclear matter in this regime changes with density from spherical ("meatballs"), over nuclear slabs ("lasagna"), to voids embedded in nuclear matter ("Swiss cheese"), and finally, when the density is so high that the nucleons touch, it forms a uniform nuclear fluid ("sauce"). This regime is therefore often referred to as "nuclear pasta."

5.7.3 *Outer core*

The outer core begins where nuclei cease to exist and matter consists of a homogeneous neutron–proton fluid. This occurs near a density of 10^{14} g cm^{-3}. The outer core, of course, contains also negatively charged leptons, electrons, and muons to

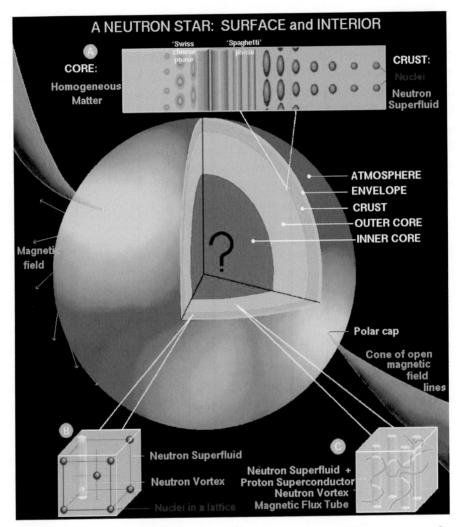

Figure 5.10   Structure of the various regions of a neutron star. Figure courtesy of Dany Page.

neutralize the charges of the protons. The core contains about 99% of the mass of a neutron star. Again, the neutrons are likely to be superfluid, and the protons may be superconducting, which means that they do not exhibit any electrical resistance.

5.7.4   Inner core

We had mentioned previously that at high densities baryons other than neutrons and protons could appear. If the Fermi energies of the nucleons are large enough, nucleons can be transformed into hyperons via weak interactions; see Table 5.1. To conserve charge, this generally also produces leptons. Such a process can be energetically favorable as it transforms a high-Fermi-energy nucleon into a low-energy hyperon. Relieving the Fermi pressure of the nucleons, this process softens the equation of state and lowers the maximum neutron star mass. The importance of this effect, however, is difficult to quantify because of our incomplete knowledge of the coupling strengths of the various particles with each other. Therefore, the densities where they start to populate the neutron star or the neutron star mass for which hyperons appear in the centers are not accurately known.

There are further physically plausible possibilities that are still unsettled and therefore referred to as *exotic*:

- Condensation of pions and/or kaons: it has been argued that pions and kaons could be produced inside neutron stars. Ignoring for a moment the strong interaction between pions and nucleons, the condition for the reaction $n \rightarrow p + \pi^-$ is that the electron chemical potential $\bar{\mu}_e = \bar{\mu}_n - \bar{\mu}_p$ exceeds the pion rest mass of about 139 MeV. Being bosons, the condensation of pions or kaons would offer the interesting possibility of Bose–Einstein condensates in the cores of neutron stars. Such condensates would have two effects: they would substantially soften the equation of state, and they would lead to an enhanced neutrino cooling. Again, quantitative predictions are hampered by our incomplete knowledge of the "in-matter" properties of the involved matter constituents and their interactions.
- Quark matter: baryons are built up of quarks. The latter ones possess a property usually called *asymptotic freedom*: the interaction becomes the weaker the closer the particles are. This is also the reason why quarks are confined in nucleons: separating them would increase their attractive interaction. The densities in the core, however, may be high enough that the quarks are essentially free. This means that as nucleons lose their individuality, they dissolve into a soup of colorless quark matter. The universe went through a similar phase of deconfined quark matter during the first moments after the Big Bang. If the high pressure in neutron star cores can really dissolve hadrons into quark matter, hybrid stars with cores made of quark matter and more conventional neutron star matter around it should exist. It has also been speculated that strange quark matter could be the absolute ground state of the strong interaction, rather than nucleons and nuclei. In this case, matter would finally decay into strange quark matter, and the observed "neutron stars" might be strange quark stars rather than the conventional neutron stars described earlier.

Figure 5.11 shows the mass–radius relationship derived with various modern equations of state.

Obviously the radius of a neutron star should be larger than its Schwarzschild radius, $R > R_S = 2GM/c^2$, so regions where $R < R_S$ can be excluded in the

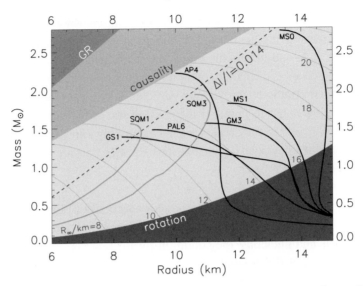

Figure 5.11   Mass–radius relationship of several modern equations of state together with various constraints. Figure from Lattimer, J. and Prakash, M. (2004). *Science*, **304**, 536.

diagram (General Relativity, GR). Moreover, the equation of state should be causal; that is, the speed of sound should not exceed the speed of light. This condition provides the constraint[11] $R < 3GM/c^2$ (causality). A lower limit on the mass at a given radius comes from rotation (see below): if the mass is below a given threshold, the star would start mass shedding at its equator. The remaining gray area in Fig. 5.11 is the allowed region.

5.8   Observational tests of the neutron star structure

Obviously, how our theoretical ideas about neutron star interiors can be compared to the compact objects that exist out in space is an important question. As examples of observational constraints on the neutron star structure, we briefly address rotation periods, cooling, and gravitational waves.

5.8.1   *Rotation periods*

We had seen in Section 5.3 that the critical frequency at which a Newtonian, spherical star starts to shed mass is given by

$$\omega_c = \sqrt{\frac{GM}{R^3}};$$

(5.56)

[11] Lattimer, J. et al. (1990). *The Astrophysical Journal*, **355**, 241.

see Eq. (5.5). The shortest possible rotation frequency is then given by

$$\tau_c = \sqrt{\frac{4\pi^2 R^3}{GM}} = 0.461 \text{ ms } \left(\frac{R}{10 \text{ km}}\right)^{3/2} \left(\frac{1.4 \text{ M}_\odot}{M}\right)^{1/2}. \tag{5.57}$$

This result is approximate in the sense that it does not account for a possible deformation of the star due to centrifugal forces, and it also does not account for General Relativistic effects (such as stronger gravity – see Eq. (5.35) – and the so-called frame-dragging effect). Nevertheless, it provides a reasonable estimate for the minimum spin period as a function of mass and radius.

Assume, for example, that we had a rapidly spinning neutron star in a relativistic binary system. As we show in Chapter 6, for such systems it is possible to obtain accurate, individual masses. In that case, the radius, and therefore the equation of state, could be constrained by Eq. (5.57). An extremely short rotation period could only be realized by a particularly compact star. This would point to a very soft equation of state and possibly exotic material in the stellar core.

5.8.2   *Cooling of compact stars*

As we have seen in Chapter 4, neutron stars are born in a Type II supernova explosion. The newborn protoneutron star is very hot, with temperatures in excess of 10^{11} K. Compared to old neutron stars, it is still relatively lepton rich (where the leptons are mainly electrons and neutrinos). In the first few seconds of its existence, such a protoneutron star radiates away about 10^{53} erg in neutrinos, and after about one day, the temperatures have dropped to less than 10^{10} K. Neutrino emission dominates the cooling processes for the first 10^5 years, after which the surface temperatures have reached $\sim 10^6$ K, and then photon cooling takes over.

Because the temperature evolution of a neutron star depends sensitively on the physical processes involved, the hope is that observations of neutron star temperatures can provide insights into the neutron star interior. It is, however, not a trivial task to infer the surface temperatures from observations because photons may be produced in thermal and nonthermal ways. For example, we have seen that newborn neutron stars possess strong surface magnetic fields. Charged particles moving along the field lines produce a strong, nonthermal synchrotron component (see Section 3.8.2), which has to be disentangled from the thermal emission component. Once the nonthermal photon component is known and subtracted, the effective surface temperature, T_s, can be defined using the Stefan–Boltzmann law:

$$L_\gamma = 4\pi R^2 \sigma_{SB} T_s^4, \tag{5.58}$$

where R is the neutron star radius and σ_{SB} the Stefan–Boltzmann constant. The produced photons have to leave the gravitational potential well of the neutron star

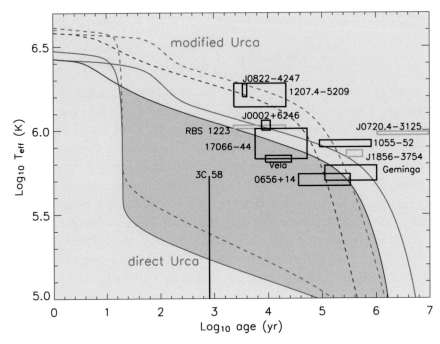

Figure 5.12   Cooling curves and comparison with observed neutron stars. The boxes indicate the errors of the observation; the theoretical curves have been obtained for various assumptions about the neutron star interior (e.g., superfluidity or not). The upper four curves correspond to slow cooling via the modified URCA processes; the lower two curves are for direct URCA. See the text for more details. Figure from Lattimer, J. and Prakash, M. (2004). *Science*, **304**, 536.

to reach an observer. Therefore, their energy is redshifted (see the box on *redshift* in Chapter 4), and the observer infers a lower temperature. For reasons of simplicity, we will ignore such General Relativistic effects in the following discussion. Typical surface temperatures of old neutron stars are $\sim 5 \cdot 10^5$ K; therefore the photon energies are $\sim k_B T \sim 50$ eV, and the emission is mainly in the UV/soft X-ray band. Cooling observations of neutron stars have been obtained by several X-ray satellite missions: EINSTEIN (1978–1981), EXOSAT (1983–1986), ROSAT (1990–1998), CHANDRA, and XMM-NEWTON (both since 1999).

To investigate neutron star cooling by neutrinos, it has to be known whether the neutron star is opaque or transparent to the neutrinos. The neutrinos are typically produced at energies around $k_B T$. If we take elastic scattering off nucleons, $\nu_i + N \rightarrow \nu_i + N$, with a cross section

$$\sigma = \frac{1}{4}\sigma_0 \left(\frac{E_{\nu_i}}{m_e c^2}\right)^2 \qquad \text{with } \sigma_0 = 1.76 \cdot 10^{-44} \text{ cm}^2 \qquad (5.59)$$

as the dominant interaction, we can estimate the mean free path (see Eq. [3.41]) as

$$l = \frac{1}{n\sigma} \approx 2500 \, \text{km} \left(\frac{R^3}{12 \, \text{km}} \right) \left(\frac{1.4M_\odot}{M} \right) \left(\frac{10^9 \text{K}}{T} \right)^2, \qquad (5.60)$$

where we have used the average density of the neutron star, $\bar{\rho} = 4\pi M/3R^3$ and $E_{\nu_i} \approx k_B T$, for this estimate. The neutron star is transparent for neutrinos if $l \gg R$ and opaque if $l \ll R$. Because of the quadratic dependence on the neutrino energy (temperature), a newborn neutron star with $T \sim 10^{11}$ K is very opaque to a neutrino, but at $T \sim 10^9$ it is practically transparent.

The internal composition of the star is decisive for the possible neutrino reactions and the rates at which they occur. Therefore, if the age of a neutron star is known (for example, from its spin properties, $\tau_{\text{dipole}} = P/2\dot{P}$ – see Eq. (5.17) – or better from dynamical information such as space velocity together with birthplace of the neutron star) and its surface temperature can be measured, we can compare it to theoretical cooling curves such as Fig. 5.12.

If we consider a conservative neutron star composition of neutrons, protons, and electrons/muons, the cooling proceeds mainly via two processes, the direct and the modified URCA process.[12] The *direct URCA processes*[13] proceed via

$$n \rightarrow p + e^- + \bar{\nu}_e, \qquad (5.61)$$
$$p + e^- \rightarrow n + \nu_e, \qquad (5.62)$$

which obviously conserve the baryon and the lepton number. The neutrinos leave the neutron star and carry away energy. This process, however, requires a relatively large proton fraction ($Y_p \geq 0.1$). Usually, the proton fraction is highest (>0.1) at high densities, then drops down to values less than 0.01 toward the edge of the outer core and, in the crust, rises again close to the iron value of $Y_p = 26/56 \approx 0.46$. Thus, how efficiently a neutron star can cool via the URCA process depends on how much of its mass is at high Y_p/high density. Again, the stiffness of the equation of state plays an important role because it determines how compact the neutron star is.

If the neutron star has a rather low mass, or if the equation of state is particularly stiff, the densities inside the star will be too low to have Y_p values in excess of the direct URCA-required 0.1. In this case, the star could still cool via the *modified URCA processes*:

$$n + (n, p) \rightarrow p + (n, p) + e^- + \bar{\nu}_e, \qquad (5.63)$$
$$p + (n, p) \rightarrow n + (n, p) + e^+ + \nu_e. \qquad (5.64)$$

---

[12] This process was named by George Gamov after a casino in Rio de Janeiro. The neutrinos produced by this process are a constant sink of energy, just as the casino is a constant sink of money.
[13] These are just β-decay and electron capture.

The role of the additional, "spectator" nucleons is to ensure momentum and energy conservation. They can make processes possible that would otherwise be forbidden because they would not conserve energy or momentum. This process, however, is much slower than the direct reaction. Therefore, if a neutron star cools via modified URCA reactions, it will, after the same amount of time, be hotter than a more massive neutron star that can cool via direct URCA processes.

More exotic matter constituents such as a pion or kaon condensates or quark matter can dramatically increase the cooling rates in neutron stars. Such stars would even at a relatively young age be rather cold. However, to date it is still difficult to obtain reliable numbers for such cooling scenarios, because they are rather sensitive to the poorly constrained in-medium properties of the constituents in the neutron star core.

Figure 5.12 shows examples of theoretical cooling curves for various assumptions together with neutron stars whose ages and temperatures are known. The determination of the ages and temperature both have their subtleties, but it generally seems that the modified URCA processes are preferred.

5.8.3   *Gravitational waves*

As discussed previously, the appearance of new particle species in neutron stars has the tendency to soften the equation of state, which, in turn, leads to more compact stars. Having accurate radii for neutron stars of known mass could severely constrain the high-density equation of state.

As is discussed in some detail in Chapter 6, there are several known binary systems that contain two neutron stars. As the stars revolve around their common center of mass, the system loses energy and angular momentum to gravitational waves, and therefore both components slowly drift inward.

At the time of writing, several ground-based gravitational wave detectors are reaching their design sensitivity. Some of them are laser interferometers that try to detect gravitational waves that pass through the L-shaped detector arms, thereby stretch/squeeze the arms differently, and produce a different interferometric pattern. Among these detector projects are the German–British collaboration GEO600, the American LIGO, the Japanese TAMA300, and the Italian–French VIRGO. They are most sensitive to frequencies from about 10 to 1000 Hz. A binary system emits gravitational waves at a frequency of twice its orbital frequency. Therefore, inspiraling double neutron star binaries become visible for the detectors several minutes before the final coalescence; see Exercise 6.6, in Chapter 6. The highest frequency is reached at the last orbit when the neutron star surfaces begin to touch. For the simplest case, we can assume a system with two equal masses (as one can see, for example, from Fig. 5.4, this is a close approximation to reality),

Newtonian gravity, and negligible tidal deformations. Under these assumptions, the peak gravitational wave frequency is

$$\nu_{GW,\text{peak}} = 2 \cdot \nu_{\text{orb}}(a = 2R_*) = \sqrt{\frac{GM}{8\pi^2 R^3}}.$$ (5.65)

Thus, if we know the mass of the system, the peak frequency of the gravitational wave signal tells us about the neutron star radii; see Chapter 6.

Gravitational waves

A detailed treatment of gravitational waves is beyond the scope of this undergraduate book, but good introductions to the topic can be found in many textbooks; see, for example, *A First Course in General Relativity* by Schutz; *Gravitation* by Misner, Thorne, and Wheeler; *Gravitation and Cosmology* by Weinberg. We give only the basic ideas and some useful order of magnitude estimates.

If you think of space-time as a smooth background, gravitational waves can be thought of as ripples on this background (very much as waves on the ocean surface) that travel at the speed of light and carry away energy and linear and angular momentum from the source. Contrary to electromagnetic radiation, gravitational radiation is the lowest order quadrupole radiation. Gravitational waves are transverse in the sense that they stretch/squeeze space perpendicular to their propagation direction. They can be thought of as superpositions of two polarization modes, *plus* and *cross*. The effect of these modes on a ring of dust particles is illustrated in Figure 5.13: the plus mode first stretches the x-direction while squeezing the y-direction and then stretches the y-direction while squeezing the x-direction. The cross pattern is rotated by 45 degrees. This stretching is *locally undetectable*, because, for example, also the ruler to measure the distance change would be stretched/sqeezed. If lengths in different directions are compared, a detection is possible.

Similar to the electromagnetic radiation field, the gravitational wave amplitude h (either h_+ or $h_\times$) falls off $\propto 1/d$, Eqs. (3.103) and (3.104), where d is the distance to the source. Its physical interpretation is the fractional change in length induced by the illustrated streching and squeezing:

$$h \sim \frac{\delta L}{L}.$$ (5.66)

As an order of magnitude estimate, we can calculate the amplitude from temporal changes of the quadrupole moment of the source, Q:

$$h \sim \left(\frac{G}{c^4}\right) \cdot \frac{\ddot{Q}}{d}.$$ (5.67)

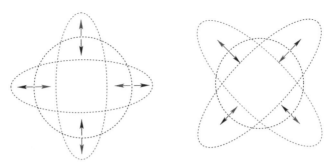

Figure 5.13   Illustration of the effect of the +- and ×-polarization mode of gravitational waves on a ring of dust particles.

It is important to realize that G/c^4 is a tremenduously small number: $8.3 \cdot 10^{-50}$ (cgs-units). For a rough order of magnitude estimate we can use

$$Q \sim MR^2 \qquad \text{and} \qquad \ddot{Q} \sim Mv^2, \tag{5.68}$$

where M, R, and v are typical masses, radii, and velocities of the source. Using typical numbers, $M = 10^6$ g, $v = 10^3$ cm/s, and distance $d = 1$ km, we find that on Earth we can generate amplitudes of $\sim 10^{-43}$. For relativistic stellar objects, $M = 10^{33}$ g, $v \approx c$, and distance $d = 10^9$ light-years, the amplitude, h, is $\sim 10^{-22}$. That means that a detector that is to measure such an amplitude faces the challenge of resolving fractional changes in length of 10^{-22}, or less than a proton radius if the arm length, L, if the detector is 1 km.

5.9 Magnetars

Observations show that typical young neutron stars host a magnetic field of the order 10^{12} G (see Fig. 5.3). Such field strengths are the natural outcome if a normal star is compressed to the size of a neutron star; see Exercise 5.4. Under special circumstances, the magnetic field can be amplified further by *dynamo action*.[14] In such cases, much higher field strengths between 10^{14} and 10^{15} G result. Such highly magnetized neutron stars are called *magnetars*.

Field strengths of this size are greater than the so-called quantum-critical magnetic field where quantum-mechanical effects become important. The critical field strength is reached once the Larmor radius of an electron $r_L = v m_e c / e B$ (see Eq. [2.22]) becomes comparable to its reduced de Broglie wavelength, $\lambda_{dB} = \hbar / m_e v$. Here, m_e is the electron mass and v the velocity perpendicular to the magnetic field. Equating λ_{dB} and r_L and solving for the magnetic field yields

---

[14] In a dynamo, the motion of a conducting fluid in a magnetic field acts to regenerate the magnetic field. This occurs, for example, in the Earth's core and inside the Sun.

$B = (v/c)^2 B_{QC}$, where

$$B_{QC} = \frac{m_e^2 c^3}{\hbar e} = 4.4 \cdot 10^{13} \text{G}. \tag{5.69}$$

At such extreme field strengths, several quantum-mechanical effects occur. For example, vacuum is said to become birefringent: photons propagate at different speeds depending on their polarization. This is similar to the double refraction occuring in a calcite crystal, where a ray of light is decomposed into two rays. Another quantum-electrodynamical effect in the presence of strong fields is that photons can split or merge with each other. If the electron gyroradius, r_L, is much smaller than the photon wavelength, it is hard to excite electron oscillations, and therefore photon electron scattering is suppressed. The strong fields also have a strong influence on the electrons inside atoms in the magnetar atmosphere: as the electron motion perpendicular to the magnetic field is largely suppressed, atoms are squeezed into needle-like shapes.

But what decides whether a core-collapse supernova produces a normal neutron star or a magnetar? The decisive quantity is the stellar rotation speed. As we have seen in Chapter 4, a star in the mass range $8\,M_\odot < M < 25\,M_\odot$ produces a protoneutron star with densities in excess of 10^{14} g cm^{-3} and temperatures higher than 10^{11} K. Under these conditions, the newborn neutron star radiates away most of the gravitational binding energy gained in the collapse in the form of neutrinos. Because its center is very opaque to neutrinos, they are essentially trapped inside the neutron star. Near the surface of the neutron star, the neutrino mean free paths become long enough so that they can escape from the neutron star. As a result, most neutrinos are emitted from regions close to the neutron star surface. The neutrinos carry away entropy and lepton-number and thus build up entropy- and lepton-number gradients. A generalization of the analysis of the Schwarzschild criterion for convection (see Section 2.9) shows that not only entropy gradients but also lepton-number gradients drive convection.[15] As a result, parcels of hotter neutron star matter rise while cooler ones sink down. Neutron star matter is expected to be a good electrical conductor; therefore, its magnetic field is "frozen" in the fluid. This has the effect that the magnetic field lines are dragged around with the fluid. If the star rotates fast enough, convection and differential rotation can conspire to transform kinetic energy into magnetic field energy. The criterion for such dynamo action is that the rotation period, P, is shorter than the convective overturn time, $\tau_{conv} \approx 10$ ms[16]. If the initial star does not rotate rapidly enough, magnetic fields will only be amplified by compression, and a pulsar will be formed. In the rarer

---

[15] This type of convection is known as Ledoux convection.
[16] This can be expressed in terms of the so-called Rossby number: $Ro \equiv \tau_{rot}/\tau_{conv}$ has to be smaller than unity.

cases where the progenitor star is a very rapid rotator, the field, in addition, is amplified by dynamo action and a magnetar is produced.

Is there any evidence that neutron stars with such extreme magnetic fields really exist? There are two candidates in the bestiary of astrophysics that are thought to be magnetars: the *soft gamma repeaters*, or SGRs for short, and *anomalous X-ray pulsars* (AXPs). We briefly describe these two classes of objects in the following subsections.

5.9.1   *Soft gamma repeaters (SGRs)*

SGRs are X-ray stars that emit bright, repeating flashes of soft gamma rays. The fact that a single source emits several such flashes means that the source survives the outburst. This is different in gamma-ray bursts, which seem to be catastrophic, "once-only" events; see Chapter 7. The term *soft* refers to the energy *per photon*, which is lower than in the case of a gamma-ray burst. It does not refer to the emitted energy in such a burst, which can be very large.

SGRs are peculiar in comparison to ordinary pulsars in several ways:

- they rotate very slowly with periods of 5–8 s
- they spin down very rapidly ($\dot{P} \approx 7 \cdot 10^{-11}$ s s^{-1}), and

These properties find a natural interpretation in the magnetar model. Even if the magnetar is born with a very short rotation period, it reaches rotation periods of seconds within a very short time because of the magnetic braking mechanism described in Section 5.4. By specifying Eq. (5.19) for the dipole case (braking index $n = 3$) and solving for ω, we find

$$\omega(t) = \frac{\omega_0}{\sqrt{1 + 2K\omega_0^2 t}}, \qquad (5.70)$$

where ω_0 is the angular frequency at birth, K the constant from Eq. (5.18) specified for the dipole case, and t the time that has passed since the magnetar formation. When we insert typical numbers ($B_p = 10^{15}$ G, $R = 15$ km, $\sin \alpha = 1$, $M = 1.4\ M_\odot$, initial period $P_0 = 1$ ms), we see that such a magnetar spins down in the first 300 yrs of its life to spin periods of about 5 s. Because this time is very short by astronomical standards, the chances to observe a magnetar in its rapid spin-down phase are vanishingly small, but this interpretation is in good agreement with observed SGRs. For example, the soft gamma repeater SGR 0526-66 is observed in the supernova remnant N49 in the Large Magellanic Cloud. The supernova remnant has an age of about 10^4 yr, and the SGR has a spin period of 8 s. If we equate the dipole age $\tau_{\text{dipole}} = P/2\dot{P}$ (see Eq. [5.17]) to the age of the supernova remnant, we find $\dot{P} \approx 1.3 \cdot 10^{-11}$ s s^{-1} and, using Eq. (5.23), a magnetic field strength of $\sim 10^{14}$ G.

If neutron star kicks are related to magnetic fields, one would expect particularly large birth kicks for magnetars. As we have seen in Exercise 5.3 of this chapter, a small anisotropy is enough to impart a large kick velocity to the neutron star. A large uniform magnetic field, for example, alters the local inter-actions between neutrinos and the neutron star material and thus leads to an aniso-tropic emission. An anisotropy in the emitted momentum of $\delta p/p \sim 3\%$ is enough to produce a neutron star kick of about 1000 km s^{-1}. Inserting typical magnetar numbers into the detailed calculations shows that the anisotropy of the required magnitude can be reproduced.

Giant flares

From time to time, SGRs undergo giant outbursts that release about 1000 times more energy than normal outbursts. The very first one has been observed from the object in the supernova remnant N49 on 5 March 1979. This object is now called SGR 0526-66, because of its Galactic coordinates: 05 hours, 26 minutes right ascension, −66 degrees declination. Two Russian Venus space probes, *Venera 11* and *12*, detected them first. Previously they had counted on average about 100 counts per second, but now they were suddenly swamped by gamma rays. Within milliseconds, the count rate reached 40 000 per second and then went off scale. Seconds later, this burst of radiation also reached eight other space-based detectors. As the burst faded over the next three minutes, it revealed an oscillation with a period of about 8 s. Over the subsequent four days, more than a dozen further bursts were recorded from the same position in the sky.

The most energetic burst to date was detected by 15 satellites on December 27, 2004. It was produced by SGR 1806-20 located in our own Galaxy. The burst released a total energy of $2 \cdot 10^{46}$ erg. This is more energy than our Sun would radiate in 150 000 years! The observed temporal structure of the burst (see Fig. 5.14) is typical for giant flares:

- most of the energy is released in a short, subsecond burst. This first hard spike rises on a timescale of about 1 ms.
- the rest of the energy is released in a softer tail that exhibits dozens of pulsations with periods of several seconds that are probably the imprint of the SGR's rotation period.

The first spike of the December 27 burst was orders of magnitude brighter than a typical supernova with $\sim 10^{43}$ erg s^{-1}: $2 \cdot 10^{46}$ erg released in 0.5 s. This exceeds the Eddington luminosity[17] of a neutron star, $L_{\text{Edd}} \approx 2 \cdot 10^{38}$ erg s^{-1}, by orders

---

[17] The Eddington luminosity is explained in Section 8.3.3.

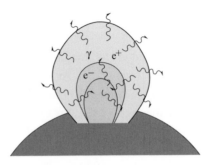

Figure 5.14   Sketch of a fireball of electron positron pairs and photons trapped by closed field lines of the magnetar.

of magnitude. Only blazars (see Chapter 8) and gamma-ray bursts (see Chapter 7) have a higher output in high-energy photons. The polar field strength of SGR 1806-20 can be estimated from its period and the period derivative (see Eq. [5.23]): $B \approx 3.2 \cdot 10^{19} G \, (P\dot{P}/1s)^{1/2} \approx 1.6 \cdot 10^{15}$ G. We can estimate the energy in the magnetic field:

$$E_{mag} \sim u_{mag} \cdot V \sim \frac{B^2}{8\pi} \cdot \frac{4}{3}\pi R_*^3 \sim 10^{48} \text{erg}, \qquad (5.71)$$

where R_* is the radius of the star. We see that the energy released in the burst is still only a few percentages of the huge magnetic energy reservoir.

What could be the mechanism behind such a giant flare? The statistical properties of such flares have been found to be similar to those of earthquakes; a "magnetar-quake" is therefore regarded as a plausible explanation. Such a re-arrangement of the neutron star crust drives strong dissipative electric currents above the star that energize particles that are trapped in the exterior magnetic field. At the same time, the magnetic field rearranges itself to a state of lower energy. This releases a dense cloud of electron–positron pairs and photons, a so-called fireball, which contains only a very small number of neutrons and protons. If too many baryons were contained in the fireball, the fireball energy would have gone into kinetic energy of the baryons rather than into hard photons (for a detailed discussion of this point, we refer to Chapter 7). Parts of this fireball can be trapped by closed loops of the strong magnetic field and produce a trapped fireball that is forced to rotate with the neutron star as sketched in Fig. 5.14. Most of the fireball is opaque to photons, but photons in the outer layers can leak out. Such an "evaporating" fireball attached to the rotating magnetar is thought to produce the emission that is observed after the first hard spike shown in Fig. 5.15.

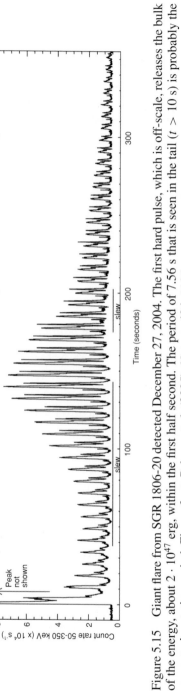

Figure 5.15 Giant flare from SGR 1806-20 detected December 27, 2004. The first hard pulse, which is off-scale, releases the bulk of the energy, about $2 \cdot 10^{47}$ erg, within the first half second. The period of 7.56 s that is seen in the tail ($t > 10$ s) is probably the magnetar's rotation period. Figure from Palmer et al. (2005). *Nature*, **434**, 1107.

5.9.2   Anomalous X-ray pulsars (AXPs)

The first AXP was observed 1981 in the center of the supernova remnant CTB 109. It was a neutron star that emitted a pulsed X-ray signal. To date, four AXPs and two more "candidate" objects that are probably AXPs are known. Their spin periods are all in the range between 5 and 12 s, and they emit pulsed X-rays with a fairly constant, high luminosity of $\sim 10^{35} - 10^{36}$ erg s^{-1}. For comparison, the photon luminosity of the Sun is $L_\odot = 4 \cdot 10^{33}$ erg s^{-1}. The observed X-ray luminosity is too large to be produced by dipole radiation from such a slowly spinning object; compare to Eq. (5.13). There is no evidence for AXPs being members of binary systems or for ongoing accretion processes. Their anomalous property is the mechanism that powers their X-ray emission. Most plausibly, this is the dissipation of a strong magnetic field.

Instead of associating AXPs with magnetars, some models propose a neutron star accreting from a fossil disk acquired in the course of the supernova explosion or during a common envelope phase. However, observations in the optical and near infrared part of the spectrum put severe constraints on such a possible remnant disk, so probably AXPs are also magnetars.

5.9.3   Dissolving boundaries?

Short after their discovery, SGRs and AXPs were considered to be two different kinds of objects and, besides that, to be very different from normal pulsars. In the past few years, several observations have been made that indicate that these once-clear boundaries between the different neutron star classes start to blur. Pulsars have been observed whose magnetic fields as derived from the dipole model, see Eq. (5.23), $B \approx 3.2 \cdot 10^{19}$G $(P\dot{P}/1\text{s})^{1/2}$, exceed the quantum critical limit B_{QC}, thereby narrowing the gap between pulsars and magnetars. On the other side, AXPs have been observed to burst in a way similar to as SGRs. This convergence in the observed properties indicates that they are probably just slightly different manifestations of the same object, a neutron star with a huge, decaying magnetic field.

5.10   Exercises

5.1   Pulse dispersion

The dispersion relation for electromagnetic waves is given by

$$\omega^2 = \frac{c^2 k^2}{\epsilon},$$ (5.72)

where ϵ is the dielectric constant of the medium through which the wave propagates. For many astrophysical plasmas, the permittivity can be approximated as $\epsilon \approx 1 - \omega_p^2/\omega^2$, where $\omega_p = 4\pi n_e e^2/m_e$ is the plasma frequency. Work out the group velocity $v_{\text{group}} = \frac{\partial\omega}{\partial k}$ and the travel time

$$t_{\text{travel}} = \frac{d}{c} + \frac{2\pi e^2}{m_e \omega^2}\text{DM}, \tag{5.73}$$

where d/c is the light travel time at very high (i.e., optical frequencies), and DM is the dispersion measure defined as

$$\text{DM} = \int_0^d n_e \, ds. \tag{5.74}$$

5.2 Order-of-magnitude estimate: Rotation period
Imagine our Sun collapses to a radius of 10 km. What would the final spin period be, assuming homogeneous spheres as initial and end states and conservation of angular momentum?

5.3 Order-of-magnitude estimate: Kick velocity
Assume that a newborn neutron star radiates 10^{53} erg in neutrinos. Show that an asymmetry in the emission as small as 1% (say, the neutron star radiates slightly more neutrinos in positive than in negative z-direction) is enough to impart a kick of several 100 km s^{-1} to the neutron star (assume a mass of 1.4 $M_\odot$). Make use of the fact that neutrinos are ultrarelativistic particles.

5.4 Order-of-magnitude estimate: Magnetic field
Assume a star of the size and magnetic field of the Sun collapses to a radius of 10 km. Estimate the magnetic field of the resulting object, assuming magnetic flux conservation.

5.5 Numerical exercise: Stellar structure
Write a little program (Fortran, C, MATLAB, Mathematica, or whatever you prefer) that integrates the stellar structure equations for a white dwarf. Proceed in the following way:
- Use a nonrelativistic, completely degenerate Fermi gas as equation of state: $P = 1/20(3/\pi)^{2/3}(h^2/m_e)n_e^{5/3}$.
- Chose units ("code units") so that the quantities your code deals with are of order unity.
- Discretise the stellar structure equations; that is, use an equidistant 1D radial grid with grid size Δr and approximate derivatives as $\frac{dP}{dr}(r_i) \approx [P(r_i + \Delta r) - P(r_i)]/\Delta r$.
- Start with a guess value for the density at $r = 0$ and integrate outward until the pressure becomes negative: this is the surface of your star.
- Check that your solution is converged; that is, if you take half the grid spacing, the solution should not change significantly.
- When you are sure that all previous steps are correct, perform an iteration over the guess for the central density until you have reached the desired mass.

- Run through masses from about 0.2 to about 1.4 $M_\odot$ and plot the obtained stellar radius as function of mass (mass–radius relationship). It should look approximately like Fig. 4.4.

5.6 **Fermi gas**

So far, we have assumed that the temperature of the Fermi gas is zero. Repeat the derivation of the Fermi gas for the case that $T \neq 0$. In this case you have to take the proper Fermi–Dirac distribution,

$$f(p) = \frac{1}{e^{(E(p) - \bar{\mu}/kT)} + 1},$$
(5.75)

into account and integrate the momenta from zero to infinity.

5.11   Further reading

Duncan, R. C. (2000). Physics in ultra-strong magnetic fields. AIP Conference Series, Vol. 256. Melville, NY: American Institute of Physics.

Duncan, R. C. and Thompson, C. (1992). Formation of very strongly magnetised neutron stars: Implication for gamma-ray bursts. *Astrophysical Journal*, **392**, L9.

Glendenning, N. K. (2000). *Compact Stars*. Berlin: Springer.

Kouveliotou, C., Duncan, R. C., and Thompson, C. (2003). Magnetars. *Scientific American*, February, 24–41.

Lattimer, J. and Prakash, M. (2004). The physics of neutron stars. *Science*, **304**, 536.

Misner, C. W., Thorne, K. S., and Wheeler, J. A. (1973). *Gravitation*. New York: Freeman and Company.

Oppenheimer, J. R. and Volkoff, M. (1939). On massive neutron cores. *Physical Review*, **554**, 374–381.

Padmanabhan, T. (2001). *Theoretical Astrophysics II: Stars and Stellar Systems*. Cambridge: Cambridge University Press.

Perkins, D. H. (2000). *Introduction to High-Energy Physics*. Oxford: University of Oxford.

Shapiro, S. and Teukolsky, S. (1983). *Black Holes, White Dwarfs and Neutron Stars*. Berlin: Wiley.

Schutz, B. F. (1989). *A First Course in General Relativity*. Cambridge: Cambridge University Press.

Weinberg, S. (1973). *Gravitation and Cosmology*. New York: Wiley.

Woods, P. M. and Thompson, C. (2006). Soft gamma repeaters and anomalous X-ray pulsars: Magnetar candidates. In *Compact Stellar X-ray Sources*, ed. W. H. G. Lewin and M. van der Klis. Cambridge Astrophysics Series, No. 39. Cambridge, UK: Cambridge University Press, pp. 547–586.

Yakovlev, D. G. and Pethick, C. J. (2004). Neutron star cooling. In *Annual Review of Astronomy and Astrophysics*, Vol. 42. Palo Alto, CA: Annual Reviews, 169–210.

Web resources

A tutorial about the physics of pulsars may be found at http://www.jb.man.ac.uk/research/pulsar/.

Lorimer, D. R. (2005). Binary and millisecond pulsars at the new millennium. Living Reviews: http://relativity.livingreviews.org/Articles/lrr-2005-7/.

An introduction to magnetars and SGRs can be found on Robert Duncan's Web page: http://solomon.as.utexas.edu/duncan/magnetar.html.

6

Compact binary systems

6.1 Introduction

About half of all stars are found in systems consisting of two or more stars, so-called binary or multiple stellar systems. Binary systems that contain a compact object – either a white dwarf, a neutron star, or a black hole – are called compact binary systems. Obviously, there are many possible combinations, as each compact star can have any other type of star as companion. Each such combination has its own peculiarities, and there is an entire branch of modern astrophysics that deals with these systems. Here, we cannot give an overview of all the fascinating objects and the often bewildering zoo of compact binary creatures that are known today. Instead, we discuss only a few such systems, some of which are probably important in the context of gamma-ray bursts; see Chapter 7.

First, we give a brief overview of the dynamics in binary systems and then focus on our selected set of binaries. We start with the discussion of low-mass (LMXB) and high-mass X-ray binaries (HMXB), and then turn our attention to systems that consist either of two neutron stars or a neutron star with a low-mass black hole companion. A broad collection of review articles dealing with different aspects of compact binary systems can be found in Lewin and van der Klis (2006); more information on accreting systems may be found in Frank, King, and Raine (2002).

6.2 Dynamics in a binary system

6.2.1 Orbital dynamics of two Newtonian point masses

Stellar binary systems in which the mutual separation is much larger than the stellar radii can be approximated very well as systems of two point masses. The problem of two point masses that interact gravitationally is a classical problem of theoretical physics and one of the few that can be solved analytically. The theoretical derivation of the laws that Kepler had found empirically was one of the great triumphs of Newtonian mechanics. In the following, we give a concise summary of the two-body problem. For further details we refer to standard textbooks of classical mechanics.[1]

Let us briefly outline the strategy. We start by reducing the general two-body problem to the problem of one ficticious body in a central potential. In such a potential, angular momentum is conserved and, as a result, the motion is restricted to a plane. In this plane, we introduce polar coordinates r and φ. The temporal change of φ is given by angular momentum conservation, and the change of r can be derived from the conservation of energy. By combining these two conserved quantities into suitable new parameters, one finds that the solutions are conic sections. From this result, Kepler's laws are derived.

Reduction to a one-body problem

Consider two point masses with masses m_1 and m_2 located at positions $\vec{r}_1$ and $\vec{r}_2$. Their common center of mass is at the position

$$\vec{R}_{cm} = \frac{m_1 \vec{r}_1 + m_2 \vec{r}_2}{m_1 + m_2} = \frac{m_1 \vec{r}_1 + m_2 \vec{r}_2}{M}, \tag{6.1}$$

where we have introduced the total mass M. The force that body 2 exerts onto body 1, $\vec{F}_{21}$, is

$$m_1 \ddot{\vec{r}}_1 = -G \frac{m_1 m_2}{|\vec{r}_1 - \vec{r}_2|^2} \hat{e}_r \equiv \vec{F}_{21}, \tag{6.2}$$

where $\hat{e}_r$ is the unit vector pointing from body 2 to body 1. It is given by

$$\hat{e}_r = \frac{\vec{r}_1 - \vec{r}_2}{|\vec{r}_1 - \vec{r}_2|} = \frac{\vec{r}}{r}, \tag{6.3}$$

where we have introduced the relative vector $\vec{r}$. As a result of Newton's third law ("actio = reactio"), the force exerted by body 1 onto body 2 has the same magnitude but opposite direction:

$$m_2 \ddot{\vec{r}}_2 = G \frac{m_1 m_2}{r^2} \hat{e}_r \equiv \vec{F}_{12} = -\vec{F}_{21}. \tag{6.4}$$

Therefore, the acceleration on the center of mass vanishes:

$$\ddot{\vec{R}}_{cm} = \frac{m_1 \ddot{\vec{r}}_1 + m_2 \ddot{\vec{r}}_2}{M} = \frac{\vec{F}_{21} + \vec{F}_{12}}{M} = 0. \tag{6.5}$$

As $m_i \ddot{\vec{r}}_i = d\vec{p}_i / dt$ is just the rate of change of momentum, Eq. (6.5) is just a statement of the conservation of the total momentum: the center of mass is not accelerated.

---

[1] E.g., Landau and Lifshitz (1976). *Mechanics*, Vol. 1 of Course of Theoretical Physics, Pergamon Press.

We can now express the position vectors of the two particles in terms of $\vec{r}$ and $\vec{R}_{cm}$:

$$\vec{r}_1 = \vec{R}_{cm} + \frac{m_2}{M}\vec{r} \quad \text{and} \quad \vec{r}_2 = \vec{R}_{cm} - \frac{m_1}{M}\vec{r}. \tag{6.6}$$

Taking the second time derivative of Eq. (6.6) and multiplying the first equation with m_1 and the second with m_2 yields

$$m_1\ddot{\vec{r}}_1 = m_1\left(\ddot{\vec{R}}_{cm} + \frac{m_2}{M}\ddot{\vec{r}}\right) = \frac{m_1 m_2}{M}\ddot{\vec{r}} \tag{6.7}$$

and

$$m_2\ddot{\vec{r}}_2 = m_2\left(\ddot{\vec{R}}_{cm} - \frac{m_1}{M}\ddot{\vec{r}}\right) = -\frac{m_1 m_2}{M}\ddot{\vec{r}}, \tag{6.8}$$

where we have used Eq. (6.5). If we now introduce the *reduced mass*

$$\mu \equiv \frac{m_1 m_2}{M}, \tag{6.9}$$

we have $m_1\ddot{\vec{r}}_1 = \mu\ddot{\vec{r}}$, or, combined with Eq. (6.2),

$$\mu\ddot{\vec{r}} = -\frac{G\mu M}{r^2}\hat{e}_r. \tag{6.10}$$

Similarly, Eqs. (6.4) and (6.8) yield

$$-\mu\ddot{\vec{r}} = \frac{G\mu M}{r^2}\hat{e}_r. \tag{6.11}$$

Thus, the two equations of motion have been reduced to a single one, and we have to deal with only an *effective one-body problem*. One calculates the motion of a fictitious body of mass μ in a central force field rather than that of the two real bodies of masses m_1 and m_2 under their mutual influence. Once this one-body problem has been solved, we can go back to the motion of the real bodies via Eq. (6.6).

Kepler problem

Let us now consider the motion of such a fictitious body. The force on it (see Eq. [6.10]) is given by

$$\vec{F}(\vec{r}) = -\frac{G\mu M}{r^2}\hat{e}_r = F(r)\cdot\hat{e}_r. \tag{6.12}$$

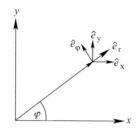

Figure 6.1   Polar coordinates in the orbital plane.

Here $\vec{r}$ is the position vector of our fictitious body. The equations for the *Kepler*[2] *problem* can be obtained by replacing μ by a planet mass m and M by the stellar mass.

   In such a central force field, the angular momentum is conserved and therefore the motion is restricted to a plane; see Exercise 6.1. It is convenient to introduce polar coordinates in this plane given by

$$\vec{r}(t) = \begin{bmatrix} x(t) \\ y(t) \end{bmatrix} = r(t) \begin{bmatrix} \cos\varphi(t) \\ \sin\varphi(t) \end{bmatrix}. \tag{6.13}$$

From Fig. 6.1, one sees that the base vectors in the Cartesian coordinate system, $\hat{e}_x$ and $\hat{e}_y$, are related to those of the polar coordinate system, $\hat{e}_r$ and $\hat{e}_\varphi$, by $\hat{e}_r = \cos\varphi\,\hat{e}_x + \sin\varphi\,\hat{e}_y$ and $\hat{e}_\varphi = -\sin\varphi\,\hat{e}_x + \cos\varphi\,\hat{e}_y$. By working out the cross product explicitly, the angular momentum becomes

$$\vec{L} = \vec{r} \times \mu\vec{v} = \mu r^2\dot{\varphi} \cdot \hat{e}_z \equiv L \cdot \hat{e}_z, \tag{6.14}$$

where $\dot{\varphi} = d\varphi/dt$, $\hat{e}_z$ is the unit vector in the z-direction, and L is the constant magnitude of the angular momentum. Solving this equation for $\dot{\varphi}$ yields the *azimuthal equation*

$$\dot{\varphi} = \frac{d\varphi}{dt} = \frac{L}{\mu r^2}. \tag{6.15}$$

So, the conservation of angular momentum has provided us with an equation for the evolution of the azimuthal angle.

   Similarly, conservation of energy yields an equation for the radial evolution. For the kinetic energy, we need to express the square of the velocity in polar coordinates

$$\vec{v}^2 = \dot{x}(t)^2 + \dot{y}(t)^2 = \dot{r}^2 + r^2\dot{\varphi}^2, \tag{6.16}$$

---

[2] One object is much heavier than the other and can be considered fixed in space, such as a planet moving around the Sun. After Johannes Kepler (1571–1630), German mathematician and astronomer.

where we have just calculated the time derivatives of the position vector components (see Eq. [6.13]) in a straightforward way. The conserved, total energy is

$$E = \frac{\mu}{2}\vec{v}^2 - \frac{G\mu M}{r} = \frac{\mu}{2}\dot{r}^2 + \frac{\mu}{2}r^2\dot{\varphi}^2 - \frac{G\mu M}{r} \equiv \frac{\mu}{2}\dot{r}^2 + U_{\text{eff}}(r), \qquad (6.17)$$

where we have used Eqs. (6.15) and (6.16) and have introduced the effective potential

$$U_{\text{eff}}(r) = -\frac{G\mu M}{r} + \frac{L^2}{2\mu r^2} = -\frac{\alpha}{r} + \frac{\beta}{r^2}, \qquad (6.18)$$

with $\alpha = G\mu M$ and $\beta = L^2/2\mu$. The second term in this effective potential is usually called *centrifugal potential*. As the particle approaches the force center, the centrifugal potential increases faster, $\propto r^{-2}$, than the gravitational potential decreases, $\propto -r^{-1}$; therefore, at some point the particle hits a centrifugal barrier and returns to larger radii again: point particles with nonzero angular momentum cannot collide.

Equation (6.17) can be solved for $\dot{r} = dr/dt$ to give

$$\frac{dr}{dt} = \sqrt{\frac{2}{\mu}\left(E + \frac{\alpha}{r}\right) - \frac{L^2}{\mu^2 r^2}}, \qquad (6.19)$$

where we have chosen the positive sign when taking the square root. This is now our second, the *radial equation*.

Equation (6.15) can be solved for $dt = (\mu r^2/L)d\varphi$ and used in dr/dt, Eq. (6.19), to give

$$\frac{dr}{d\varphi} = \frac{\mu r^2}{L} \cdot \frac{dr}{dt} = \frac{\mu r^2}{L}\sqrt{\frac{2}{\mu}\left(E + \frac{\alpha}{r}\right) - \frac{L^2}{\mu^2 r^2}}. \qquad (6.20)$$

One can now combine the two conserved quantities, the energy E and the angular momentum L, into two new parameters that make the subsequent equations particularly simple:

$$p = \frac{L^2}{\mu\alpha}, \qquad (6.21)$$

often called the parameter of the orbit, and

$$e = \sqrt{1 + \frac{2EL^2}{\mu\alpha^2}}, \qquad (6.22)$$

which is the numerical eccentricity of the orbit. With these two parameters Eq. (6.20) can be integrated analytically to give the radius in terms of the azimuthal angle as

$$r = \frac{p}{1 + e \cos \varphi}. \tag{6.23}$$

This is the equation for conic sections, whose solutions can be classified according to the eccentricity e:

- *circles* for $e = 0$,
- *ellipses* for $0 < e < 1$,
- *parabolae* for $e = 1$, and
- *hyperbolae* for $e > 1$.

These solutions include *Kepler's first law*: the orbit of each planet is an ellipse with the Sun at one focus. As the values of the cosine vary between -1 and 1, the maximum and minimum separation from the force center are given by (see Eq. [6.23]), $r_{max} = p/(1 - e)$ and $r_{min} = p/(1 + e)$, and their average is the semimajor axis

$$a = \frac{1}{2}(r_{max} + r_{min}) = \frac{1}{2}\left[\frac{p(1 + e)}{(1 - e)(1 + e)} + \frac{p(1 - e)}{(1 - e)(1 + e)}\right] \tag{6.24}$$

$$= \frac{p}{1 - e^2} = -\frac{G\mu M}{2E}. \tag{6.25}$$

For ellipses, the semiminor axis, b, is related to semimajor axis, a, via

$$b = a\sqrt{1 - e^2} = \frac{p}{\sqrt{1 - e^2}}. \tag{6.26}$$

From Eq. (6.22), we see that the bound orbits, circles and ellipses, have a negative total energy. By solving Eq. (6.22) for the energy

$$E = \frac{\mu \alpha^2}{2L^2}(e^2 - 1), \tag{6.27}$$

one sees that a circle, $e = 0$, is the orbit that has the minimal energy.

Kepler's second law, which says that a line joining the planet and the star sweeps out equal areas, dS, during equal intervals of time, dt, is just a direct consequence of the conservation of angular momentum. Mathematically, this can be stated as

$$\frac{dS}{dt} = \frac{L}{2\mu} = \text{const}; \tag{6.28}$$

see Exercise 6.2.

Kepler's third law can then be found by calculating the surface area, S, of an ellipse. Using Eq. (6.28),

$$S = \int_0^P \frac{dS}{dt} dt = \frac{L}{2\mu} P,$$ (6.29)

where P is the orbital period. However, the area enclosed by an ellipse is equal to πab; therefore we have

$$P = \frac{2\mu}{L} S = 2\pi \frac{\mu}{L} ab = 2\pi \frac{\mu}{L} a^2 \sqrt{1 - e^2},$$ (6.30)

where we have used the relation between the semiminor and semimajor axes, Eq. (6.26). If we insert the definitions of α and e, Eq. (6.22), and use Eq. (6.25), this reduces to

$$P = 2\pi \sqrt{\frac{a^3}{GM}}$$ (6.31)

or, by $\omega = 2\pi/P$, to

$$\omega_{\mathrm{K}} = \sqrt{\frac{GM}{a^3}}.$$ (6.32)

This is a statement of Kepler's third law: *the squares of the orbital periods are proportional to the cubes of the semimajor axes.*

We have seen before that circles are orbits of minimum energy. This implies that a binary system that loses orbital energy, say by tidal interactions, tends to circularize. This is also true for the emission of gravitational waves: it efficiently circularizes eccentric orbits; see Section 6.4.4. Most old binary systems are therefore close to circular.

Interestingly, there are not many potentials for which the resulting orbits are closed; this is only the case for the harmonic oscillator and the gravitational potential. This has interesting consequences: if the potential deviates from a $1/r$ behavior, the orbits no longer close. Instead, the point of closest approach, for the Sun called *perihelion* or for a general star called *periastron*, moves by an angle $\Delta\varphi$ per orbit. Such deviations can be produced, for example, by higher multipole moments in the Sun's gravitational potential, $\Phi = \tilde{\Phi}_1/r + \tilde{\Phi}_2/r^2 + O(1/r^3)$, or by corrections to the Newtonian potential by General Relativity; see Section 6.4.

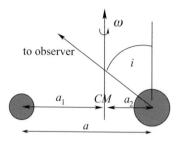

Figure 6.2   Geometry of observed binary system. The orbital plane is seen on its edge.

6.2.2   The mass function

Consider a binary system of masses m_1 and m_2 with a separation a. If we use the center of mass as the origin, we have the following relations (see Fig. 6.2):

$$a_1 + a_2 = a \qquad \text{and} \qquad m_1 a_1 - m_2 a_2 = 0, \tag{6.33}$$

or

$$\frac{a_1}{a} = \frac{m_2}{m_1 + m_2} = \frac{m_2}{M}, \tag{6.34}$$

where we have introduced the total mass M. If spectral lines of one of the stars can be measured, it is called a *single-line spectroscopic binary*. In this case it is possible to measure the velocity projected onto the line of sight, v_1, via the Doppler effect. From Fig. 6.2, we see that

$$v_1 = v_{1,\text{true}} \sin i = \left(\frac{2\pi}{P}\right) a_1 \sin i, \tag{6.35}$$

where P is the orbital period. If the radial velocity is plotted as a function of time, v_1 and P can be directly read off: P is the orbital period and v_1 is half of the radial velocity amplitude. The two can be combined into the *mass function* f_1:

$$f_1 \equiv \frac{P v_1^3}{2\pi G}. \tag{6.36}$$

The mass function of the *observed* object gives us a handle on the *unseen* object. If we use Eq. (6.35), we can write

$$f_1 = \frac{P}{2\pi G} \left(\frac{2\pi}{P}\right)^3 (a_1 \sin i)^3 = \omega_K^2 \frac{(a_1 \sin i)^3}{G}. \tag{6.37}$$

With Kepler's third law (see Eq. [6.32]),

$$\omega_K^2 = \left(\frac{2\pi}{P}\right)^2 = \frac{GM}{a^3}, \tag{6.38}$$

we find with the help of Eq. (6.34)

$$f_1 = \frac{(m_2 \sin i)^3}{M^2}. \tag{6.39}$$

Clearly, f_1 has the dimension of a mass, but what is its meaning? This becomes clearer if we introduce the mass ratio $q = m_1/m_2$. With $M/m_2 = 1 + q$, the mass function can be written as

$$f_1 = \frac{m_2(\sin i)^3}{(1 + q)^2}. \tag{6.40}$$

As $(\sin i)^3 \leq 1$ and $(1 + q)^{-2} < 1$, *the mass function of the observed star gives a strict lower limit on the mass of the unseen companion star*: $f_1 < m_2$. For example, assume we have a binary system with an unseen compact object and a normal star whose spectral lines can be measured (single-line spectroscopic binary). If the measurement of v_1 and P yields a mass larger than the upper limit on the maximum neutron star mass, $f_1 > 3.2\ M_\odot$ (see Section 5.6.2 in Chapter 5), the unseen companion has to be a black hole. This property of the mass function is used in Section 6.3.3 to identify black hole candidates.

If it is possible to determine the mass functions of both binary components (double-line spectroscopic binary),

$$f_1(m_1, m_2, i) = \frac{(m_2 \sin i)^3}{M^2} \quad \text{and} \quad f_2(m_1, m_2, i) = \frac{(m_1 \sin i)^3}{M^2}, \tag{6.41}$$

then we can obtain the mass ratio, q, of the binary $q = (f_2/f_1)^{1/3}$. If relativistic effects can be measured unambiguously, for example, in some double neutron star systems (DNSs; see Section 6.4), the *individual* component masses can be determined.

6.2.3   *Orbital evolution with gravitational wave emission*

The theory of General Relativity predicts that a system with a time-varying quadrupole moment emits gravitational waves. As we explain in Section 6.4.4, the emission of gravitational waves has been observed for some neutron star binary systems. These observations agree very well with the predictions from General Relativity. The energy and angular momentum carried away in the gravitational waves is removed from the binary orbit. As a result, the orbit decays, which means that the stars spiral toward each other. At large separations, this decay occurs in a quasistatic, "secular" way, so that the effect from the emission of gravitational waves can be considered as a small perturbation to the motion predicted by Newtonian theory. In this limit, the accelerations can be derived from the potential

$$\Phi = \Phi_N + \Phi_{GW}, \tag{6.42}$$

where Φ_N is the normal Newtonian potential and Φ_{GW} incorporates the effect from the gravitational wave emission, the so-called radiation reaction.[3] Without going through the explicit calculations, we sketch how the effect on the binary orbit, in particular on the semimajor axis a and the eccentricity e, can be calculated.[4] One can think of the effect of gravitational waves acting on the orbit as some kind of drag force that drives the system toward coalescence. The radiation reaction force on a mass element $dm = \rho\, d^3x$ is then given by

$$d\vec{F}_{GW} = -\vec{\nabla}\Phi_{GW}\,\rho d^3x. \tag{6.43}$$

The rate of change of energy is then, as usual, the product of the acting force and the velocity. This has to be integrated over the whole source, so the total energy loss is

$$\frac{dE}{dt} = \int \vec{v} \cdot d\vec{F}_{GW} = -\int \vec{v}(\vec{r}) \cdot \vec{\nabla}\Phi_{GW}\,\rho d^3x. \tag{6.44}$$

For the angular momentum loss, one proceeds in a similar way by again integrating over the source

$$\frac{d\vec{L}}{dt} = \text{radiation reaction torque} = \int \vec{r} \times d\vec{F}_{GW}. \tag{6.45}$$

Because we are interested in the evolution of the eccentricity, e, and the semimajor axis, a, and we have to relate e and a to the energy and angular momentum,

$$e = \sqrt{1 + \frac{2EL^2M}{G^2m_1^3m_2^3}} \tag{6.46}$$

and

$$a = -\frac{Gm_1m_2}{2E} \tag{6.47}$$

(see Eqs. [6.22] and [6.25]). If we take the time derivatives of a and e, we have da/dt and de/dt as a function of dE/dt and dL/dt, which, in principle, we know how to calculate via Eqs. (6.44) and (6.45). If one inserts the explicit form of Φ_{GW} and performs the calculations, one finds, as an average over several orbital

---

[3] For the explicit form of Φ_{GW}, see Burke (1971). *Journal of Mathematical Physics*, **12**, 402.
[4] The explicit steps may be found in Lightman et al. (1975).

revolutions,

$$\frac{da}{dt} = -\frac{64}{5} \frac{G^3 m_1 m_2 M}{a^3 c^5} \frac{(1 + 73e^2/24 + 37e^4/96)}{(1 - e^2)^{7/2}} \qquad (6.48)$$

and

$$\frac{de}{dt} = -\frac{304}{15} \frac{G^3 m_1 m_2 M}{a^4 c^5} e \frac{(1 + 121e^2/304)}{(1 - e^2)^{5/2}} < 0. \qquad (6.49)$$

This shows that the rates of change depend very sensitively on the orbital eccentricity. The change of eccentricity is always smaller than zero, as is apparent from Eq. (6.49). This means that *the emission of gravitational waves has the tendency to circularize orbits.* In a sense, eccentricity is "radiated away"; the lost energy and angular momentum are carried away by the gravitational waves. Prior to the final coalescence, typical neutron star binary orbits are close to circular.

If we assume that Eq. (6.48) is valid until the stars come into contact, we find for the inspiral time of a circular ($e = 0$) orbit

$$\tau_{insp} = \frac{5}{256} \frac{c^5}{G^3} \frac{a_0^4}{M^2 \mu}, \qquad (6.50)$$

where a_0 is the current separation, M the total mass, and μ the reduced mass (see Exercise 6.4). If the initial orbit is eccentric, the inspiral time is more complicated, but a good approximation[5] is given by

$$\tau_{insp} = 10^7 \text{yr} \ P_{orb,h}^{8/3} \left(\frac{M}{M_\odot}\right)^{-2/3} \left(\frac{\mu}{M_\odot}\right)^{-1} (1 - e^2)^{7/2}, \qquad (6.51)$$

where $P_{orb,h}$ is the current orbital period in hours. As an example of how sensitive the inspiral time is to the eccentricity, we show the inspiral times as a function of eccentricity for two orbital periods in Fig. 6.3: 2.45 hr is the period of the binary pulsar PSR J0737, and 7.75 is the period of PSR 1913+16. A large eccentricity can easily reduce the inspiral time by several orders of magnitude.

6.2.4   Roche geometry

For most binary systems, using two point masses in circular orbits around their common center of mass is a very good approximation. It is convenient to use a coordinate system that corotates with the binary in which both stars lie on the x-axis and the common center of mass is at the origin. A test particle in such a noninertial frame feels the force it would feel in an inertial frame, $\vec{f}$, but on top

---

[5] Lorimer (2005). Living Reviews: http://relativity.livingreviews.org/Articles/lrr-2005-7/.

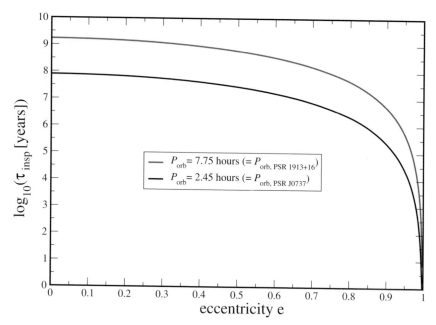

Figure 6.3 Inspiral time as a function of eccentricity for two orbital periods: 2.45 is the orbital period of the binary pulsar PSR J0737, and 7.75 is the period of PSR 1913+16.

also the centrifugal, $\vec{f}_{\text{cen}}$, and the Coriolis forces, $\vec{f}_{\text{cor}}$:

$$\vec{f}' = \vec{f} + \vec{f}_{\text{cen}} + \vec{f}_{\text{cor}} = \vec{f} - m\vec{\omega}_K \times (\vec{\omega}_K \times \vec{r}') - 2m\vec{\omega}_K \times \vec{v}'. \tag{6.52}$$

Here m is the particle mass and $\vec{r}'$ and $\vec{v}'$ are position and velocity of the particle as measured in the corotating frame. The combination of gravitational and centrifugal forces can be calculated as the gradient of the *Roche potential*:

$$\Phi_{\text{Roche}}(\vec{r}) = -\frac{Gm_1}{|\vec{r} - \vec{r}_1|} - \frac{Gm_2}{|\vec{r} - \vec{r}_2|} - \frac{1}{2}(\vec{\omega}_K \times \vec{r})^2, \tag{6.53}$$

where $\vec{r}_1$ and $\vec{r}_2$ are the position vectors of the two stars. The geometry of the Roche potential for a binary system with a mass ratio of $q = m_2/m_1 = 0.2$ is shown in Fig. 6.4. In the close vicinity of each star, the influence of the companion star is negligible. Therefore, the equipotential lines are practically circular; see the left panel. With increasing distance from the stars, the equipotential lines become deformed but the enclosed space can still be uniquely attributed to one star. There is a critical, figure-eight-shaped equipotential surface beyond which both stars are enclosed. The two volumes inside this critical surface are joined at the point marked with L_1, as shown in Fig. 6.4, and are referred to as the *Roche lobes*. Far away from the binary, the equipotential surfaces are again spherically symmetric and coincide with those of a point mass with $M = m_1 + m_2$ located at the origin.

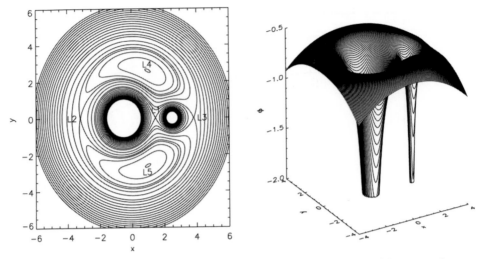

Figure 6.4   Geometry of the Roche potential of a binary system with mass ratio
$q = m_1/m_2 = 0.2$: contours in orbital plane (left); values of the potential (z-axis)
over the orbital plane (right).

At the so-called *Lagrange points*, the forces on a test particle from both stars
cancel out. The Roche lobes join at the inner Lagrange point, L_1, which lies on
the saddle between the potential wells, as shown in the right panel of Fig. 6.4. On
the line joining the two stars there are two more Lagrange points, L_2 and L_3. The
Lagrange points L_4 and L_5 form equilateral triangles with the stars.

6.2.5   Mass transfer and accretion

Close binary systems can transfer mass from one component to the other. This can
occur, for example, via strong winds as in the case of high-mass X-ray binaries
HMXBs (see below) or via the so-called Roche lobe overflow. The latter can occur
if one of the stars expands while the binary separation remains fixed. In this case,
the size of the star becomes comparable at some point to the size of the Roche lobe.
Now matter can flow over to the companion star via Lagrange point L_1, as illustrated
in Fig. 6.4. The same can happen if, for some reason, the mutual separation shrinks
to the point that the stellar radius is comparable to the Roche lobe in size. This
may happen, for example, as a result of gravitational wave emission. These two
different modes of accretion are discussed later in more detail in the context of
low-mass (LMXB) and high-mass X-ray binaries.

Generally, mass transfer does not only change the masses but also transfers
angular momentum. Hence, it can change the orbital separation. Assume, for example, that mass is transferred from the lighter to the heavier component, say from
a low-mass star to a black hole. If momentum is conserved, the center of mass

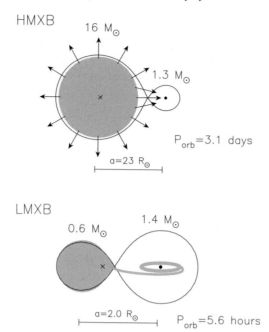

Figure 6.5 Examples of a high-mass (HMXB) and low-mass X-ray binary (LMXB) are shown. In the HMXB case, the compact object is fed by a strong wind; in the LMXB case, accretion occurs because of Roche lobe overflow. From Tauris and van den Heuvel (2006).

cannot be accelerated as we had seen in Eq. (6.5). Therefore, if the more massive component is becoming even more massive, the low-mass component has to move further out to keep the center of mass fixed. However, if the mass transfer proceeds in the opposite direction (if the more massive star transfers mass to the less massive star), this can have the effect of reducing the orbital separation. Under certain circumstances, this can lead to unstable mass transfer.[6] These two different modes of accretion are illustrated in Fig. 6.5 and discussed in more detail in the following sections.

6.2.6 *Effect of a supernova explosion*

As we had seen in Chapter 4, at the end of their lives massive stars die in a supernova explosion. If such a supernova goes off in a binary system, the binary is most likely disrupted. In reality, the process can have various complicated consequences. For example, asymmetries in the explosion can impose large velocities on the remaining

---

[6] For a further discussion of mass-transfer stability, we refer the interested reader to Frank, King, and Raine (2002).

neutron star (see Chapter 5) or the supernova explosion can ablate mass from the second star and so forth. Here, we consider only a very simple model in which we assume that the exploding star does not receive a kick. Also, we assume that the explosion occurs instantaneously, which means that the ejecta reach the companion star on a timescale that is short in comparison with the orbital timescale and that the related impact is negligible.

Consider a circular binary system with masses m_1 and m_2 and a separation of a. Its total energy is given by

$$E_{\text{tot}} = \frac{1}{2}\mu(\omega_K a)^2 - \frac{Gm_1m_2}{a} = -\frac{Gm_1m_2}{2a}, \qquad (6.54)$$

where we have used Eq. (6.32). Assume that star 1 explodes in a supernova and ejects the mass $\Delta m = m_1 - m_r$, where m_r is the mass of the remnant (either a neutron star or black hole). After the explosion, the relative velocity between star 2 and the remnant is still close to the original value, $v = \sqrt{G(m_1 + m_2)/a}$. So the total energy of the binary after the explosion is

$$E'_{\text{tot}} = \frac{1}{2}\frac{m_r m_2}{m_r + m_2}v^2 - \frac{Gm_r m_2}{a} = \frac{Gm_r m_2}{2a} \cdot \left(\frac{m_1 + m_2}{m_r + m_2} - 2\right). \qquad (6.55)$$

A positive value of the expression in the bracket corresponds to an unbound system; see the discussion of the Kepler problem. The expression in the bracket becomes larger than zero if

$$m_r < \frac{m_1 - m_2}{2}, \qquad (6.56)$$

or, using $m_r = m_1 - \Delta m$, we find that the mass loss has to be smaller than

$$\Delta m_{\text{crit}} = \frac{m_1 + m_2}{2} \qquad (6.57)$$

for the system to remain bound. From this follows that – at least in this simple model – *the binary system only survives if less than half of its total mass is ejected*. This immediately implies that a disruption of the binary does not occur if the less massive star explodes. Generally, a high companion mass makes the survival of the binary more probable. Additional kicks again tend to disrupt the system more easily, but in particular cases, where the kicks are just in the right direction, it may also save the binary from being disrupted.

6.2.7   Common envelope evolution

If mass is transferred to the accretor more rapidly than the latter one can accept it, a hot envelope forms around the accretor. If this envelope becomes larger than the size of the Roche lobe, the Roche-lobe approximation is not useful any-more. Instead, it is more appropriate to consider the motion of the binary inside a *common envelope* (CE). This is likely to happen if the more massive compo-nent evolves off the main sequence and becomes a red giant star or if the mass transfer becomes dynamically unstable. Once inside the CE, the envelope exerts a drag force on the orbiting stars and thereby extracts energy at the expense of the orbit. This leads to a rapid reduction of the orbital separation. The energy ex-tracted from the orbit is deposited in the envelope as thermal energy. Depending on how fast the inspiral occurs, the inspiralling star can either merge with the other core or, if energy is deposited fast enough into the envelope, the latter one can become unbound once its gravitational binding energy has been deposited. In the latter case, the outcome is a much closer and more tightly bound binary system.

Reliable quantitative calculations of CE evolution, however, are difficult and represent a major uncertainty in the evolution of close binary systems.

6.3   X-ray binary systems

X-ray binaries are thought to harbor either a neutron star or a black hole that has a noncompact companion star close enough to transfer mass. According to accretion theory (see Chapter 8), the radiation emitted from an accreting compact star is predominantly X-ray radiation. Of course, also white dwarfs can be accretors, but because their radii, R, are much larger, the corresponding accretion luminosities, which are $\propto 1/R$ (see Eq. [8.6]), are much smaller, and therefore only systems relatively close to Earth can be observed. Here, we focus on binaries that contain either a neutron star or a black hole.

A few hundred X-ray binaries are estimated to exist in our Galaxy. The comparison with the estimated huge number of stellar mass black holes, $\sim 10^8$, shows that these binaries are very rare creatures. This is a direct con-sequence of the large number of improbable steps that are necessary for their formation, such as the survival of the supernova explosion, as described in Section 6.2.6.

More than 90% of the Galactic X-ray sources belong to either high-mass X-ray (HMXBs) or low-mass X-ray binaries (LMXBs). In both cases, the ac-creting star can be either a neutron star or a black hole. In the HMXB case, the donor star is a young and massive ($>10\ M_\odot$) O/B star with strong winds, whereas in the LMXB case, the donor is a slowly evolving low-mass ($\lesssim 1.4\ M_\odot$) main-sequence star. Therefore, they belong to different stellar populations: HMXBs with their short-lived massive companion stars are found close to star-forming regions, in particular close to spiral arms in the galactic plane, whereas LMXBs are usually

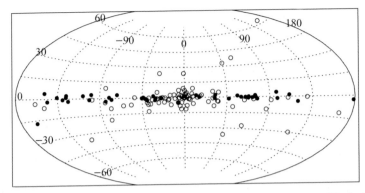

Figure 6.6   Distribution of LMXB (open circles) and HMXB (filled circles) within the Galaxy. From Grimm, Gilfanov, and Sunyaev (2003). *Chinese Journal of Astronomy and Astrophysics*, **3**, Supplement, 257–269.

old systems that have had the chance to move away from their birthplaces. This is reflected in their spatial distribution within the Galaxy, as shown in Fig. 6.6.

X-ray binaries with accreting neutron stars are necessary stages in the evolution toward millisecond and binary pulsars. Pulsars are important tools to probe fundamental physics: the highest observed spin frequencies probe the neutron star equation of state (see Chapter 5), and binary pulsars are formidable laboratories for General Relativity, as we discuss later. X-ray binaries also produce neutron star–white dwarf binaries in which the neutron star masses may be much closer to their (unknown) upper mass limit, which is discussed in Chapter 5.

The nomenclature of X-ray binaries is somewhat involved. Often objects are named according to "constellation + X + number," such as, for example, LMC X-3, the third discovered X-ray source in the Large Magellanic Cloud. Some objects are also named according to the satellite mission that discovered them, such as RX J1940.1-1025, discovered by the ROSAT X-ray all-sky survey.

Black holes

It was speculated as early as in the eighteenth century, by the French mathematician Laplace and by Cambridge professor John Mitchell, that objects may exist in the universe whose gravity is so strong that not even light can escape. In 1915, shortly after Einstein formulated his theory of General Relativity, German astronomer Karl Schwarzschild found a solution to Einstein's field equations that describes a nonrotating black hole. Remarkably, Schwarzschild completed this calculation while serving as a frontline soldier in World War I. Shortly after finding this solution, Karl Schwarzschild died of an illness.

A detailed discussion of black holes is beyond the scope of this book, but good introductions to black holes can be found in many textbooks; see, for example, *A First Course in General Relativity* by Schutz (1989), *Black Holes* by Raine and Thomas (2005), or, for more detail, *Gravitation* by Misner, Thorne, and Wheeler (1973). We restrict ourselves to give a brief summary of those black-hole properties that are relevant to our discussion here.

Black holes in equilibrium are fully descibed by three quantities: mass, angular momentum, and electric charge. Apart from this they are featureless. All other properties are radiated away, either electromagnetically or via gravitational waves. The astrophysical black holes that we discuss here are unlikely to carry any significant electric charge because a separation of charges on a macroscopic scale would cost tremendous amounts of energy. Thus, we are left with mass and angular momentum. Black holes without angular momentum, or spin, are called *Schwarzschild black holes*. They are spherically symmetric and characterized by a radius, which solely depends on its mass and is called the Schwarzschild radius:

$$R_{\text{eh}} = R_{\text{S}} = \frac{2GM}{c^2},\tag{6.58}$$

The Schwarzschild radius defines a one-way boundary that separates the inside of the black hole from the rest of the universe. This boundary, which is not made of any material, is called the event horizon and is located at a radial coordinate R_{eh}. All events that occur inside the boundary are forever hidden from us. Once inside, particles and light rays can never escape to the outside. Therefore, the inside of a black hole is disconnected from the rest of the universe in both space as well as in time.

Circular orbits of test particles become unstable inside the *innermost stable circular orbit* at

$$R_{\text{ISCO}} = \frac{6GM}{c^2},\tag{6.59}$$

inside of which they fall into the hole within about one orbital period.

The geometry of space-time around spinning black holes is described by the Kerr metric found by New Zealander Roy Kerr. Kerr showed that there is a region around a rotating black hole, called the ergoregion, that drags space and time around with the rotating black hole, rather like a vortex. Particles falling toward the black hole within the ergosphere are forced to rotate faster and thereby gain energy. Because they are still outside the event horizon, they may escape the black hole. The net process is that the rotating black hole can emit energetic particles at the cost of its own total energy. The possibility of extracting spin energy from a rotating black hole was first proposed by the British mathematician Roger Penrose in 1969 and is called the Penrose process. Orbits around Kerr black holes are generally complex. If we drop an object onto a Kerr black hole, it would spiral around the black hole's spin axis as it falls. The only way to fall in a straight line into the black hole is to fall along the axis of rotation. Traveling along any other line, an object spirals down a cone whose apex is at the black hole's center and whose axis is the black hole's axis of rotation.

The black hole spin angular momentum, J, cannot become arbitrarily large but must be smaller than GM^2/c. Therefore, the black-hole spin is often characterized by the dimensionless quantity $0 \le a = cJ/GM^2 < 1$. The angular momentum shifts the position of both the event horizon and the innermost stable circular orbit. For a maximally ($a = 1$) spinning black hole, both of these orbits move closer to the hole:

$$R_{\text{eh}} = R_{\text{ISCO}} = \frac{GM}{c^2}. \tag{6.60}$$

It is often useful to know whether General Relativity has to be used or whether Newtonian gravity is a good enough approximation. General relativistic effects only occur close to the hole, at distances comparable to the Schwarzschild radius. Therefore, if all interesting length scales are much larger than GM/c^2, it is safe to use Newtonian gravity.

6.3.1   Low-mass X-ray binaries

Observations

LMXBs are mainly found in the bulge and the globular clusters of our Galaxy. These regions harbor predominantly old stars, suggesting that LMXBs belong to an old stellar population. Relative to the number of stars, LMXBs occur more often in globular clusters than elsewhere. This probably points to additional formation channels due to multiple star encounters that require large stellar number densities to occur at a reasonable rate. The orbital periods of LMXBs range from a few minutes up to several days. Usually, the companion star is very difficult to observe, and it can be seen only in systems with a large orbital separation. In all other cases, the spectrum is dominated by the spectrum of an accretion disk. The optical luminosity, L_{opt}, is generally much lower than the energy output rate in X-rays, L_{X}, $L_{\text{opt}}/L_{\text{X}} \ll 0.1$. The typical photon energies in LMXBs are $k_B T \lesssim 10$ keV, and they are usually lower than in the HMXB case; the spectrum is said to be softer.

Accretion mechanism

Accretion in LMXBs is driven by the Roche-lobe overflow of the noncompact star: matter passes the saddle point at L_1, see Fig. 6.4, and settles into an accretion disk around the more massive, compact companion. Consider a binary system that contains a neutron star or black hole and a low-mass ($\lesssim 1.4\, M_\odot$) main-sequence star. Generally, such main-sequence stars only have very small mass loss via winds; therefore accretion cannot be driven by winds. Mass can only be transferred once the star expands because of the nuclear reactions in its interior and fills its Roche lobe. Because of the low mass of the noncompact star, its nuclear evolution timescale is long, and accretion in LMXBs can go on for hundreds of millions years.

Material that passes L_1 has a specific angular momentum with respect to the accreting compact object of $l \approx bv_p = b^2\omega_K$, where b is the distance of the accretor from L_1, v_p is the velocity perpendicular to the line joining the two stars, and ω_K is the orbital frequency of the binary system. The overflowing material cannot fall straight onto the accretor, but instead it settles into an orbit that is determined by the angular momentum of the accreted material. The circular radius that corresponds to this angular momentum is called the *circularization radius*. If we assume that the material moves on nearly Keplerian orbits, the velocity is $v = r\omega_K = r\sqrt{GM/r^3} \propto r^{-1/2}$, where we have made use of Eq. (6.32). If adjacent orbits couple to each other via viscous processes, the faster inner orbits lose angular momentum to the slower outer orbits and, thus, there is a net angular momentum transport outward. This leads to a spreading of the material into an *accretion disk*. The viscous processes heat up the disk material, which then predominantly radiates in X-rays; see Section 8.3.3. If matter is continuously flowing toward the compact object, the accretion stream has to impact on the disk; see Fig. 6.7, where it produces a hot spot.

Evolution

Obviously, as stars evolve, a LMXB is only a snapshot in the life of a binary system. So how did such a system form and how will it end up? An interesting observation is that many observed systems have a smaller orbital separation than the radius of the giant star that must have produced the compact object. This means that the orbit must have shrunk considerably, probably during a common envelope phase; see Section 6.2.7.

An evolutionary scenario together with typical numbers for the masses, orbital periods, and ages is shown in Fig. 6.8. Initially, on the zero age main-sequence (ZAMS), the system consists of a wide binary with a large mass ratio. As the more massive component evolves into the giant stage, its size becomes comparable to its Rochelobe, and mass transfer via the Lagrange point L_1 (see Fig. 6.4) sets in. As the star evolves further, it engulfs the lower mass star and forms a common envelope system. Because of dissipation while moving through the envelope, the orbital separation shrinks, and once enough energy has been deposited, the common envelope is ejected. From the more massive star, only the helium core survives. Later, the helium core explodes in a core-collapse supernova (see Chapter 4) and produces the compact component, either a neutron star or a black hole. If the binary survives this explosion (see Section 6.2.6), the further evolution is driven by the low-mass star: as it evolves, it fills its Rochelobe and transfers mass to the compact component. This is the LMXB stage. As the low-mass star evolves only

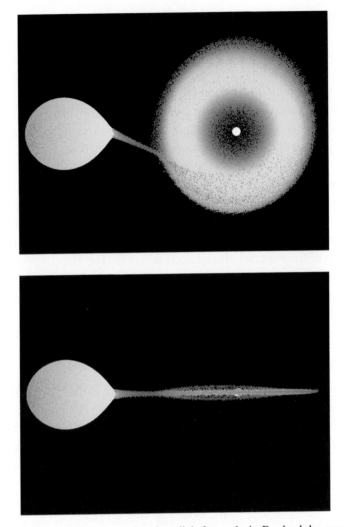

Figure 6.7   Illustration of an accretion disk formed via Roche-lobe overflow of a nondegenerate star (left) onto a compact object (right). The upper panel shows the xy-plane, the lower panel the xz-plane. The accretion stream is impacting on the disk (upper panel); parts of the material are overflowing and streaming along the disk surface (lower panel). Figure is modeled after a smoothed particle hydrodynamics simulation by Roland Speith.

very gradually, the binary can remain in this stage for a very long time: typical lifetimes range from 10^7 to 10^9 yr.

Magnetic fields

Typically, neutron stars are born with magnetic fields of $\sim 10^{12}\,G$; see Fig. 5.3 in Chapter 5. During the long accretion phase in LMXBs, the magnetic field is buried

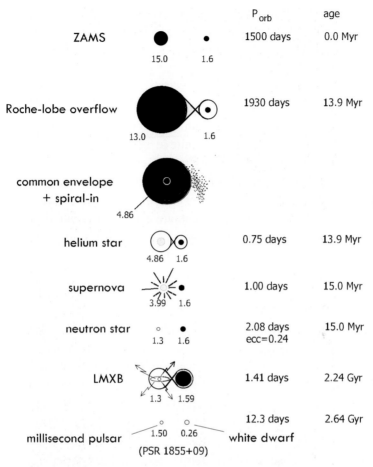

			P_{orb}	age
ZAMS			1500 days	0.0 Myr
	15.0	1.6		
Roche-lobe overflow			1930 days	13.9 Myr
	13.0	1.6		
common envelope + spiral-in				
	4.86			
helium star			0.75 days	13.9 Myr
	4.86	1.6		
supernova			1.00 days	15.0 Myr
	3.99	1.6		
neutron star			2.08 days ecc=0.24	15.0 Myr
	1.3	1.6		
LMXB			1.41 days	2.24 Gyr
	1.3	1.59		
			12.3 days	2.64 Gyr
millisecond pulsar (PSR 1855+09)	1.50	0.26 white dwarf		

Figure 6.8 Sketch of the evolution of a binary system that at some stage is observable as an LMXB. Typical numbers for the masses orbital periods, and age are indicated on the right. From Tauris and van den Heuvel (2006).

and is thought to decay to field strengths that are not important for the accretion dynamics, $B \sim 10^9 \, G$. This is in contrast to the neutron stars in HMXBs that still have a strong magnetic field, which influences the inflowing mass and leads to "pulsations" in the X-ray lightcurves. This is explained in Section 6.3.2.

Spin-up

Let us consider the case where the accretor is a neutron star. During the long accretion phase, the low-mass star can transfer a substantial amount of mass and angular momentum to the neutron star. In this way, the latter one can be spun up to periods of milliseconds. Such millisecond pulsars seem old in terms of their (low) magnetic field (see Chapter 5) but young in terms of their short rotation period. Therefore, they are also called *recycled* pulsars. After a long accretion phase, such a binary becomes a neutron star white dwarf system. Consistent with these evolutionary ideas, neutron stars in such binaries seem to have consistently higher masses than the standard value of 1.35 $M_\odot$; see Chapter 5.

X-ray bursts

Material that is accreted on the weakly magnetized surface of a neutron star consists of mostly hydrogen that burns into helium. Under the conditions at the neutron star surface, helium is degenerate and can ignite in a so-called *pycnonuclear reaction.*[7] Once ignition starts, runaway burning processes set in, which result in a rapid (∼1 s) increase of the X-ray luminosity. As the material cools, the luminosity settles back into its original state on timescales of tens of seconds. Such thermonuclear explosions on the surface of neutron stars are called *X-ray bursts*. In some cases, much longer (∼hours) outbursts have been observed. These so-called *superbursts* are believed to be the result of explosive carbon burning in the deeper layers of the accreted material.

Transients

Many LMXBs show transient behavior, which means that they undergo long periods of very low accretion rates with correspondingly low luminosities ($L_{low} \sim 10^{30}$–10^{32} ergs/s) that are interrupted by accretion periods with luminosities close to the Eddington limit ($L_{high} \sim 10^{38}$ ergs/s) (see Section 8.3.4). The mechanism behind this transient behavior is thought to be related to a disk instability: the accreted material is initially cold and therefore has a low viscosity. The accretion rate is low, and matter piles up in the disk. Once a critical amount of mass has piled up, ionization sets in, the viscosity rises sharply, and that leads to large accretion rates and luminosities.[8] During periods of low accretion rates, the system is relatively clean, which makes it easier to observe the companion star.

6.3.2   High-mass X-ray binaries

Observations

As indicated by their name, high-mass X-ray binaries contain massive ("early type," O or B) companion stars with masses beyond 10 $M_\odot$. The luminosities are very high, $\geq 10^5\ L_\odot$, more characteristic of stars of mass $\geq 20\ M_\odot$. As these stars have only a very limited lifetime, they still reside close to their birthplaces. Therefore, HMXBs are found near star-forming regions in the Galactic disk. Currently, there are about 130 known HMXB systems in our Galaxy. About half of them show

---

[7] Nuclear reactions occuring under degenerate conditions, from the Greek *pycnos* for "dense."
[8] For more information on disk instabilities, see Frank, King, and Raine (2002).

pulsations with periods of 10–300 s. Their orbital periods range from a few hours to several hundred days. The spectra of HMXBs are harder than those of LMXBs; typical photon energies are $k_B T \gtrsim 15$ keV. Well-known examples are Cen X-3 and Cyg X-1.

Accretion mechanism

Consider a massive O/B star as a companion. With their high luminosities and the large opacities due to metal lines, such massive stars blow out their dilute outer envelopes at rates as large as 10^{-6}–10^{-4} $M_\odot$ per yr. For comparison: the mass loss in the solar wind is only about $2 \cdot 10^{-14}$ $M_\odot$ per yr. If such stars are members of a compact binary system, some fraction of the wind material is accreted by the compact object. This wind accretion is similar to the accretion of a star from the interstellar medium, called *Bondi–Hoyle accretion*.

Let us make some simple estimates of such an accretion process. For simplicity, we assume that the wind velocity v_w is constant. Typically, it is around the escape velocity from the star

$$v_w \approx v_{\rm esc} \approx \sqrt{\frac{2GM_*}{R_*}}, \tag{6.61}$$

where M_* and R_* are the mass and the radius of the star. Such wind velocities are very large (up to several thousand km/s), much larger than the sound velocity (few 10 km/s) in the gas. Therefore, one can expect shocks to form; in particlular, a bow shock forms around the compact object, similar to the bow shock around the Earth in the solar wind. If we assume that the wind is in a stationary state, all the mass that enters in a time dt a shell at radius r has to leave it at $r + dr$. Therefore, the mass loss rate is

$$\frac{dM}{dt} = \dot{M} = 4\pi r^2 \rho(r) v(r), \tag{6.62}$$

where ρ is the density and r the distance from the center of the star. Or, if we insert Eq. (6.61) into Eq. (6.62), we find

$$\rho(r) = \frac{\dot{M}}{4\pi r^2} \sqrt{\frac{R_*}{2GM_*}}. \tag{6.63}$$

The compact object has to orbit through the wind from the massive star, as shown in Fig. 6.9. It experiences a relative velocity

$$v_{\rm rel} = \sqrt{v_w^2 + v_c^2}, \tag{6.64}$$

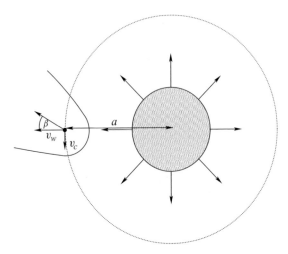

Figure 6.9   Accretion onto a compact object (left) via the wind from a massive star.

where v_c is the orbital velocity of the compact object. The angle β between the wind velocity and the resulting velocity as seen by the compact object is given by $\tan \beta = v_c/v_w$. For simplicity, we assume that the companion star is massive enough so the center of mass practically coincides with the geometric center of the massive star. In this case, $v_c = \omega_K a$, where a is the binary separation; see Fig. 6.9.

Matter streaming toward the compact object is accreted, if its total energy is negative, $mv^2/2 - GmM_c/r < 0$, where M_c is the mass of the compact object, or if we solve for the critical radius, where the total energy is zero, we find the accretion radius

$$R_{\rm acc} = \frac{2GM_c}{v_{\rm rel}^2}. \tag{6.65}$$

Now, we can work out what fraction of the wind mass is accreted onto the compact object. For this estimate, we assume that the orbital velocity is much smaller than the wind velocity, so that Eq. (6.64) becomes $v_{\rm rel} \approx v_w$. The compact object accretes everything contained in the cylinder with radius $R_{\rm acc}$ streaming toward the compact object:

$$\dot{M}_c = \pi R_{\rm acc}^2 \rho v_w = 4\pi \frac{(GM_c)^2}{v_w^3} \rho, \tag{6.66}$$

where we have inserted Eq. (6.65). This equation looks very similar to the Bondi–Hoyle rate for the accretion of zero angular momentum matter:

$$\dot{M} = 4\pi \lambda \frac{(GM)^2}{(c_s^2 + v^2)^{3/2}} \rho_\infty, \tag{6.67}$$

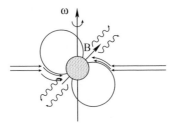

Figure 6.10   Close to a magnetized neutron star, inflowing matter can be forced
to corotate with the magnetic field lines and is channeled toward the magnetic
poles. In general, the magnetic and the rotation axes are not aligned. As a result,
the X-ray emission from the pulsar appears to pulsate with the spin period of the
neutron star.

where ρ_∞ is the matter density far away from the accretor, c_s is the sound velocity,
and λ is a factor of order unity.

The fraction η of accreted material is then

$$\eta = \frac{\dot{M}_c}{\dot{M}} = \frac{\pi R_{\text{acc}}^2 \rho v_w}{4\pi r^2 \rho(r) v(r)} = \frac{G^2 M_c^2}{a^2 v_w^4}, \tag{6.68}$$

where we have used Eqs. (6.62), (6.65), and (6.66). If we again insert the escape
velocity for the wind speed, we obtain

$$\eta = \frac{1}{4} \left(\frac{M_c}{M_*} \right)^2 \left(\frac{R_*}{a} \right)^2. \tag{6.69}$$

By inserting typical numbers, one finds that a fraction of 10^{-4} to 10^{-3} is accreted
by the compact object. This implies that *wind accretion is very inefficient* and only
works for very strong winds from very massive stars.

Contrary to the LMXB case, the formation of an HMXB requires *two* stars above
the threshold for a supernova explosion, $\sim 8\ M_\odot$. The initially less massive star
can have a somewhat lower mass as long as it later on gains enough mass from its
companion to undergo a supernova explosion. The fate of such a system is usually
very sensitive to its initial conditions, such as mass and orbital separation. In the
next section, we discuss an evolutionary channel involving an HMXB system that
can produce either a double neutron star or a neutron star–black hole system.

Magnetic accretion and pulsations

As mentioned earlier, HMXBs are young objects. If they contain a neutron star, it
has still a magnetic field close to the strength at birth, $\sim 10^{12}\ G$ (see Fig. 5.3). Close
to the star, the magnetic field can force the infalling material to corotate with the
magnetic field lines. In this way, the matter is channeled toward the magnetic poles

where X-rays are produced. This is illustrated in Fig. 6.10. The magnetic field becomes dominant when the magnetic energy density, $u_{mag} = B^2/8\pi$, becomes comparable to the kinetic energy density of the wind, $1/2\rho v^2$.

In general, the magnetic and the rotation axis are not aligned. An observer therefore sees an X-ray luminosity modulated by the rotation period of the neutron star. Such pulsations are commonly observed in HMXBs. Such *X-ray pulsars* are distinguished from the *anomalous X-ray pulsars* (AXPs) whose emission is powered by the decay of an ultrastrong magnetic field; see Chapter 5, Section 5.9.

Let us assume a magnetic field of dipolar geometry. We had seen from Eq. (5.9) that the magnetic moment, $\vec{m}$, can be expressed via the field at the poles, $\vec{B}_p$:

$$\vec{m} = \frac{\vec{B}_p R^3}{2}. \tag{6.70}$$

If we insert this into the equation for the dipole magnetic field, Eq. (5.7), and assume the remaining angle terms from the scalar products to be of order unity, we have

$$B(r) \approx B_p \left(\frac{R}{r}\right)^3, \tag{6.71}$$

where R is the neutron star radius and r the distance from its center. The magnetic field becomes important for the accretion process if

$$\frac{1}{2}\rho v^2 \approx \frac{B(r)^2}{8\pi} = \frac{B_p^2}{8\pi}\left(\frac{R}{r}\right)^6. \tag{6.72}$$

We can now solve Eq. (6.62) for ρv, multiply by an additional velocity, and insert again the escape velocity, Eq. (6.61), to find

$$\rho v^2 = \frac{\dot{M}}{4\pi r^2}\sqrt{\frac{2GM_*}{R_*}}. \tag{6.73}$$

Inserting this into Eq. (6.72) yields the radius where the magnetic field of the neutron star in an HMXB becomes important,

$$R_{mag} = B_p^{1/2} R^{3/2} R_*^{1/8} \dot{M}^{-1/4} (2GM_*)^{-1/8}, \tag{6.74}$$

which for typical numbers is several hundred neutron star radii. Therefore, the magnetic field dominates the accretion process out to large radii.

The properties of LMXBs and HMXBs are summarized in Table 6.1.

Table 6.1   *Summary of the differences between LMXB and HMXB.*

Property	LMXBs	HMXBs
Accreting object	Low B-field neutron star or black hole	High B-field neutron star or black hole
Companion	Low-mass star, $L_{opt}/L_X \ll 0.1$	High-mass star, $L_{opt}/L_X > 1$
Stellar population	Old: $> 10^9$ yr	Young: $< 10^7$ yr
Mechanism	Roche-lobe overflow, disk accretion	Wind, Bondi–Hoyle accretion
Angular momentum of accreted material	High	Low
Accretion timescale	10^7–10^9 yr	10^5 yr
Variability	Often transient behavior, X-ray bursts	Regular X-ray pulsations, no X-ray bursts
X-ray spectra	Soft, $k_B T \lesssim 10$ keV	Hard, $k_B T \gtrsim 15$ keV
System status	Clean in low state; dirty in high state	Dirty

6.3.3   *Evidence for stellar-mass black holes*

There is good evidence for two classes of black holes: stellar-mass black holes with several solar masses, which we describe later, and supermassive black holes that are found in the centers of galaxies, which are described in Chapter 8. There are some hints that a third class of black holes, *intermediate-mass black holes* with masses between hundreds and thousands of solar masses may exist. However, the evidence is substantially weaker than for the other two classes.

It is believed that there are about $3 \cdot 10^8$ stellar-mass black holes in our Galaxy. Because these black holes do not radiate themselves, they are difficult to identify. It was shown in Section 6.2.2 that the mass function gives a strict lower limit on the mass of the unseen companion. In Chapter 5, we had seen that the maximum mass of a neutron star is not well known, but that under plausible assumptions it cannot lie beyond 3.2 $M_\odot$; see Section 5.6.2. Therefore, binary systems with mass functions beyond this mass limit must contain a black hole. Such systems are usually called *black-hole candidates*: candidates, because a *direct* proof of, for example, the existence of a event horizon is not available. To date there are 18 black-hole candidates known in our Galaxy, both in

Table 6.2   *Black hole candidates in binaries.*

Name	Type	D (kpc)	P_{orb} (hr)	f ($M_\odot$)	Est. mass ($M_\odot$)
V518 Per	L,T	2.6 ± 0.7	5.1	1.19 ± 0.02	3.2–13.2
LMC X–3	H,P	50 ± 2.3	40.9	2.3 ± 0.3	5.9–9.2
LMC X–1	H,P	50 ± 2.3	101.5	0.14 ± 0.05	4.0–10.0
V616 Mon	L,T	1.2 ± 0.1	7.8	2.72 ± 0.06	3.3–12.9
MM Vel	L,T	5.0 ± 1.3	6.8	3.17 ± 0.12	6.3–8.0
KV UMa	L,T	1.8 ± 0.5	4.1	6.1 ± 0.3	6.5–7.2
GU Mus	L,T	5 ± 1.3	10.4	3.01 ± 0.15	6.5–8.2
IL Lupi	L,T	7.5 ± 0.5	26.8	0.25 ± 0.01	7.4–11.4
V381 Nor	L,T	5.3 ± 2.3	37.0	6.86 ± 0.71	8.4–10.8
V1033 Sco	L,T	3.2 ± 0.2	62.9	2.73 ± 0.09	6.0–6.6
V821 Ara	L,T	4	42.1	>2.0	–
V2107 Oph	L,T	8 ± 2	12.5	4.86 ± 0.13	5.6–8.3
V4641 Sgr	L,T	7.4–12.3	67.6	3.13 ± 0.13	6.8–7.4
V406 Vul	L,T	11	9.2	7.4 ± 1.1	7.6–12
V1487 Aql	L,T	11–12	804.0	9.5 ± 3.0	10.0–18.0
Cyg X–1	H,P	2.0 ± 0.1	134.4	0.244 ± 0.005	6.9–13.2
QZ Vul	L,T	2.7 ± 0.7	8.3	5.01 ± 0.12	7.1–7.8
V404 Cyg	L,T	2.2–3.7	155.3	6.08 ± 0.06	10.1–13.4

L: LMXB; H: HMXB; P: persistent; T: transient. After J. E. McClintock and R. A. Remillard (2006).

HMXBs and LMXBs. Some properties of these black-hole candidates are given in Table 6.2.

Black-hole candidates in LMXBs usually show transient behavior. They show outbursts that are usually referred to as *soft X-ray transients*. For HMBXs, there are also sources that are constantly seen such as Cyg X-1, LMC X-1, and LMC X-3, which are called *persistent black-hole candidates*. Some black-hole candidates show features such as superluminal jets (see Chapter 8) that are known from quasars and are therefore sometimes called *microquasars*.

6.4   Double neutron star systems

6.4.1   *History*

In 1973, just a few years after the discovery of the first pulsar, Russel Hulse and Joseph Taylor started a systematic pulsar search at the Arecibo radio telescope in Puerto Rico. In summer 1974, they discovered a pulsar whose pulse frequency periodically shifted back and forth, a fact that they attributed to the pulsar being a member of a binary system. It soon became clear that the companion star is also

Table 6.3   *Properties of the binary pulsar PSR
1913+16.*

Property	PSR 1913+16
Orbital period	7.75 hr
Pulsar period	59 ms
Masses	1.4414 $M_\odot$ and 1.3867 $M_\odot$
Eccentricity	0.617
Periastron advance	4.227 deg/yr
Distance to system	5.90 kpc
Inspiral time	$3 \cdot 10^8$ yrs

Numbers from Weisberg and Taylor (2005).

a neutron star orbiting the pulsar in less than 8 hr. This binary system, called PSR 1913+16, was immediately recognized as a precious laboratory for gravitational physics. Several general relativistic effects have been observed, the most important one being the inspiral of the binary components toward each other in excellent agreement with General Relativity's prediction from gravitational wave emission. In 1993, Hulse and Taylor were awarded the Nobel Prize in physics for their discovery.

In 2003, a binary neutron star system, J0737-3039 A+B, was discovered where *both* neutron stars are pulsars. With a period of only 2.45 hr, its orbit is even tighter, and therefore general relativistic effects accumulate even faster than for PSR 1913+16.

6.4.2   *Fundamental importance*

The importance of the discovery and continued observation of PSR 1913+16 can hardly be overrated. It is an important result of classical mechanics that potentials $\propto 1/r$ lead to bound orbits in the form of closed ellipses.[9] Deviations from a $1/r$-behavior lead to a slow motion of the periastron, an effect that has been measured in the Solar System for the planet Mercury (perihelion shift). Such deviations may come from deviations from Newton's law of gravity, but also, for example, from higher order multipole moments of the gravitational potential that can be induced by tidal interactions between the stars. The current orbital separation of PSR 1913+16, however, is large enough that tidal effects are completely negligible, and the stars can be excellently approximated as point masses. However, the orbit is tight enough and the eccentricity, $e = 0.62$, is large enough so that general relativistic effects

---

[9] See any textbook on classical mechanics, e.g., Landau and Lifshitz (1976). *Mechanics, Course of Theoretical Physics*. Vol. 1 of Pergamon Press.

can build up to a measurable size within a few years. Moreover, the system is clean in the sense that there are no further complicating effects such as winds, and the deviations from a Newtonian point mass binary are exclusively due to relativistic effects (rather than, say, Newtonian tidal interaction). It is a lucky coincidence that the pulsar emission cone sweeps across the Earth, otherwise the system would have gone unobserved. PSR 1913+16 can be considered a clock that moves at a velocity of $10^{-3}c$ through the strong gravitational field of a companion star. Nature has provided us with an excellent laboratory for gravity.

The mere existence of such a double neutron star system sets constraints on the evolution theory of stars in binary systems: at least in some cases, the formation of black holes can be avoided, either by a restricted amount of accreted material in a common envelope phase (see scenario I next section) or by avoiding neutron stars in a common envelope phase in the first place (see scenario II in the next section). The knowledge of the orbital parameters of PSR 1913+16 provides us with invaluable information: i) accurate masses of both neutron stars (see Section 6.4.4) and ii) the proof that gravitational waves do exist (see Section 6.4.4). The mass of the pulsar in PSR 1913+16, $1.4414 \pm 0.0002 \, M_\odot$, is currently the highest *well-known* neutron star mass and therefore provides an accurate lower limit to the maximum possible neutron star mass. To date PSR1913+16 is the tightest constraint on theories of the strong interaction: nuclear equations of state that cannot stabilize this mass can be ruled out immediately. Moreover, the orbit is decaying in excellent agreement with the gravitational wave predictions derived from General Relativity (see Section 6.4.4). The measured cumulated shift in periastron time agrees with the theoretical prediction to better than 0.2%, an amazing accuracy for an object that is more than 15 000 light-years away.

6.4.3   *Formation channels*

DNSs are rare, as the initial binary system has to survive two subsequent supernova explosions (see Section 6.2.6). Most systems are disrupted; some fraction, however, survives as a highly eccentric DNS. At the time of writing, eight double neutron stars are known, among them are PSR 1913+16 and J0737-3039 A+B, two highly relativistic systems.

How do such binary systems form? Generally, there are several paths that can lead to the formation DNSs. We present only two of them. It is useful to review first some mass thresholds that determine the fate of the star. If a star has an initial mass of less than $M_{SN} \approx 8 \, M_\odot$, it will not undergo a core-collapse supernova but rather end its life as a white dwarf. If the mass is more than M_{SN} but less than M_{BH} ($\approx 25 \, M_\odot$), the supernova explosion leaves a neutron star behind. At masses more than M_{BH}, a black hole will form, either via fallback if the star is lower than M_{Coll}

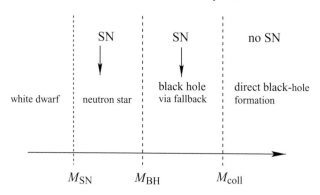

Figure 6.11   Mass thresholds that decide the final fate of the star. The numerical values are $M_{SN} \approx 8\ M_\odot$, $M_{BH} \approx 25\ M_\odot$, and $M_{coll} \approx 40\ M_\odot$.

($\approx 40\ M_\odot$) or directly if it is more than this threshold. This situation is summarized in Fig. 6.11.

The standard scenario for the formation of a DNS starts out from a close, massive binary system: both stars have masses between M_{SN} and M_{BH}, and their separation is less than 1 AU. The heavier of both stars, the so-called primary, evolves faster and bloats up when evolving toward a giant phase. At this stage, it can overflow its Roche lobe and transfer mass to its lower mass companion (the secondary). At the end of its lifetime, the primary explodes as a core-collapse supernova and forms a neutron star. If the system survives the explosion, it consists of a neutron star together with a main-sequence star. If the main-sequence star transfers mass to the neutron star, the system may be observable as an HMXB system. As the secondary evolves further and bloats up, it engulfs the neutron star in a *common envelope phase*; see Section 6.2.7. In this phase, the evolved core of the secondary star and the neutron star revolve around each other inside the hydrogen envelope of the secondary. They lose orbital energy, which, in turn, is deposited in the envelope. This is thought to finally eject the envelope, leaving behind a close binary system consisting of the naked core of the secondary (He-star) and the neutron star. The He-star itself will at some stage produce a neutron star in a supernova, and if the system also survives this second explosion, a close DNS is left behind.

What happens to the neutron star as it moves through the common envelope is critical for the subsequent evolution: if it accretes too much mass, it may collapse to become a black hole and the described scenario would produce – maybe in most cases – a neutron star and a (low-mass) black hole rather than a DNS. Such neutron star–black-hole systems may be important as progenitors of the subclass of short gamma-ray bursts; see Section 7.5.2.

The problematic issue of a neutron star spiralling through a common envelope may be avoided altogether if the masses of the stars in the initial system are nearly, say, within 5%, the same. In this case, the secondary leaves the main sequence directly after the primary and before the primary can explode. Consequently, there are two He cores (rather than a He-core and a neutron star) orbiting inside a common hydrogen envelope. After the dissipation of orbital energy has unbound the common hydrogen envelope, a close system containing two He-cores is left behind. If the binary system survives the two subsequent supernova (SN) explosions, it produces a close DNS. Contrary to the standard scenario, this alternative does not lead to an X-ray binary phase. These two evolutionary paths are sketched in Fig. 6.12.

6.4.4   Relativistic effects and determination of the individual masses

Masses

A Newtonian point mass binary system can be described with *five Keplerian parameters*: (i) the orbital period of the binary system, P_b, (ii) the projected (pulsar) semimajor axis, $a_p \sin i$, (iii) the orbital eccentricity, e, (iv) the longitude of periastron, ω (see Fig. 6.13; this quantity ω is not to be confused with the orbital frequency, which we denoted ω_K), and (v) the epoch of periastron passage, T_0.

The description of relativistic effects requires up to *five post-Keplerian parameters*. In any theory of gravity, they can be written as functions of both masses and the Keplerian parameters. If we assume that General Relativity is the correct theory, the post-Keplerian parameters are (i) the periastron advance, $\dot{\omega}$ (this is the analogon to the famous perihelion shift of Mercury),

$$\dot{\omega} = 3 \left(\frac{P_b}{2\pi} \right)^{-5/3} \frac{(T_\odot M)^{2/3}}{1 - e^2}, \tag{6.75}$$

(ii) a parameter that describes the combined effect of gravitational redshift (see, the box on redshift) and time dilation,

$$\gamma = e \left(\frac{P_b}{2\pi} \right)^{1/3} T_\odot^{2/3} M^{-4/3} M_c (M_p + 2M_c), \tag{6.76}$$

(iii) the change in the orbital period

$$\dot{P_b} = -\frac{192\pi}{5} \left(\frac{P_b}{2\pi} \right)^{-5/3} \left(1 + \frac{73}{24} e^2 + \frac{37}{96} e^4 \right) (1 - e^2)^{-7/2} T_\odot^{5/3} M_p M_c M^{-1/3}, \tag{6.77}$$

and two more parameters that describe the so-called Shapiro-delay effect[10]: r (for range) and s (for shape).

---

[10] Gravitational time delay effect predicted by Irwin Shapiro in 1964: because of general relativistic effects, a (radar) signal passing near massive objects will take slightly longer than predicted by Newtonian theory.

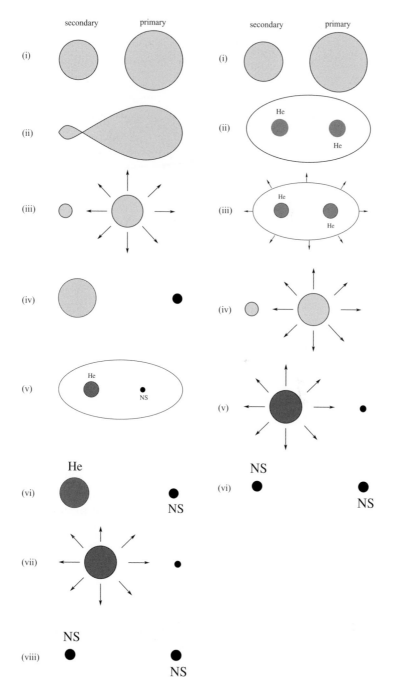

Figure 6.12 Two evolutionary paths to the formation of a close double neutron star system. Left: standard channel; right: alternative to avoid a neutron star that has to go through a common envelope. See text for details.

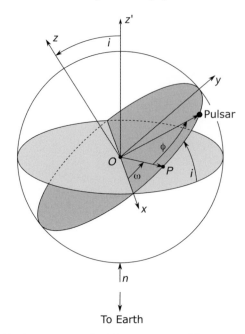

Figure 6.13   Orbital geometry of a binary system. Figure courtesy of Matthias
Hoeft, after Shapiro and Teukolsky (1983).

Here the subscripts p and c refer to pulsar and companion, respectively; $M = M_p + M_c$; and a characteristic timescale, $T_\odot = GM_\odot/c^3 = 4.9255\mu$ s, has been introduced.

As the two masses are at this stage the only unknowns, the measurement of any two of the post-Keplerian parameters allows the determination of the *individual stellar masses*. If more than two of these parameters can be measured, the system is overdetermined. This allows for self-consistency checks of the solutions: plot the companion mass as determined from the post-Keplerian parameter as a function of the pulsar mass. If the same is done for another post-Keplerian parameter, the intersection of the two curves yields the masses of both the pulsar and its companion. In Fig. 6.14, this is done for periastron advance $\dot{\omega}$ and the parameter γ; see Eq. (6.76). If the other measured post-Keplerian parameters are plotted in the same diagram, they should (ideally) all pass through the intersection point of the previous curves. The degree to which this is fulfilled serves as an error indication for the individual masses. The values for the masses of the binary pulsar PSR 1913+16 and its companion are

$$M_p = 1.4414 \pm 0.0002 \, M_\odot \qquad \text{and} \qquad M_c = 1.3867 \pm 0.0002 \, M_\odot. \qquad (6.78)$$

These are the most accurately known neutron star masses to date.

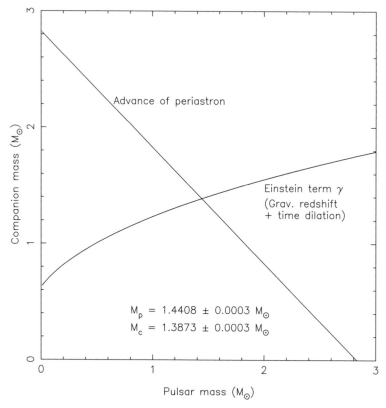

Figure 6.14   Mass–mass diagram for the binary pulsar PSR1913+16 using the post-Keplerian parameters periastron advance and the combined redshift plus time dilation parameter γ. From Weisberg and Taylor (2003). Note that in the meantime the best values for the masses have changed slightly.

Gravitational radiation

The linearized field equations of Einstein's theory of General Relativity permit wave-like solutions; see the box on gravitational waves in Chapter 5. They are in many respects similar to electromagnetic waves; for example, they are *transverse* in the sense that the wave amplitude oscillates in a direction that is perpendicular to the propagation direction of the wave. They also propagate at the speed of light, the amplitude drops of like $1/r$ (like the radiative part of the electromagnetic field; see Eq. [3.100] in Chapter 3), and they transport energy and (angular) momentum.

According to theory, a double neutron star system such as PSR 1913+16 should emit gravitational waves, and indeed it does. The system constantly loses energy and angular momentum and consequently slowly spirals in as discussed in Section 6.2.3. The orbit of PSR 1913+16 shrinks by 3.2 mm (the separation between the two components is about a solar radius, $\sim 7 \cdot 10^{10}$ cm) per orbit, and the current gravitational wave luminosity is $L_{GW} = 0.6 \cdot 10^{33}$ erg/s, about 16% of the solar

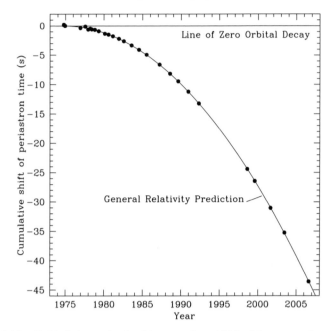

Figure 6.15   Orbital decay in the binary pulsar 1913+16 system demonstrated as an increasing orbital phase shift for periastron passages with time. The general relativistic prediction due to the emission of gravitational radiation is shown by the parabola. Figure courtesy of Weisberg, Taylor and Nice, to appear in (2007).

luminosity! Using Kepler's third law, $P_K = 2\pi \sqrt{a^3/GM}$, for a simple estimate, we see that as the orbit shrinks, the orbital period becomes shorter. So in this sense, the system is braked and as a result speeds up! The periastron is reached slightly earlier after each orbital revolution, and this effect builds up over time. This effect has been measured very accurately for PSR 1913+16.

Figure 6.15 shows the measured cumulative shift of periastron time (filled circles) and compares it to a purely Newtonian orbit that does not decay (horizontal line). The prediction from General Relativity is shown as the solid parabola. For this (indirect) demonstration of the existence of gravitational waves, Hulse and Taylor were awarded the Nobel Prize for physics in 1993.

The ultimate fate of such a double neutron star system is a violent collision that can possibly produce a short gamma-ray burst. This is explained in more detail in Section 7.5.

6.5   Exercises

6.1 Central force field

For a central force field, $\vec{F}(\vec{r}) = F(r) \cdot \hat{e}_r$, show that
(i) the angular momentum $\vec{L} = \vec{r} \times \vec{p}$ is conserved, $d\vec{L}/dt = 0$, and therefore (ii) the orbital motion is restricted to a plane that is perpendicular to the angular momentum vector.

6.2 **Kepler's second law**

Use the conservation of angular momentum and the geometric interpretation of the absolute value of the vector product to show Kepler's second law: $dS/dt = L/2\mu$, where dS is the infinitesimal surface area swept out in time dt.

6.3 **Order of magnitude estimate: Black holes in the Galaxy**

Starting from the supernova rate, estimate the number of black holes in our Galaxy.

6.4 **Inspiral time**

Assume that Eq. (6.48) remains valid until the stars come into contact. Show that the inspiral time from the current orbital separation a_0 for a circular binary orbit is given by

$$\tau_{\text{insp}} = \frac{5}{256} \frac{c^5}{G^3} \frac{a_0^4}{M^2 \mu}, \qquad (6.79)$$

where M is the total and μ the reduced mass.

6.5 **Pulsar: A binary system?**

In principle, the observed periodicity could also come from the orbital motion of a binary system of neutron stars.

(i) Calculate the separation a_0 of a circular neutron star binary system with two neutron stars of 1.4 $M_\odot$ each that has an orbital period of $P = 1.5$ ms.

(ii) Such a system will lose energy via gravitational radiation at a rate of $L_{\text{GW}} = (32/5)(G^4/c^5)(M^3\mu^2/a^5)$, where M is the total and μ the reduced mass of the system. Relate this energy emission to the rate of change of the separation, da/dt. Show that the lifetime of such a binary, that is, the time it takes to spiral in from the current position, a_0, until the surfaces of the neutron stars touch (assume a radius of 12 km) is given by Eq. (6.79).

6.6 **Gravitational wave detection**

Use Eq. (6.79) to calculate how long a neutron star binary will be visible in the frequency window from 10 to 1000 Hz. Assume for the binary system 2×1.4 $M_\odot$, a circular orbit, and radii of 12 km.

6.6 Further reading

Bondi, H. (1952). On spherically symmetrical accretion. *Monthly Notices of the Royal Astronomical Society*, **112**, 195.

Burke, W. L. (1971). Gravitational radiation damping of slowly moving systems calculated using matched asymptotic expansions. *Journal of Mathematical Physics*, **12**, 401.

Frank, J., King, A. R. and Raine, D. (2002). *Accretion Power in Astrophysics*. Cambridge: Cambridge University Press.

Landau, L. D. and Lifshitz, E. M. (1976). *Mechanics*, Vol. 1 of *Course of Theoretical Physics*. Oxford: Pergamon Press.

Lewin, W. H. G. and van der Klis, M. (Eds.) (2006). *Compact Stellar X-ray Sources*. Cambridge: Cambridge University Press.

Lightman, A. P., Press, W. H., Price, R. H., and Teukolsky, S. A. (1975). *Problem Book in Relativity and Gravitation*. Princeton, NJ: Princeton University Press.

McClintock, J. E. and Remillard, R. A. (2006). Black hole binaries. In *Compact Stellar X-ray Sources*. Cambridge: Cambridge University Press.

Narayan, R. (2005). Black holes in astrophysics. *New Journal of Physics*, **7**, 1999.

Padmanabhan, T. (2001). *Theoretical Astrophysics II: Stars and Stellar Systems*. Cambridge: Cambridge University Press.

Raine, D. and Thomas, E. (2005). *Black Holes*. London: Imperial College Press.

Shapiro, S. and Teukolsky, S. A. (1983). *Black Holes, White Dwarfs and Neutron Stars*. Wiley: New York.

Tauris, T. T. and van den Heuvel, E. (2006). Formation and evolution of compact stellar X-ray sources. In *Compact Stellar X-ray Sources*, ed. W. H. G. Lewin and M. van der Klis. Cambridge: Cambridge University Press.

Weisberg, J. M. and Taylor, J. H. (2005). Relativistic binary pulsar B1913+16: Thirty years of observations and analysis. In *Binary Radio Pulsars*, ASP conference series, ed. F. A. Rasio and I. H. Stairs. San Francisco: Astronomical Society of the Pacific.

Web resources

Lorimer, D. R. (2005). Binary and millisecond pulsars at the new millennium. *Living Reviews*: http://relativity.livingreviews.org/Articles/lrr-2005-7/.

Stairs, I. H. (2003). Testing general relativity with pulsar timing, http://relativity.livingreviews.org/Articles/lrr-2003-5/index.html.

7

Gamma-ray bursts

7.1 Introduction

Since their discovery in 1967, gamma-ray bursts (GRBs) have been described by
superlatives: the brightest explosions since the Big Bang, the biggest mystery, or
the toughest nut to crack in modern astronomy. Contrary to other discoveries that
were made at about the same time, say pulsars (Chapter 5) or active galactic nuclei
(Chapter 8), GRBs were not explained convincingly within a few years. Instead,
they have persistently resisted a convincing explanation for decades. It seems that
we are now close to answering the question about what objects cause GRBs, but
many questions about the physics of the GRB explosions are still waiting for
convincing answers. Here we summarize our understanding of GRBs as of the time
of writing in summer 2006.

GRBs were discovered accidentally during the time of the cold war. In 1963,
the United States of America launched the VELA[1] satellite system to monitor a
nuclear test ban treaty that outlawed the tests of atomic bombs underwater, in the
atmosphere, and in space. A nuclear explosion in space produces neither a huge
amount of visible radiation nor sound. Instead, it produces an intense flash of
X-rays, gamma rays, and neutrons. These explosion products would be absorbed
by the Earth's atmosphere, so that an atomic explosion in space can be detected
only be in space. The VELA satellites were launched in extremely high-altitude
orbits, about 100 000 km above the Earth's atmosphere, to be outside the Van
Allen radiation belt[2] and thus to guarantee a low level of detector noise. The
high orbits had the additional effect that a possible atomic test explosion behind
the Moon (this was considered a serious possibility) could have been detected.
Gamma rays are notoriously difficult to focus by lenses or mirrors; gamma-ray

---

[1] Vela is Spanish for *watchman*.
[2] The Van Allen radiation belt is a torus of energetic charged particles around Earth, trapped by Earth's magnetic field.

detectors merely absorb the photon energy but usually cannot tell from which direction the radiation arrived. It is, however, possible to measure accurately the arrival times of the photons. If several satellites detect the same event, the direction to the explosion can be reconstructed from the satellite positions and the respective arrival times. Similar to the case of a radio telescope, where the baseline determines the obtainable resolution, this method is the more accurate the larger the distance between the satellites, which is again an argument for the high-altitude satellite orbits.

On July 2, 1967, the VELA satellites detected an unexpected flash of gamma rays. The signal expected from an atomic bomb explosion consists of a very short ($\sim$1 ms) spike followed by a much more prolonged, less intense, and gradually fading radiation signal. The detected explosion, however, looked very different: it lacked the initial spike but instead showed two distinct peaks in its lightcurve. To the Los Alamos scientists, led by Ray Klebesadel, who analyzed the data, it was clear that this was an explosion but very likely not one associated with an atomic bomb, therefore no urgent action had to be taken. It was only in 1973 that the first scientific results on the discovery of 16 "gamma-ray bursts of cosmic origin" were published in a scientific journal.[3]

7.2   Observed properties

Because it is both exciting and instructive, we follow roughly the chronological order in describing the GRB observations.

The discovery of gamma-ray bursts triggered furious research activity and brought to light a flurry of often rather exotic theoretical models. They included magnetic flares, cosmic strings, neutron star glitches, thermonuclear explosions, and meteoritic impacts on neutron stars, to name just a few. There were probably times when more models existed than actually observed bursts. At the Texas conference in 1974, only 1 year after the first scientific publication on this subject, Malvin Ruderman gave a review on the existing GRB theories. He summarized the situation: "The only feature that all but one (and perhaps all) of the very many proposed models have in common is that they will not be the explanation of GRBs." His favorite mechanism, matter accretion onto a black hole, is still a key ingredient in most of today's models.

The observed GRB lightcurves vary substantially on timescales as short as a millisecond, $\Delta t \sim 1$ ms. Therefore, from straightforward causality arguments (see Section 8.2.3), it was concluded that the source of the explosion most probably

---

[3] Klebesadel, R. W., et al. (1973). *Astrophysical Journal*, **182**, L85.

involves a compact object:

$$D \leq c \cdot \Delta t \approx 300 \text{ km}. \qquad (7.1)$$

A compact object of this scale could only be a neutron star or a black hole. By the end of the 1980s, a remarkable consensus had grown that gamma-ray bursts are caused by neutron stars in our own Galaxy. As the distribution of neutron stars is concentrated around the plane of our Galaxy, see Fig. 5.2, the expectation was that future satellite experiments with higher positional accuracy than the VELA satellites would confirm the suspected concentration of the sources in the Galactic plane.

7.2.1   The Burst and Transient Source Experiment (BATSE)

In April 1991, the Compton Gamma-Ray Observatory (CGRO) was launched, which carried, among others, the Burst and Transient Source Experiment (BATSE). The observed distribution of the GRB sources on the sky (see Fig. 7.1) came as a big surprise: contrary to expectations, gamma-ray bursts are distributed completely isotropically; they occur with the same probability in each direction of the sky.

The BATSE results also showed a large diversity in the behavior of gamma-ray bursts. Some examples of lightcurves detected by BATSE are shown in Fig. 7.2. GRBs form a very inhomogeneous class of events: some exhibit long times of

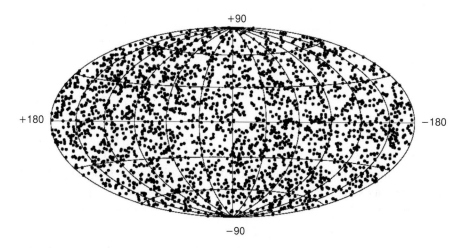

Figure 7.1   Distribution of 2704 gamma-ray bursts on the sky as found by BATSE. The Galactic plane corresponds to the central horizontal line. Courtesy of the Gamma-Ray Astronomy Team at the National Space Science and Technology Center (NSSTC); see http://gammaray.nsstc.nasa.gov/.

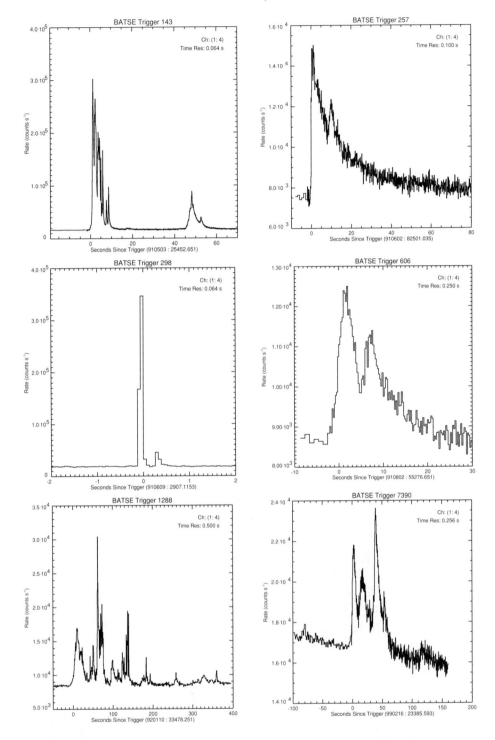

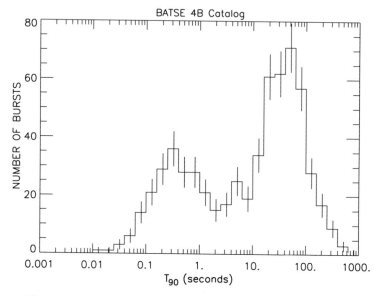

Figure 7.3   Distribution of the duration of the gamma-ray bursts detected by BATSE. The duration is measured by the parameter T_{90}, which is the time over which a burst emits from 5% of its total measured counts to 95%. Courtesy of the Gamma-Ray Astronomy Team at the National Space Science and Technology Center (NSSTC); see http://gammaray.nsstc.nasa.gov/.

quiescence between two phases of activities (first panel), some are so-called FREDS ("Fast Rise Exponential Decay"; second panel), some are just isolated spikes of subsecond duration (panel 3), while others are long sequences of erratic subbursts (panels 4–6).

As a measure for the duration of a GRB, one usually uses a quantity denoted by T_{90}. It measures the duration of the time interval during which 90% of the total counts have been detected. The start of the T_{90} interval is defined by the time at which 5% of the total counts have been detected; the end is defined by the time at which 95% of the total counts have been detected. As is apparent from the BATSE observations shown in Fig. 7.3, the bursts fall into two classes: short ones with typical durations of ~ 0.3 s and long ones with a typical duration of about 30 s. Only about 25% of the bursts belong to the short category.

Figure 7.2   Gamma-ray bursts show a large diversity in their behavior. The figures show the number of counts per second as a function of time as measured by BATSE. Note the different scales on the abscissae. Courtesy of the Gamma-Ray Astronomy Team at the National Space Science and Technology Center (NSSTC); see http://gammaray.nsstc.nasa.gov/.

This immediately raised the question whether there are two different types of central engines or whether the same source can act in two modes.

7.2.2   Spectra

Not only the durations but also the spectra show differences: the spectra of the short GRBs are systematically *harder* than long ones. Harder means that they contain a larger fraction of high-energy photons. BATSE collected the GRB photons in four different energy channels: channel 1 for photons with energies between 20 and 50 keV, channel 2 for 50–100 keV, channel 3 for 100–300 keV photons, and everything greater than energies of 300 keV was collected in channel 4. The hardness of a burst is then quantified by the *hardness ratio*, which is defined as the ratio of the photon counts in channels 3 and 2. It measures the relative occurrence of high-energy photons and is consequently a simple characterization of the photon spectrum. A plot of the hardness ratio versus burst duration is shown in Fig. 7.4: although there is a smooth transition between both classes, short bursts are consistently harder than long ones. This suggests two different central engines, a conjecture that has received a convincing confirmation by the first afterglow observations of short bursts in the year 2005; see Section 7.2.9.

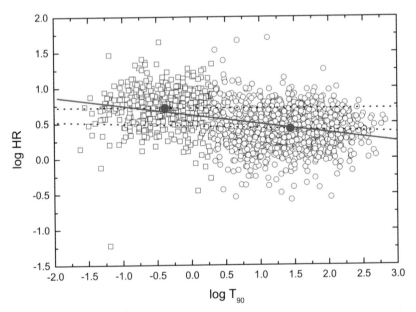

Figure 7.4   Plot of the hardness ratio (photon counts in the 100 to 300-keV channel divided by the photon counts in the 50 to 100-keV channel) versus burst duration, T_{90}; see text. The lines are fits to different burst samples. Figure taken from Qin et al. (2000). *Publications of the Astronomical Society of Japan*, **52**, 759.

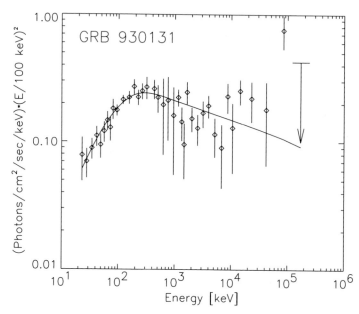

Figure 7.5   Example of a gamma-ray burst spectrum. Shown is the specific flux as a function of energy measured in keV; the line is a fit with two smoothly joined power laws. Figure taken from Bromm and Schaefer (1999). *Astrophysical Journal*, **520**, 661.

The spectra of GRBs are good examples of nonthermal spectra, which were sketched in Fig. 3.6. They can be fitted with two power laws that are smoothly joined at the so-called break energy. Typical break energies of GRBs are in the range of a few hundred keV. One example of a GRB spectrum is shown in Fig. 7.5. In contrast to a blackbody spectrum, the high-energy part does not drop off exponentially. Instead, it declines slowly in a *high-energy tail*. Some GRB spectra extend up to several GeV ($1\ \mathrm{GeV} = 10^9\ \mathrm{eV}$). These nonthermal spectra, together with variations on very short timescales, set tight constraints on the properties of the outflow that produces the radiation of GRBs. In particular, these two observations can be reconciled with each other only if highly relativistic motion is involved. This is explained in Section 7.3.1.

7.2.3   *Source distribution and distance scale*

Although the isotropic distribution clearly showed that previous ideas about the location of GRB sources had not been correct, it did not say anything about the distance scale to the bursting sources: they could be either spherically distributed in a restricted volume around the observer (see upper-right panel in Fig. 7.6) or

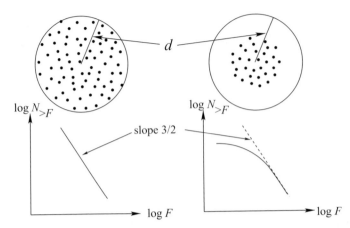

Figure 7.6   Illustration of two isotropic distributions. On the left side, the number density remains constant out to the largest observable distance d, and therefore $\log N - \log F$ has a slope of $-3/2$, and the distribution is said to be homogeneous. On the right side, the sources are distributed to distances shorter than the largest observable ones; one sees the "edge of the distribution." Therefore, the ($\log N - \log F$) curve remains below the $-3/2$ line for low values of F.

have a constant source number density out to the largest detectable distances (see upper-left panel in Fig. 7.6). The latter case is called a *homogeneous* distribution.

Imagine for a moment a (Euclidean) space filled with sources of a constant number density. In this case, the number of sources increases proportionally to the volume or the cube of the distance d:

$$N \propto d^3. \tag{7.2}$$

Because of the inverse square law, Eq. (3.9), the fluxes, F, from the sources are smaller for more distant sources:

$$F \propto d^{-2}. \tag{7.3}$$

Combining Eqs. (7.2) and (7.3) yields

$$N \propto F^{-3/2}. \tag{7.4}$$

If we plot $\log(N_{>F})$ versus $\log(F)$, where $N_{>F}$ is the number of sources with a flux larger than F, we find a straight line with a slope of $-3/2$ as shown in the lower-left panel of Fig. 7.6. If d extends beyond the edge of the source distribution, the counted number of sources at the faint end is smaller than predicted by the $-3/2$ line, such as shown in the bottom-right panel of Fig. 7.6. This turnover at the faint end is a clear indication that we see the edge of the source distribution.

In principle, the source distribution can be any combination of (un-)isotropic and (in-)homogeneous. The BATSE experiment found that GRB sources are isotropic (see Fig. 7.1) and inhomogeneous (see lower-right panel in Fig. 7.6). This means that we see an isotropic distribution out to some large distance; beyond this distance, the number density drops. Such an inhomogeneous and isotropic distribution is typical for cosmological sources such as galaxies, for example.

The BATSE observations definitely ruled out a *disk* population in our Galaxy, but not an extended spherical distribution, a *halo*, around our Galaxy. Earth is located in a spiral arm of our Galaxy at a distance of about 8 kpc from the center. Because a possible halo would be centered around the Galaxy center, its radius would have to be very large to still appear isotropic as seen from Earth.

Still, this did not settle the distance issue. There was a passionate controversy between adherents of a cosmological origin and advocates of an extended halo of neutron stars around our Galaxy. In commemoration of the great debate in 1920 in which Harlow Shapley and Heber Curtis[4] debated over the scale of the Universe, whether the observed spiral nebulae were simply nearby gas clouds or galaxies like our own, in 1995 a second great debate was held in New York to discuss the distance scale of gamma-ray bursts.[5] A democratic vote at the end of the debate gave about equal numbers for the Galactic and the cosmological hypotheses.

7.2.4 Energies and rates

The distance to the bursting source determines how much energy is required to explain the observations. We start from the simplest assumption that the central engine emits equal amounts of energy in all directions. If the typical time-integrated energy flux, the fluence, is $f \sim 10^{-6}$ erg cm^{-2}, the total involved energy is

$$E_{\text{iso}} = 4\pi d^2 f \sim \begin{cases} 2 \cdot 10^{40} \text{ erg}, & \text{for } d = 15 \text{ kpc} \\ 2 \cdot 10^{41} \text{ erg}, & \text{for } d = 50 \text{ kpc} \\ 2 \cdot 10^{51} \text{ erg}, & \text{for } d = 5 \text{ Gpc} \end{cases} . \tag{7.5}$$

The radius of our Galaxy is about 15 kpc, 50 kpc may be the radius to some Galactic halo, and the 5 Gpc corresponds approximately to a cosmological redshift $z = \Delta\lambda/\lambda = 1$; see the redshift box in Chapter 4. Energies derived under the assumption of isotropic emission are called *isotropized energies*, E_{iso}.

Closely related to the distance and energy issue is the question at which *rate* GRBs actually occur. BATSE observed about one burst per day. If the bursts were Galactic, it would be a relatively common event with a rate of about one per day

---

[4] Harlow Shapley (1885–1972): American astronomer and writer. Heber Doust Curtis (1872–1942): American astronomer.

[5] The main protagonists were Don Lamb, who advocated the Galactic origin, and Bogdan Paczyński, who argued that the observed isotropic and inhomogeneous distribution was typical for a cosmological origin. The debate was moderated by the Astronomer Royal, Sir Martin Rees.

per galaxy and a large but not tremendous amount of energy. If, however, they were cosmological, they would be extremely rare and require a very exotic central engine to release such a huge amount of energy to still be detectable close to Earth. The BATSE data suggest a total ($=$ long $+$ short) GRB rate of about 10^{-7} events per year per galaxy.

Modifications for the energy estimates and the rates for the case that the emission is cone shaped rather than isotropic are discussed in Section 7.2.8.

7.2.5   *Soft gamma repeaters and GRBs*

The detection of a GRB in a wavelength band different from gamma rays was thought to be the key to an understanding of the nature of the bursts, but all efforts to point other instruments to the positions given by gamma-ray instruments turned out to be fruitless. The only exception was an energetic flash of gamma rays detected on March 5, 1979. It came from a neighboring galaxy, the Large Magellanic Cloud, and could be associated with a supernova renmant. After some controversy, it turned out not be a GRB at all but rather a *giant flare* of a Soft Gamma Repeater (SGR) (see Section 5.9.1), one of the manifestations of an ultramagnetized neutron star (magnetar). We know that such SGR outbursts can, in contrast to GRBs, repeat, which means that the same object can launch several such outbursts without being destroyed. For more information on SGRs and magnetars, we refer to Chapter 5.

7.2.6   *Finally: A smoking gun . . .*

The final breakthrough that settled the question about the distance to the bursting source came after the Italian–Dutch X-ray satellite BeppoSAX (see Appendix A) had been launched in 1996. On February 28, 1997, BeppoSAX detected an after-glow[6] in the X-ray band that was obviously related to a gamma-ray burst, called GRB 970228 (in the nomenclatura of GRBs, the first two numbers give the year, the second two, the month and the final two, the day). Twenty-one hours later, a rapidly fading optical afterglow was detected at the same position (within the positional accuracy of the BeppoSAX detection) of GRB 970228. This afterglow seemed to be related to a faint galaxy, suggesting that the burst occurred at a distance far from our own Galaxy. Later, also radio afterglows were detected, and finally, in May 1997, a redshifted spectral line was identified for the burst GRB 970508 that indicated a cosmological distance at a redshift (see box in Chapter 4) of at least $z = 0.835$. This settled the long debate about the distance scale of GRBs, so at

---

[6] The term *afterglow* is used for delayed, lower energy emission of the burst.

least long GRBs are coming from distances far outside of our Galaxy and are thus caused by an extremely rare event that releases a tremendous amount of energy.

7.2.7 Connection between long GRBs and supernovae

The emission of gamma rays from the initial stages of supernovae had already been predicted in 1968.[7] Consequently, the first investigation of GRBs searched for "gamma-ray fluxes near the times of appearance of supernovae," but "these searches proved uniformly fruitless."[8] It took until the first discoveries of GRB afterglows in 1997 to obtain observational hints on a connection between long GRBs and supernovae. First, only indirect hints were available: the host galaxies of the observed bursts showed signs of active star formation and thus suggested a connection to massive, short-lived stars. This is similar to the case of normal supernovae (see Chapter 4): Type II supernovae occur close to star-forming regions; they are massive and young but already dying stars. By contrast, Type Ia supernovae occur in all kinds of galaxies, also in those where star formation has ceased long time ago. Therefore, Type Ia supernovae must come from an old stellar population, and it is widely believed that white dwarfs are the progenitors. Similarly, the observation of long GRBs in galaxies with star formation suggests a young progenitor population.

Supernova 1998bw and GRB 980425

On April 25, 1998, BEPPOSAX detected GRB 980425 and at about the same position a peculiar, very energetic Type Ib/c supernova named SN 1998bw. This supernova had a redshift of $z = 0.0085$. Because the positional errors were large, the two observations could not be conclusively linked to each other, but a connection was very plausible. Compared with others, the GRB was very weak, with an energy E_{iso} of only $\sim 10^{48}$ erg, whereas the supernova of Type Ic was particularly bright and rapidly expanding. The kinetic explosion energy is estimated to be $2 \cdot 10^{52}$ erg, about one order of magnitude larger than for a normal core-collapse supernova. Such particularly energetic supernovae are sometimes called *hypernovae*.[9]

After GRB 980425, "bumps" have been detected in the late lightcurves of several other GRBs. They appeared about 20 days after the burst and were in shape and color roughly comparable to SN 1998bw, again suggesting a connection between (at least some) long GRBs with supernovae, but all this was only circumstantial evidence.

---

[7] Colgate, S. A. (1968). *Canadian Journal of Physics*, **46**, 467.
[8] Klebesadel, R. W., et al. (1973). *Astrophysical Journal*, **182**, L85.
[9] Paczynski, B. (1998). *Astrophysical Journal*, **494**, 45.

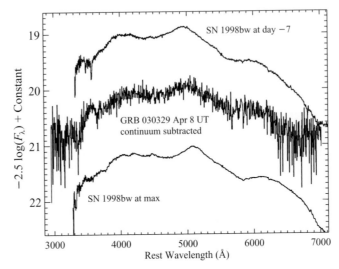

Figure 7.7   Comparison of the spectra of two supernovae that were related to a GRB: SN 2003dh/GRB 030329 and SN 1998bw/GRB 980425. Figure taken from Stanek et al. (2003). *Astrophysical Journal*, **591**, 17.

Supernova SN 2003dh and GRB 030329

The case was closed only in 2003, when the connection between long GRBs and supernovae was put on a sound basis. On March 29, 2003, the extremely bright GRB 030329 was detected. After 6 days, again a supernova became visible as a bump in the lightcurve. The spectrum of this supernova, SN 2003dh, was very similar to the peculiar spectrum of SN 1998bw; see Fig. 7.7. Since the discovery of this unambiguous connection, it is generally accepted that at least some long GRBs are related to supernovae.

7.2.8   Collimated outflows

The estimates for the involved energy stated in Eq. (7.5) assumed that the radiation is emitted isotropically. If the source is instead emitting into two cones that cover a solid angle $\Delta\Omega$ (see Fig. 7.8), the involved energy is reduced by a large factor:

$$E_{\text{true}} = E_{\text{iso}} \left(\frac{\Delta\Omega}{4\pi} \right). \tag{7.6}$$

$\Delta\Omega$ is given by

$$\Delta\Omega = 2 \cdot \int_0^{2\pi} \int_0^{\theta_j} \sin\theta \, d\theta \, d\varphi = 4\pi(1 - \cos\theta_j) \approx 4\pi \left(\frac{\theta_j^2}{2} \right), \tag{7.7}$$

Figure 7.8   Sketch of outflow that is collimated into two cones of opening half-angle θ_j.

where in the last step we have expanded the cosine for the case of small angles θ_j.
Therefore, the true emitted energy is only

$$E_{\text{true}} \approx E_{\text{iso}} \left(\frac{\theta_j^2}{2} \right) \approx \frac{E_{\text{iso}}}{65} \left(\frac{\theta_j}{10°} \right)^2 . \tag{7.8}$$

The quantity $(4\pi/\Delta\Omega)$ is often referred to as *beaming factor*, f_b. If the radiation
is emitted into a small solid angle $\Delta\Omega$, f_b is large, and the source is said to be
strongly beamed.

If GRBs produce collimated emission with a beaming factor f_b, their emission
would be directed into only a $1/f_b$ fraction of the whole sky. This implies that most
of the GRBs would go unnoticed unless their radiation cones by accident point
toward Earth. As a result, the true rate has to be larger by a factor f_b.

The first observational evidence for collimated outflow came from the particularly bright GRB 990123, a burst detected at a redshift of $z = 1.6$. This implied
an isotropic energy of $4.5 \cdot 10^{54}$ erg in gamma rays, that is, more than the energy
equivalent of a solar mass, $1\ M_\odot\ c^2 = 1.8 \cdot 10^{54}$ erg. Every stellar mass object
would face serious problems if challenged with transforming a good fraction of its
rest mass into gamma rays, but with a small enough collimation angle θ, this energy
could be reduced by large factors to energies that a stellar mass object can provide.

More evidence for collimated outflows comes from *achromatic breaks* in GRB
lightcurves.[10] Imagine an outflow that is collimated into a cone with an opening
angle θ_j[11] and that moves relativistically with a Lorentz factor γ toward the observer.

---

[10] Note, however, that there are also alternative explanations of such breaks.
[11] It is important to distinguish between collimated outflow, i.e., matter outflowing into only a fraction of the full
solid angle, and relativistic beaming. The latter is purely special relativistic effect (see Section 1.3.5) and has
nothing to do with the geometry of the outflow.

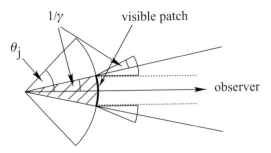

Figure 7.9   A jet with physical opening angle θ_j moves with Lorentz factor γ toward an observer. As the radiation from each emitting patch is beamed into a forward cone with an opening angle $1/\gamma$, the observer sees only a small fraction, a cone with opening angle $1/\gamma$, of the emitting surface.

If each radiating patch on the surface of the cone radiates isotropically in its rest frame, for the observer it looks as if the radiation was beamed in the forward direction into a cone of opening angle $1/\gamma$; see Section 1.3.5 and Exercises 1.2 and 1.3 in Chapter 1. This situation is sketched in Fig. 7.9. The observer sees only the patch inside an angle of $1/\gamma$ with his line of sight; the other parts are "beamed away" from him. As the jet is moving through interstellar material, it is constantly slowed down; that is, its Lorentz factor is decreasing. As it slows down, the fraction of the real jet that the observer can see is increasing. Once $1/\gamma > \theta_j$, the observer will suddenly realize that parts of the sphere he thought he was seeing are missing and that he is only seeing a cone. This causes the lightcurve in all wavelengths to drop faster than before. Such a sudden steepening of the lightcurve that is independent of the wavelength is called an *achromatic break*.

   Such achromatic breaks have been observed in several bursts, for example in the lightcurve shown in Fig. 7.10. From the time at which this jet break occurs, the opening angle can be deduced.[12] Typical values for long bursts are spread around opening angles of 4 degrees. Taking collimation into account, the large spread in isotropized energies (see upper panel in Fig. 7.11) reduces to a relatively narrow range of true emitted energies around 10^{51} erg (see lower panel in Fig. 7.11). The current estimates for the beaming angles suggest that there are at least 1000 GRBs/day in the universe (see Exercise 7.1).

7.2.9   The afterglows of short GRBs

The detection of the first *afterglows of short GRBs* had to wait until 2005. On November 20, 2004, the SWIFT satellite was launched. To capture the rapidly fading afterglows of short GRBs early on, SWIFT was constructed to slew very

---

[12] Frail, D., et al. (2001). *Astrophysical Journal*, **562**, 55.

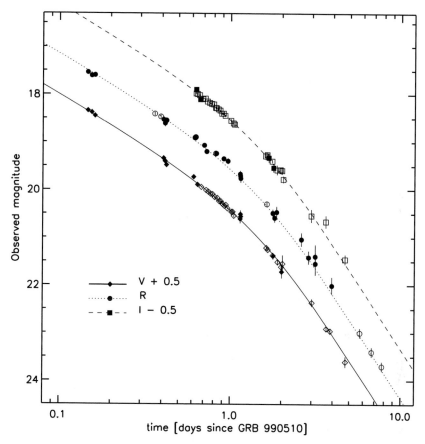

Figure 7.10   Observation of an achromatic break in the lightcurve of GRB
990510. Figure taken from Harrison, F.A., et al. (1999). *Astrophysical Journal*,
523, 121.

quickly to a given burst location. On May 9, 2005, the first afterglow of a short
GRB was detected,[13] and several more followed quickly. Not really surprisingly,
short bursts also turned out to occur at cosmological distances, far outside our
Galaxy. These observations showed further differences to long bursts:

- Short bursts occur in all types of galaxies and also in elliptical galaxies without
  any star formation. Similar to the case of Type Ia supernovae (Chapter 4), this
  points to an old progenitor population.
- They have systematically lower redshifts than long bursts.
- The involved energies are smaller by several orders of magnitude.

Figure 7.12 shows a comparison of redshift, isotropic energy and X-ray luminosity
between long and short bursts.

[13] Bloom, J., et al. (2006). *Astrophysical Journal*, **638**, 354.

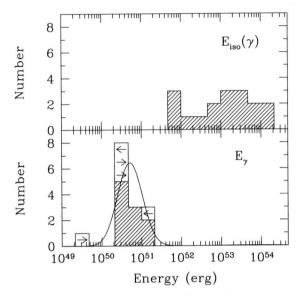

Figure 7.11   Energy estimates of some long GRBs. The upper panel shows the derived γ-ray energies if isotropic emission is assumed; the lower panel shows the energies corrected for collimation. Figure taken from Frail, D., et al. (2001). *Astrophysical Journal*, **562**, 55.

Several observations showed a puzzling X-ray activity a long time after the bursts occurred. At the time of writing, this is not yet properly understood.

7.2.10   Cosmological redshifts

The GRB redshifts measured by BATSE, all for long bursts, were typically at around $z \sim 1$. Because of its faster slewing, SWIFT could detect afterglows that otherwise would have gone unnoticed. This increased the number of high redshift bursts and put the median observed redshift of long bursts to ~ 2 as is shown in Fig. 7.12. At the time of writing, the record holder in terms of redshift is GRB 050904, with a measured redshift of $z = 6.29$. Short bursts have consistently smaller redshifts than long ones.

Some of the milestones of gamma-ray burst observations are summarized in Table 7.1. The properties of long and short bursts are compared in Table 7.2.

7.3   Constraints on the central engine

7.3.1   The compactness problem

Originally, the so-called compactness problem was used as an argument against a cosmological origin of GRBs. It arises because of the combination of large energies, short-time variability, and the observed nonthermal spectrum. The short-time

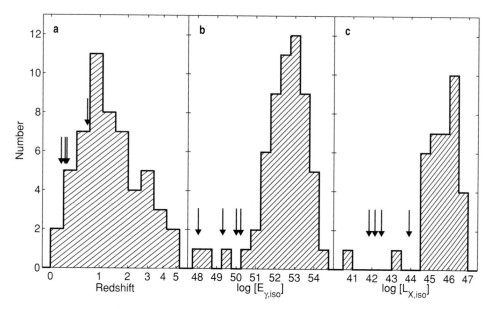

Figure 7.12 Comparison of the properties of short and long GRBs; the histograms refer to long GRBs, the arrows to the first set of observed short GRB afterglows. Short GRBs seem to come from closer, i.e., they have smaller redshifts, and the involved energies seem to be much lower than for long bursts. Figure taken from Fox, D., et al. (2005). *Nature*, **437**, 845.

variability implies a small size of the source, but a compact source with a huge luminosity is opaque to its own radiation because of the formation of a large number of electron–positron pairs. Therefore, it should produce a thermal spectrum (Chapter 3), which is not what is observed.

Let us make an order-of-magnitude estimate for the optical depth a photon would encounter. This is a rather crude estimate, but as it turns out, the optical depth is *huge*. Even if we change some of the poorly known numbers by large factors, the result remains the same: the source is optically thick to its photons! From $d\tau = n\sigma ds$ (see Eq. [3.46]), we estimate

$$\tau \sim n_e \sigma_T D, \tag{7.9}$$

where $\sigma_T = 6.65 \cdot 10^{-25}$ cm^2 is the Thomson cross section, n_e the number density of free electrons, and D the size of the source.

Assuming that the lightcurve varies substantially on a timescale Δt, straightforward causality arguments (see Section 8.2.3) suggest a source size given by

$$D \leq c \cdot \Delta t. \tag{7.10}$$

Table 7.1   *Some milestones in gamma-ray-burst science.*

Date/burst	Milestone
July 2, 1967	First-ever gamma-ray burst detection
March 5, 1979	An unusual gamma-ray transient is localized in a supernova remnant in the Large Magellanic Cloud; it is often referred to as just "the March 5 event"; after a decade of controversy, the event turns out to have come from a soft gamma repeater (SGR) rather than being a proper GRB
April 5, 1991	NASA launches the Compton Gamma-Ray Observatory (CGRO). Among the experiments is the Burst and Transient Source Experiment (BATSE); unexpectedly, BATSE finds an isotropic source distribution
April 30, 1996	Italian–Dutch BeppoSAX satellite is launched
February 28, 1997 (GRB 97228)	First-ever afterglow is detected by BeppoSAX
May 8, 1997 (GRB 970508)	First measurement of the cosmological redshift of a GRB; this ends a long debate about the distance scale: (at least long) GRBs are definitely cosmological
April 25, 1998 (GRB 980425)	Energetic supernova SN 1998bw detected coincidently with the weak GRB 980425: this is the first hint of a possible connection between GRBs and supernovae
January 23, 1999 (GRB 990123)	For the first time, the afterglow is detected within seconds of the initial burst and detection of an optical flash; the inferred isotropic energy is $4.5 \cdot 10^{54}$ erg, which would correspond to more than one solar mass being tranformed into gamma rays. This suggests that the emission may not be isotropic but instead collimated
October 9, 2000	Launch of the HETE II satellite; it detects outbursts similar to normal GRBs, but with much softer spectra and smaller spectral breaks, so-called X-ray flashes
March 29, 2003 (GRB 030329)	Safe connection between supernova and the extremely bright GRB 030329 the supernova spectrum is similar to SN 1998bw
November 20, 2004	NASA launches the Swift satellite
May 9, 2005 (GRB 050509b)	First-ever afterglow of a short GRB detected by Swift: also short bursts occur at cosmological distances; they are not connected with supernovae
September 4, 2005 (GRB 050904)	Swift detects a GRB with redshift $z = 6.29$; this is thought to be close to the end of the "dark ages" of the universe, where the first sources of light appeared.

If we denote the fluence by f, the burst occurs at a distance d, and the emission is isotropic, the involved energy is

$$E \sim 4\pi d^2 f. \tag{7.11}$$

Table 7.2   *Comparison of the properties of long and short GRBs.*

Property	Long GRB	Short GRB
Duration	~ 30 s	~ 0.3 s
Observed rate (BATSE)	$\sim 500 \, \mathrm{yr}^{-1}$	$\sim 170 \, \mathrm{yr}^{-1}$
Variability timescale	~ 1 ms	~ 1 ms
Host galaxy	Galaxies with active star formation	Galaxies with and without star formation
Supernova?	Confirmed in some cases	Prob. not
Isotropized γ-energy $E_{\gamma,\mathrm{iso}}$	$\sim 10^{53}$ erg	$\sim 10^{50}$ erg
Median redshift	~ 2	~ 0.3
Popular model	"collapsar"	Compact binaries

With the typical observed photon energy, $\bar{E}_\gamma \sim 1$ MeV, we have a photon number density of

$$n_\gamma \sim \frac{4\pi d^2 f}{\bar{E}_\gamma D^3} \tag{7.12}$$

at the source. Because the observed photon energy is very high, a reasonable fraction of photons is energetic enough to produce electron–positron pairs in collisions. Remember that a photon needs to have at least the energy of twice the rest-mass energy of an electron (2×0.511 MeV) to produce an electron–positron pair. If we write their density as $n_e = f_e \cdot n_\gamma$ and set $D \sim c \cdot \Delta t$, the optical depth becomes

$$\tau \sim n_e \cdot \sigma_T \cdot D \sim f_e \frac{4\pi d^2 f}{\bar{E}_\gamma D^2} \sigma_T \sim f_e \frac{4\pi d^2 f}{\bar{E}_\gamma c^2 \Delta t^2} \sigma_T \tag{7.13}$$

$$\tau \sim 10^{16} f_e \left(\frac{d}{5 \text{ Gpc}}\right)^2 \left(\frac{f}{10^{-6} \text{ erg cm}^{-2}}\right) \left(\frac{1 \text{ MeV}}{\bar{E}_\gamma}\right) \left(\frac{0.01 \text{ s}}{\Delta t}\right)^2 ; \tag{7.14}$$

that is, for every reasonable f_e, the optical depth is *much* higher than the value required to be optically thin, $\tau \sim 1$. If this were true, the photons would produce large amounts of electron–positron pairs, equilibrate with the latter, and produce a thermal spectrum. This would contradict the observed, nonthermal GRB spectra see (Fig. 7.5).

So where is the flaw in this reasoning? It lies in our implicit assumption of nonrelativistic motion. Relativistic motion leads to two modifications. First, the source can be larger than our original estimate by a factor of order γ^2, where γ is the Lorentz factor (see also Exercise 1.2). Second, the photons that we observe are blueshifted, and therefore their energy in the fireball frame is lower by a factor

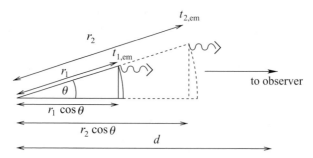

Figure 7.13   Geometry to relate time intervals in the emission frame to the observer frame.

$1/\gamma$. Hence, a much smaller fraction of photons has enough energy to produce electron–positron pairs.

Let us first relate the time intervals in the emission and in the observer frame. For simplicity consider a spherical shell that expands at a velocity v. At time $t_{1,\text{em}}$ and radius r_1, it emits a photon toward the observer, as shown in Fig. 7.13. This photon arrives at the observer at time

$$t_{1,\text{obs}} = t_{1,\text{em}} + \frac{d - r_1 \cos \theta}{c}. \tag{7.15}$$

At time $t_{2,\text{em}} = t_{1,\text{em}} + \Delta t_{\text{em}}$, the shell has reached a radius $r_2 = r_1 + \Delta t_{\text{em}} v$, and it emits a photon again. This photon reaches the observer at time

$$t_{2,\text{obs}} = t_{2,\text{em}} + \frac{d - r_2 \cos \theta}{c}. \tag{7.16}$$

The observer measures a time difference between the arrival of the two photons of

$$\Delta t_{\text{obs}} = t_{2,\text{obs}} - t_{1,\text{obs}} = \Delta t_{\text{em}}(1 - \beta \cos \theta), \tag{7.17}$$

where we have used $\beta = v/c$. If we assume that θ is small, we can set $\cos \theta \approx 1$ and then make use of the relation $1 - \beta \approx 1/2\gamma^2$ that we had derived in the context of synchrotron radiation; see Eq. (3.167). This yields the relation between the observed and the time interval in the emission frame:

$$\Delta t_{\text{obs}} = \frac{\Delta t_{\text{em}}}{2\gamma^2}. \tag{7.18}$$

Therefore, the original source-size estimate, $D \leq c\Delta t$, was wrong: the short observed variability timescale can be obtained even for large sources if only the Lorentz factor is high enough: $\Delta t = \Delta t_{\text{em}}/2\gamma^2$. As Δt enters quadratically (see Eq. [7.13]), this modification has an effect $\propto \gamma^4$ on the estimate of the optical depth.

The second effect comes from the detected photons being blueshifted: as the source is moving toward us with a Lorentz factor γ, the observed photon energy is $E_{\gamma,\text{obs}} = E_{\gamma,\text{em}} \cdot \gamma$. As an effect, only a small fraction of the photons at the high-energy end of the photon spectrum is energetic enough to produce electron–positron pairs in collisions. The value of this fraction depends on the shape of the spectrum. A detailed calculation[14] shows that for a typical spectrum the naive estimate for τ has to be corrected by a factor $\propto \gamma^{-6}$. This results in a lower limit on the Lorentz factor of about 100,[15] *so to circumvent the compactness problem, relativistic motion with Lorentz factors of hundreds toward the observer is required.* Such relativistic motion has indeed been inferred from the radio afterglows of GRBs.[16]

7.3.2   *The baryonic pollution problem*

Closely related to the compactness problem is the baryonic pollution problem. Assume we have a fireball with energy E made of photons, electron–positron pairs, and baryons (neutrons and protons) of mass M. This fireball expands under its own pressure until its thermal energy has been transformed into kinetic energy of its baryons. The Lorentz factor that can be reached is determined by the quantity $\eta \equiv E/Mc^2$:

$$\gamma_{\text{asym}} \approx \eta. \tag{7.19}$$

To achieve at least a Lorentz factor of γ, the baryonic mass M must be smaller than

$$M_\gamma = \frac{E}{\gamma c^2} = 6 \cdot 10^{-6} \, M_\odot \left(\frac{E}{10^{51} \, \text{erg}} \right) \left(\frac{100}{\gamma} \right). \tag{7.20}$$

This is a serious constraint on any GRB model: how can a stellar mass object release a huge amount of energy without polluting it with too many baryons?

7.3.3   *Requirements for a GRB-progenitor model*

These considerations impose several constraints on GRB models. Every plausible model for the central engine of a GRB has to provide more than just a lot of energy; it has to fulfill the following requirements:

• The central engine has to be able to provide a large enough *energy reservoir* to account for the observed energy. Realistically, this energy reservoir should be *much* larger than the observed photon energy as every plausible central engine is likely to emit large amounts of energy in other channels such as neutrinos or gravitational waves.

---

[14] Lithwick, Y. and Sari, R. (2001). *Astrophysical Journal*, **555**, 540.
[15] The spectral shapes of short and long bursts are somewhat different. A recent analysis of short bursts concludes that a lower limit on their Lorentz factor is about 30, see Nakar, E. (2007), astro-ph/0701748.
[16] Taylor, G. B., et al. (2005). *Astrophysical Journal*, **622**, 986.

- The model for the central engine has to explain what sets the *duration of the burst*: why is the duration of a short GRB about 0.3 s, and why about 30 s for a long burst?
- It has to provide plausible arguments as to how the resulting lightcurves can vary substantially on a subsecond timescale.
- These ideas must be consistent with the types of host galaxies in which the bursts are observed. For example, a model for a long GRB must explain why the burst should be related to galaxies with active star formation. A model for short bursts has to explain why bursts can occur in galaxies with *and* without star formation.
- Closely related is the question of an accompanying supernova explosion. For long bursts, in some cases there is convincing evidence for such a connection. For short bursts, there are observations of good quality that did not find any sign of a supernova.
- Models for GRB central engines have to explain how the required relativistic outflow is produced, in particular how the "pollution" with too many baryons is avoided.
- GRBs show a huge diversity in their behavior; see Fig. 7.2, so whatever produces the burst must depend sensitively on at least one parameter. A plausible model has to provide arguments as to why each burst looks different.

7.3.4   The general picture

Before we come to specific models of GRBs, let us first summarize the general accepted picture. An object of stellar mass undergoes a catastrophic event that leads to the formation of the GRB central engine. This engine rapidly releases a large amount of energy in a compact region of $\sim 10^6 - 10^7$ cm. This energy can be in the form of radiation, heat and/or electromagnetic field and is the driver behind the subsequent acceleration to ultra-relativistic speeds. The flow is initially optically thick and cools by adiabatic expansion until it becomes optically thin at large distances ($> 10^{13}$ cm). The initial energy does not produce the observed ("prompt") GRB emission, instead, the radiation is produced once the flow's energy is dissipated at large distances from the source. This dissipation produces highly relativistic electrons that emit the observed gamma rays by interacting with a magnetic field and/or a radiation field, for example, via synchrotron or inverse Compton emission, see Chapter 3.8. These processes dissipate only a fraction of the flow's energy, the rest is transferred at larger radii ($\sim 10^{16} - 10^{18}$ cm) into the ambient medium, where the decelerating blast wave produces electromagnetic emission at lower frequencies, the so-called afterglow.

Both the energy deposition mechanism and the composition of the resulting relativistic outflow are not known with certainty. The main candidates for the composition are a *baryonic plasma* made of protons, electrons, and likely neutrons and a strongly magnetized plasma with a large ratio of magnetic to particle energy, a so-called *Poynting-flux dominated outflow*.

In the following, we will sketch the first possibility, for the second one we refer to the existing literature.[17]

7.4 The fireball model

Whatever causes a GRB releases a large amount of energy. It is comparatively easy to identify possible energy sources, but problems arise when one considers how to transform this energy into photons of the right fluence, energy, and temporal structure. We had seen that relativistic motion is required to circumvent the compactness problem, which in turn requires a very low baryon loading.

Studies of fireball expansion, however, show that a mass in baryons as small as 10^{-8} $M_\odot$ suffices to trap the photons for long enough so that most of the fireball energy is converted into kinetic energy of the baryons before the photons can escape. This has led to the idea that the fireball is *not* responsible for producing the observed burst of gamma rays. Instead its role is to accelerate a small amount of baryons to ultrarelativistic velocities. The dissipation or "re-thermalization" of the baryon kinetic energy in shocks produces both the observed gamma-ray burst (sometimes called *prompt emission*) and the later, lower-photon-energy afterglow. This is, in a nutshell, the essence of the so-called fireball model.[18] For an in-depth discussion of the fireball model, we refer to the existing literature.[19]

Detailed studies[20] show that the time structure of the inner engine is closely imprinted on the gamma-ray lightcurve. Apart from this, the evolution of the fireball is decoupled from whatever astrophysical catastrophe brought it into being. This is a strength of the model as the prediction of the evolution of the fireball is robust. At the same time, it is also a weakness as we cannot learn much (apart from the imprinted time structure) about the properties of the central engine. In the fireball model, the central source just has to (i) provide a large ratio η of energy to rest mass so that ultrarelativistic outflow can develop, (ii) be active for the duration of the observed bursts, and (iii) to produce the time variability on subsecond timescales that is observed in the bursts.

The main stages of the fireball model are (i) energy deposition by the central engine, (ii) expansion to ultrarelativistic velocities, (iii) escape of the fireball photons in a preburst, (iv) production of the gamma-ray burst, and (v) the afterglow. These stages are described next and are summarized in Fig. 7.14.

---

[17] See, for example, Lyutikov, M. (2006). *New Journal of Physics*, **8**, 119 and references therein.
[18] Note that this picture is widely accepted but not completely free of controversy.
[19] E.g., Piran, T. (1999). *Physics Reports*, **314**, 575, *Physics Reports*, **442**, 166. Or, Meszaros, P. (2006). *Reports on Progress in Physics*, **69**, 2259. Or, Nakar, E. (2007).
[20] Kobayashi, S., et al. (1997). *Astrophysical Journal*, **490**, 92.

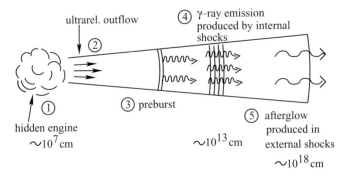

Figure 7.14   Sketch of the fireball model.

7.4.1 *Energy deposition*

The central engine deposits a large amount of energy, of the order of 10^{51} erg, in a small, relatively baryon-free volume of diameter $r_0 \sim 100$ km. How exactly this is done is not the subject of the fireball model and can be different for different central engines. The resulting fireball consists of an optically thick plasma of photons, electron-positron pairs, and baryons with $\eta = E/Mc^2 \geq 100$. This stage is marked as number 1 in Fig. 7.14.

7.4.2 *Expansion to ultrarelativistic velocities*

For a large energy-to-rest mass ratio, η, a fireball accelerates under its own pressure to ultrarelativistic velocities with Lorentz factors given by $\gamma_{\text{asym}} \approx \eta$. For too-large baryon loadings, the fireball produces mainly nonrelativistic baryons. Conservation of energy and entropy leads to the following scaling relations as a function of radius r, which we do not derive here:

$$\gamma(r) \propto r, \qquad T(r) \propto \frac{1}{r}, \qquad \text{and} \qquad n(r) \propto \frac{1}{r^3}. \tag{7.21}$$

Here, T is the temperature in the fireball frame and n the (comoving) baryon number density. The fireball has reached the asymptotic Lorentz factor at a distance

$$r_f = \eta r_0. \tag{7.22}$$

Once the asymptotic Lorentz factor has been reached, the fireball coasts with a constant Lorentz factor. At this stage, we have

$$\gamma(r) = \gamma_{\text{asym}} \approx \eta, \qquad T(r) \propto r^{-2/3}, \qquad \text{and} \qquad n(r) \propto r^{-2}. \tag{7.23}$$

The expansion phase is sketched as number 2 in Fig. 7.14.

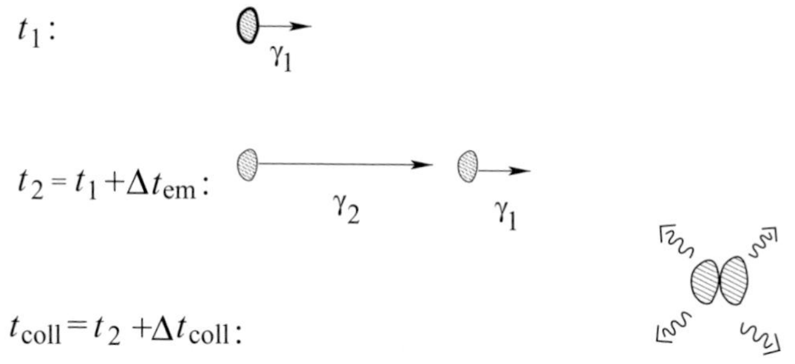

$t_1:$

$t_2 = t_1 + \Delta t_{em}:$

$t_{coll} = t_2 + \Delta t_{coll}:$

Figure 7.15   Sketch of a so-called internal shock: a blob of matter with a higher Lorentz factor γ_2 catches up with a previously ejected blob with γ_1.

7.4.3   Thermal preburst

Once the optical depth inside the fireball drops below unity, its photons escape. This leads to a preburst with thermal spectrum; see number 3 in Fig. 7.14.

7.4.4   Production of gamma rays at large distances via internal shocks

During its expansion, the fireball transforms most of its thermal energy into kinetic energy of its baryons so that the outcome of the fireball is a small amount of ultrarelativistic particles. If the outflow is not completely homogeneous but contains portions with different Lorentz factors, the faster parts of the flow can catch up with slower ones, thus generating internal shocks. This situation is sketched in Fig. 7.15. Consider a first blob that is ejected at time t_1 with a Lorentz factor γ_1. A time Δt_{em} later, at $t_2 = t_1 + \Delta t_{em}$, a second, faster blob is ejected from the engine with a Lorentz factor γ_2. The second blob will catch up with the first at a time, $t_{coll} = t_2 + \Delta t_{coll}$, determined by the condition

$$v_2 \cdot \Delta t_{coll} = \Delta t_{em} \cdot v_1 + \Delta t_{coll} \cdot v_1. \tag{7.24}$$

By solving this equation for Δt_{coll}, we find the distance where the collision occurs,

$$r_{coll} = v_2 \cdot \Delta t_{coll} = c \Delta t_{em} \frac{\beta_1 \beta_2}{\beta_2 - \beta_1}, \tag{7.25}$$

where the $\beta_i = v_i/c$. For the shock to be reasonably efficient, γ_2 should be considerably larger than β_1, so let us assume $\gamma_2 \sim 2\gamma_1 \gg 1$. If we solve the definition of the Lorentz factor γ_i for β_i, we find $\beta_i = (1 - 1/\gamma_i^2)^{1/2} \simeq 1 - 1/2\gamma_i^2$. If we neglect a numerical factor of order unity, we have $(\beta_1 \beta_2)/(\beta_2 - \beta_1) \sim \gamma_1^2$ and thus the internal shocks occur at

$$r_{coll} \sim \gamma_1^2 c \Delta t_{em} \sim 10^{13} \text{ cm} \left(\frac{\gamma_1}{200} \right)^2 \left(\frac{\Delta t_{em}}{0.01 \text{ s}} \right), \tag{7.26}$$

where we have inserted typical numerical values after the last equal sign. This means that the internal shocks and therefore the gamma-ray production occur far from the central engine.

The shocks are believed to be collisionless plasma shocks,[21] in which the ambient magnetic field is strongly amplified and the shocked electrons produce synchrotron radiation; see Section 3.8.2. Depending on the detailed conditions, also inverse Compton processes (Section 3.8.3) may become important. At this stage, the relativistic matter is far away from where it was produced and the density is low enough for the newly produced gamma rays to escape freely. This prompt emission phase is sketched as stage 4 in Fig. 7.14.

7.4.5 *Production of the afterglow by an external shock with an ambient medium*

In this stage, the outflow material dissipates its kinetic energy via shocks with the ambient medium. These shocks are called *external* as they do not occur within the outflowing material itself but rather between the outflow and external matter (either interstellar medium or a wind that was previously ejected by the progenitor). As the flow decelerates, the emission shifts to lower and lower energies, from X-rays down to the radio wave band.

The electrons in the external shock are accelerated into a power-law distribution of electron Lorentz factors. With the assumption that a fixed fraction of the energy in the shocked material goes into electrons and another fixed fraction into magnetic field energy, the spectra from synchrotron emission consist of several power-law segments.[22] These fit the observed spectra reasonably well.

7.5 The central engine

Hundreds of models for the central engines have been proposed over the years, but with increasingly better observational data, more and more models, in particular the Galactic ones, had to be abandoned. At the time of writing, there are still a few models left that have survived for good reasons, but even for these models, despite their merits, there remain many open questions. For each burst category, long-soft and short-hard, we present what is currently the most popular model. It has to be stated explicitly that it is by no means sure that this is the true central engine, and plausible alternatives exist for each category.

---

[21] A shock is called *collisionless* if its thickness is much smaller than the collisional mean free path of the gas constituents. A prominent example of such a shock is the Earth's bow shock created by the interaction of the solar wind with the Earth's magnetic field.

[22] Sari, R., Piran, T. and Narayan, R. (1998). *Astrophysical Journal*, **497**, L17.

The basic ingredients needed to launch both long and short GRBs may be rather similar: a newborn, rapidly accreting black hole surrounded by a massive disk.

7.5.1   *Collapsars*

Observations suggest that (at least some) long GRBs occur in conjunction with peculiar types of supernovae. Interestingly, after correction for collimation, the involved gamma-ray energy is not too different from the electromagnetic energy output of a normal core-collapse supernova, but while a normal supernova imparts its energy to a large mass of order $\sim 1\,M_\odot$ and thus produces nonrelativistic outflows with velocities between 5000 and 30 000 km/s, a GRB somehow manages to impart a lot of energy to a very small amount of mass and thus to produce ultrarelativistic outflow.

The "collapsar" model had been suggested *before* observational evidence for a connection of long GRBs and supernovae existed.[23] In a nutshell, this model involves a rapidly rotating, massive Wolf–Rayet star[24] that has lost its hydrogen envelope and now has a radius comparable to that of the Sun. Such stars have helium cores of more than $10\,M_\odot$ and consequently had much higher masses when they were on the main sequence. Once the central core collapses, it forms a black hole of a few solar masses that is surrounded by a thick accretion disk. During the subsequent accretion process, energy is deposited along the stellar rotation axis, for example, via annihilation of the neutrinos that are emitted from the hot accretion disk, $\nu_i + \bar{\nu}_i \rightarrow e^+ + e^-$. This energy deposition drives a jet along the original rotation axis of the star. The outflow is collimated by the pressure from the stellar mantle and attains high Lorentz factors as it breaks through the stellar surface. The gamma rays are produced at distances of many stellar radii in internal shocks between portions of the jet with different Lorentz factors. The afterglow is produced when the outflow runs into the interstellar medium or the matter previously lost as a wind.[25] Only a small fraction of supernovae produces long GRBs. According to the collapsar model, each long GRB should be accompanied by a supernova similar to the observed SN 1998bw.

Several aspects of this model deserve a closer look.

---

[23] Woosley, S. (1993). *Astrophysical Journal*, **405**, 273.

[24] Wolf–Rayet stars are very hot and massive stars that have very strong winds associated with them. Named after the French astronomers Charles Wolf (1827–1918) and Georges Rayet (1839–1906).

[25] For the interstellar medium one would expect a roughly constant density, for the simplest wind case we had seen in Eq. (6.63) that a density $\rho \propto 1/r^2$, where r is the distance from the star.

Progenitor star and accretion disk formation

The formation of a massive accretion disk around the newly formed black hole requires a fair amount of angular momentum. As outlined in Chapter 6, a black hole possesses a last orbit inside of which no stable, circular test particle orbit can exist; see the box on black holes. For a nonrotating (Schwarzschild) black hole, this orbit lies at

$$R_{ISCO} = \frac{6GM_{BH}}{c^2},$$
(7.27)

where M_{BH} is the black-hole mass. The radius of this orbit moves closer in for a rotating, Kerr black hole. For the infalling material to settle into a circular orbit of radius R, it needs to have a specific angular momentum of

$$l \approx R \cdot (R \cdot \omega_K(R)) = R^2 \sqrt{\frac{GM_{BH}}{R^3}} = \sqrt{GM_{BH}R},$$
(7.28)

where we have used Kepler's third law, Eq. (6.32). In other words, for an accretion disk to form outside the last stable orbit of a Schwarzschild black hole at R_{ISCO} the matter must have a specific angular momentum larger than

$$l_c = \frac{\sqrt{6}GM_{BH}}{c}.$$
(7.29)

This means that the progenitor stars need to retain a very high specific angular momentum in their cores up to the time they collapse. If processes that extract or redistribute angular momentum from or within the star, such as stellar winds or magnetic fields, are ignored in stellar evolution, a promising fraction of stars should be able to fulfill this constraint on the angular momentum. However, if these processes are accounted for using today's best available prescriptions, there is a shortage of viable progenitors. This possible "angular momentum crisis" may be resolved by either a better understanding of angular momentum transport during stellar evolution or perhaps by different initial conditions. Such possibilities include tapping additional angular momentum reservoirs, for example, from the orbit of a binary system. The collapsing core could have been produced in the merger of two massive stars or after mass transfer in a binary system. Another possibility are very metal-poor stars that have formed early in the history of the universe. Because of their lower content of heavy elements, such stars have a reduced opacity. Consequently, photons leak out more easily and entrain less material. Therefore, such stars lose much less mass and angular momentum. GRBs are rare events, and so it is not surprising that they require very particular initial conditions. To date, it is difficult to quantify how serious this angular momentum problem is.

Jet launch and penetration of the star

Jets occur everywhere and on many different scales in the universe, from young stellar objects, over neutron stars, to gamma-ray bursts and active galactic nuclei (Chapter 8). There is no commonly accepted theory about how they are launched. Accretion disks and magnetic fields are very plausible ingredients, but other mechanisms may also work.

The collapsar model does not specify explicitly how the jet is launched. It merely assumes that a jet forms in one way or another. In the equatorial plane, the infalling material is halted at some stage by a centrifugal barrier. Along the stellar rotation axis, there is much less centrifugal support. Here matter can fall relatively unhindered into the hole, thereby clearing a low density funnel. Energy injected into this region will, if injected at a sufficiently large rate, drive an outflow along the rotation axis. If a massive accretion disk forms around the newly born black hole, it can possibly produce neutrinos at large enough rates so that neutrino antineutrino annihilation, $v_i + \bar{v}_i \rightarrow e^+ + e^-$, can become a viable energy deposition mechanism. Another possibility is the extraction of the energy related to the spin of the black hole. Further mechanisms related to strong magnetic fields in the accretion disk could also supply the energy.

Whatever the energy deposition mechanism is, it has to keep pushing the jet for as long as it takes to cross the star, which is typically some tens of seconds. Numerical simulations of jets launched in collapsar progenitor stars show that jets not sustained for long enough are not able to break through the stellar surface at relativistic speeds. Instead they deposit their energy in the stellar mantle and do not yield the conditions required for a GRB. An example of a jet simulation is shown in Fig. 7.16. The energy deposition time approximately sets the duration of the burst. Hence, a Wolf–Rayet star with jet crossing times of a few tens of seconds is needed. If the jet is fed for long enough, it is collimated by the density structure of the star and several tens of seconds after it has been launched, breaks through the surface of the helium core as a collimated, relativistic outflow.

Production of the gamma-ray burst

The jet possesses some intrinsic variablity because of the time-dependent accretion from the disk, which is fed by the still collapsing parts of the star. In addition, a strong shear flow develops at the interface between the jet and the stellar mantle; see Fig. 7.16. Such shear flows are Kelvin–Helmholtz unstable; see Section 2.9.3. Thus perturbations grow rapidly and make the flow turbulent. These perturbations produce velocity fluctuations that are imprinted on the jet. Portions of the flow with different Lorentz factors catch up with each other at distances of several stellar radii, where they produce internal shocks. When this occurs, the flow is already

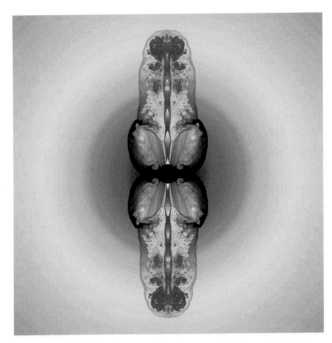

Figure 7.16   Snapshot of a computer simulation of a jet working its way out of a star. Figure courtesy of Andrew MacFadyen.

optically thin to the gamma rays produced via synchrotron and possibly inverse Compton processes.

While the jet is drilling through the stellar material, it is surrounded by a cocoon-like structure; see Fig. 7.16. When breaking through the surface, the cocoon produces an outflow that is much slower and less collimated than the jet itself. It may be that the X-ray flashes that have been observed (see Table 7.2) are cases where we miss seeing the jet and instead only detect the radiation from this cocoon.

7.5.2 Compact binary mergers

In Chapter 6, we had seen that binary systems exist that contain two neutron stars. The orbit of the first discovered and best studied system, the relativistic binary PSR 1913+16, decays because of the emission of gravitational radiation: the two neutron stars will collide in about $3 \cdot 10^8$ years from now. Two possibilities to form two such neutron stars from a massive binary system were discussed in Chapter 6. The discussion there made clear that it requires a fair amount of fine-tuning of the system parameters to produce two neutron stars in a binary system. In a good fraction of cases, the binary may instead end up as a neutron star together with a low-mass black hole. These two types of binary systems are often collectively referred to as compact binaries, and their potential as a possible central engine of GRBs was already recognized in the 1980s. For a long time, they were the best candidates for GRB central engines in general. However, since the first afterglow observations in 1997 and the clear association of long bursts with supernovae, their role has been downgraded to the central engines of short GRBs only.

In the following, we discuss how the merger of such binaries may possibly lead to a short GRB. Double neutron star systems are discussed first, then we describe in which respect neutron star–black-hole systems may deviate from the double neutron star behavior. We discuss how a burst could be launched and, finally, the expected observational signatures are summarized and linked to the recent observations.

The merger process of a double neutron star binary may be roughly divided into three phases: i) the slow ("secular") inspiral phase, ii) the merger, and iii) the final "ringdown" phase, in which the death struggle of the remnant and probably the collapse to a black hole takes place.

Inspiral

Initially, the orbit decays in a quasistatic way. The time until coalescence can, to a good approximation, be calculated from the current orbital period, $P_{orb,h}$ (in units of hours), and the orbital eccentricity, e, by[26]

$$\tau_{insp} = 10^7 \text{yr} \ P_{orb,h}^{8/3} \left(\frac{M}{M_\odot} \right)^{-2/3} \left(\frac{\mu}{M_\odot} \right)^{-1} (1 - e^2)^{7/2}. \qquad (7.30)$$

This shows the sensitivity of the inspiral to both the separation (hidden in the orbital period) and the eccentricity of the orbit. For more details of the orbital evolution under the influence of gravitational wave emission, we refer to Section 6.2.3. Generally, there can be large differences in the inspiral times for different orbital parameters. As an example: a binary system with the orbital period of PSR 1913+16, $P_{orb,h} = 7.75$, and zero eccentricity will inspiral in about $5 \cdot 10^9$ yr[27]; the same system with an eccentricity of 0.95 would coalesce within only 10^6 yr. Because of their formation via two supernovae, neutron star binaries are typically born with large eccentricities. However, as we had seen in Section 6.2.3, eccentricities are radiated away, so that the orbits will be close to circular at the time the merger occurs. The implications of the inspiral dynamics on the merger location with respect to the host galaxy is discussed below.

During the last minutes of their inspiral, compact binaries emit gravitational waves in a frequency range (10–1000 Hz). This frequency range can be detected by ground-based gravitational wave detectors. Both the frequency and the amplitude

---

[26] Lorimer, D. R. (2005). *Living Reviews in Relativity*. http://relativity.livingreviews.org/Articles/lrr-2005-7/.
[27] The real system PSR 1913+16 has an eccentricity of $e = 0.62$, therefore $\tau_{insp} = 3 \cdot 10^8$ yr.

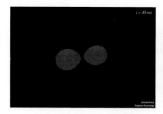
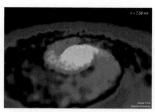
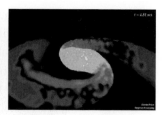

Figure 7.17   Computer simulation of the last stages of a double neutron star coalescence: shortly after contact, the stars shed mass into spiral arms that subsequently form a hot torus. The central object will probably at some stage collapse to a black hole. Figure from Price, D. and Rosswog, S. (2006). *Science*, **312**, 5774.

of the gravitational waves increase during the inspiral, and the binary is said to "chirp." A gravitational wave detection just prior to a short GRB would be the ultimate proof of the binary nature of short GRBs.

Merger

While the initial decrease in separation is tiny, PSR 1913+16 reduces its separation by 3.2 mm per orbit, and the last 100 km will take less than a second. Once the separation of the stars is a few times their radius, the neutron stars will become deformed by their mutual tidal interaction and merge within a matter of milliseconds. Once they come into contact, excess angular momentum is shedded into a pair of spiral arms that will be wrapped around the central object within milliseconds. The supermassive, hot neutron-star-like object in the center is likely to collapse at some point to a black hole. The timescale for this to happen rather uncertain. Snapshots from a computer simulation of a neutron star merger are shown in Fig. 7.17. The remnant has temperatures in excess of 10^{11} K in the central object before collapse and several times 10^{10} K in the surrounding torus. It radiates neutrinos at rates comparable to supernovae, $\sim 10^{53}$ erg s^{-1}.[28]

Ring-down

In this final phase, the central object of the remnant settles down into its stable state, which is probably a black hole of about 2.5 $M_{\odot}$.[29] This configuration is similar to the inner regions in the collapsar: a low-mass black hole surrounded by a hot and massive accretion disk.

The merger of a neutron star with a low-mass black hole initially proceeds very similarly that between two neutron stars. However, because of its larger

---

[28] The supernova neutrino signal is dominated by electron-type neutrinos from electron captures $e^- + p \to n + \nu_e$. Being much more neutron rich from the start, the remnant of a compact merger emits predominantly electron antineutrinos produced via $e^+ + n \to p + \bar{\nu}_e$.

[29] Note that energy has been lost via gravitational waves and neutrinos; this reduces the gravitational mass of the central object as discussed after the Tolman–Oppenheimer–Volkoff equations in Chapter 5.

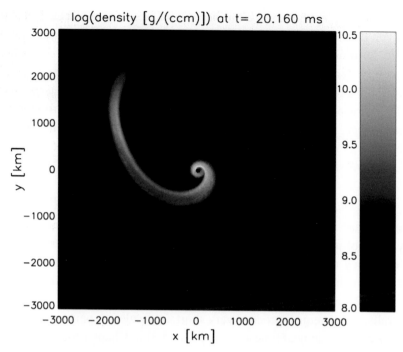

Figure 7.18   Computer simulation of a merger of a 1.4 $M_\odot$ neutron star with a 14 $M_\odot$ black hole. Once close enough, the neutron star becomes tidally disrupted. Parts of it become ejected in a tidal tail, others are accreted by the black hole. (Rosswog S. [2005], ApJ, **634**, 1202)

mass, the gravitational wave amplitude is somewhat larger. Differences may occur once the neutron star is close enough to transfer mass into the black hole. At
this stage, the evolution may be complicated by the transfer of mass and angular momentum, complications that to a lower extent may also occur for double
neutron stars with unequal masses. This may lead to a longer lived merger phase,
possibly several encounters before the neutron star is completely disrupted, but
this depends on the details of the binary system, particularly on the mass ratio
between the neutron star and the black hole. A snapshot from a computer simulation showing the disruption of a neutron star by a 14 $M_\odot$ black hole is shown in
Fig. 7.18.

Production of a GRB?

Two ultimate energy sources are available: the gravitational binding energy and
the spin of the central object. Other possible sources, such as thermal energy
or magnetic fields, derive their energy from the other two. The most promising
place to launch a fireball is, similar to the collapsar case, the region above the
poles of the newly formed central object: this region is close to the source of

energy (central object/hot torus), but because of centrifugal forces, it has a very low baryon density. Neutrinos from the hot torus can efficiently annihilate in this region to form electron–position pairs, $\nu_i + \bar{\nu}_i \rightarrow e^+ + e^-$, and – if collimated appropriately – could be able to produce the energies that have been derived from the first short GRBs with afterglows; see Table 7.2. Moreover, neutron stars are naturally endowed with strong magnetic fields; see Chapter 5. During a neutron star coalescence, chances are good that these fields can be amplified substantially, but these energy deposition mechanisms are still subject to current research with many questions still left unanswered.

Comparison with the observed properties of short GRBs

Compact binary mergers release several times 10^{53} erg of gravitational binding energy. Most of it is radiated away as gravitational waves and neutrinos, but even if the efficiency of the transformation into gamma-rays is only moderate, there should be more than enough energy to account for the observed, isotropized energy of $E_{\gamma,\mathrm{iso}} \sim 10^{50}$ erg ($\Delta\Omega/4\pi$); see Table 7.2.

Naturally, such mergers produce very short timescales. A neutron-star-like central object has a dynamical timescale (see Eq. [5.3] in Chapter 5) of

$$\tau_{\mathrm{dyn,ns}} = \frac{1}{\sqrt{G\bar{\rho}}} = 0.4 \text{ ms} \left(\frac{10^{14} \text{ g cm}^{-3}}{\bar{\rho}} \right)^{1/2}. \tag{7.31}$$

For a nonrotating black hole, the last stable circular orbit lies at $R_{\mathrm{ISCO}} = 6GM_{\mathrm{BH}}/c^2$, where M_{BH} is the black-hole mass. The shortest timescales in an accretion disk around such a black hole should be comparable to the orbital timescale outside this radius,

$$\tau_{\mathrm{dyn,bh}} = \frac{2\pi}{\omega_{\mathrm{K,ISCO}}} = \frac{2\pi\sqrt{216}}{c^3}GM_{\mathrm{BH}} \approx 1 \text{ ms} \left(\frac{M_{\mathrm{BH}}}{3 \text{ M}_\odot} \right), \tag{7.32}$$

where we neglected the test particle mass with respect to the black-hole mass and have used the Kepler frequency, $\omega_K = (GM_{\mathrm{BH}}/R^3)^{1/2}$ at R_{ISCO}; see Eq. (7.27).

The rates for such compact binary mergers are uncertain. One can try to estimate them via binary stellar evolution theory. As a cross-check, one can infer a rate from the small number of observed systems in our own Galaxy (at the time of writing, eight such systems are known) by taking into account all possible reasons why existing systems could go undetected. Both methods have relatively large uncertainties but agree on a "best-guess value" of a few times 10^{-5} mergers per year per galaxy. This has to be compared with the rate required to explain the observations. The total burst rate is around 10^{-7} per year per galaxy, about 25% of which are short bursts, so even allowing for strongly collimated outflows and the possibility that not every merger might produce a GRB, there should be more than enough mergers to be consistent with the observed GRB rate.

Compact binary systems are born with a distribution of initial separations and eccentricities. Systems that are born with very short initial separations and/or large eccentricities may die close to their birthplaces, the star-forming regions of galaxies. Wide systems such as PSR 1913+16 take a very long time until coalescence (PSR 1913+16 will merge $3 \cdot 10^8$ years from now). As we had seen when discussing neutron stars in Chapter 5, neutron stars usually receive a strong kick of several hundred kilometers at birth. As a result, a binary is likely to possess a large center of mass velocity with respect to its host galaxy. Systems with a very long inspiral time may merge in the outskirts or even outside the galaxy where they were born. This behavior is consistent with the types of galaxies in which afterglows of short GRBs have been detected: some occurred in galaxies with star formation, others in elliptical galaxies, where star formation had ceased long ago. Because of the long inspiral of (at least some of) the binary systems, mergers occur relatively late in the evolution of the universe. Therefore, we expect only moderate cosmological redshifts for these events. Indeed, the first redshift observations of short bursts have shown generally lower redshifts than those of long bursts.

So far, our understanding of GRB sources is still incomplete. The current ideas will certainly be challenged by future observations.

7.6 Exercises

7.1 Order of magnitude estimate

Assume that a typical GRB emits its gamma rays into two cones of half-opening angle $\theta_j = 4°$ as shown in Fig. 7.8. Adopt a value of two observed bursts per day to obtain a lower limit on the true the GRB rate.

7.2 Order of magnitude estimate

A spherical fireball with mass loading M expands into a constant density medium with one hydrogen atom per cm^{-3}. At which radius has it swept up a mass M? How does this radius change if the fireball is instead collimated into two cones with opening angle θ_j?

7.3 Order of magnitude estimate

Starting from the observed rates, estimate how many gamma-ray bursts must have taken place in our own Galaxy.

7.4 Blast wave

Consider a spherical and adiabatic fireball with an initial energy $E_0 = \gamma_0 M_0 c^2$. After a while, it has swept up and shocked the mass m, which has (in the observer frame) a thermal energy of $\gamma^2 mc^2$. Show that the deceleration becomes important once a mass of $m = M_0/\gamma_0$ has been swept up. Moreover, once $m \gg M_0$, show that $\gamma \propto m^{-1/2}$.

7.5 **Synchrotron emission**

Consider a relativistic shock propagating through a uniform, cold medium with particle number density n. According to the solution of Blandford and McKee,[30] the energy density behind the shock is given by $u = 4\gamma^2 n m_p c^2$, where γ is the Lorentz factor of the shocked fluid. Assume that the magnetic energy density behind the shock is a constant fraction ϵ_B of this energy density to show that the magnetic field (in the fluid frame) behind the shock is

$$B = \sqrt{32\pi m_p \epsilon_B n}\ \gamma c. \tag{7.33}$$

Calculate the characteristic, observed synchrotron frequency of a randomly oriented electron with Lorentz factor γ_e.

7.6 **Order of magnitude estimate**

To produce a GRB from a neutron star–black-hole system, an accretion disk is thought to be necessary. A simple estimate for the separation at which a star of mass M_* and radius R_* is disrupted by a black hole of mass M_{BH} is the so-called *tidal radius*

$$R_{tid} = \left(\frac{M_{BH}}{M_*}\right)^{1/3} R_*. \tag{7.34}$$

Assume that this estimate holds for a neutron star. Estimate the masses of black holes for which the neutron can be disrupted outside the last stable orbit R_{ISCO}.

7.7   **Further reading**

Blandford, R. D. and McKee, C. F. (1976). Fluid dynamics of relativistic blast waves. *Physics of Fluids*, **19**, 1130.

Hurley, K., Sari, R., and Djorgovski, S. G. (2003). Cosmic gamma-ray bursts: Their afterglows and their host galaxies. In *Compact Stellar X-Ray Sources*, ed. W. Lewin and M. van der Klis. Cambridge: Cambridge University Press.

Katz, J. I. (2002). *The Biggest Bangs*. Oxford: Oxford University Press.

Lyutikov, M. and Blandford, R. D. (2003). Gamma-Ray Bursts as electro magnetic outflows, astro-ph/0312347.

Lyutikov, M. (2006). The electromagnetic model of Gamma-Ray Bursts. *New Journal of Physics*, **8**, 119.

MacFadyen, A. I. and Woosley, S. E. (1999). Collapsars: Gamma-ray bursts and explosions in "Failed Supernovae." *Astrophysical Journal*, **524**, 262.

Mészáros, P. (2006). Gamma-ray bursts. *Reports on Progress in Physics*, **69**, 2259.

Nakar, E. (2007). Short-Hard Gamma-Ray Bursts, astro-ph/0701748.

Piran, T. (1999). Gamma-ray bursts and the fireball model. *Physics Reports*, **314**, 575.

Piran, T. (2004). The physics of gamma-ray bursts. *Reviews of Modern Physics*, **76**, 1143.

Sari, R., Piran, T., and Narayan, R. (1998). Spectra and light curves of gamma-ray burst afterglows. *Astrophysical Journal Letters*, **497**, L17.

---

[30] Blandford, R. D. and McKee, C. F. (1976). *Physics of Fluids*, **19**, 1130.

Woosley, S. E. (1993). Gamma-ray bursts from stellar mass accretion disks around black holes. *Astrophysical Journal*, **405**, 273.

Waxman, E. (2003). Gamma-ray bursts: The underlying model. *Lecture Notes in Physics*, **598**, 393–418.

Wijers, R. A. M. J., Rees, M. J., and Mészáros, P. (1997). Shocked by GRB 970228: The afterglow of a cosmological fireball. *Monthly Notices of the Royal Astronomical Society*, **288**, L51.

8

Active galactic nuclei

8.1 Introduction

8.1.1 A bit of history

The history of the discovery of active galaxies is a nice example of the often erratic, unpredictable, and fascinating ways in which revolutions in our view of the Universe come about. In many ways, the history of identifying active galaxies is parallel to the history of the discovery of the origin of gamma-ray bursts, which is described in Chapter 7. They both started off with an observation of something that nobody could explain, and then faced the distance–energy dilemma: if the sources were in our own Galaxy and thus relatively close, why had these objects not been observed earlier? However, if they were outside our Galaxy, the large distances implied huge energies. By which process could such tremendous energies be released? In the case of active galaxies, it began with the observations of cosmic radio sources. These sources were first discovered serendipitously in 1932 by Karl Jansky. Jansky was a radio engineer with the Bell Telephone Laboratories in New Jersey and was charged with finding out what caused the noise in transatlantic radio transmissions. While most of the noise was caused by thunderstorms, there was another source that did not go away even after the thunderstorms had ceased. In 1935, he identified the source of this noise. It came from the center of our Galaxy, the Milky Way. This discovery was largely ignored by astronomers and began to receive attention only in the 1940s. After World War II, many radar scientists, particularly in the United Kingdom and Australia, were looking for civil applications of their skills and turned their attention to Jansky's celestial radio sources. What was puzzling was that the most powerful radio source in the sky was not the Sun but sources that were much more distant. Consequently, the physical origin of this radio emission could not be *normal* blackbody radiation because radio frequencies are at the low-frequency end of the electromagnetic spectrum where a blackbody spectrum decreases. Thus, to still have so much power in the radio part of a blackbody spectrum, the power in other wave bands would have been so huge that it would not have gone undetected. In 1953, Russian theoreticians Vitali Lazarevich Ginzburg and Iosif Samuelovich Shklovski had independently suggested that synchrotron radiation from highly relativistic electrons was responsible for the strong radio emission from cosmic sources. Their theory was confirmed directly by the discovery of the predicted polarizations in the light of the galaxy M87. However, this discovery immediately created a problem. From the theory of synchrotron radiation, the British astrophysicist Geoffrey Burbidge computed the minimum energy contained in relativistic electrons and in the magnetic field of Cygnus A, the second brightest radio source in the sky, as $2.8 \cdot 10^{59}$ erg. Immediately, the question arose: what could be the source of so much energy?

In the early 1960s, astronomers at Jodrell Bank near Manchester in England were measuring the angular sizes of these radio sources and found that a small subset of their sources were very small, in that they were less than 1 arcsecond across. What could these sources be? Did the radio emission come from a nearby object in our own Galaxy, or did it have an extragalactic origin?

One of these radio sources was 3C48.[1] Allan Sandage from the Carnegie Institute in Pasadena and his collaborators found that its spectrum was peculiar and that its emission varied with time. At the time, they labeled the object a *radio star* because it appeared to coincide with a star. The idea that this emission could come from a galaxy, with $\sim 10^{10}$ stars that all changed their brightness so rapidly, seemed remote. Thus, these strong, point-like radio sources became known as quasistellar objects (QSO) or quasars.

The main problem at the time was to find the optical counterpart of the radio source. For this, it was crucial to determine the position of the source with great accuracy. One method involved using the Moon and measuring the exact time when the Moon occulted the source. This way, the exact position of the source can be determined to within arcseconds. A compact source from the *Third Cambridge Catalogue* that was in the Moon's path was 3C273. In 1962, C. Hazard, M. B. Mackay, and A. J. Shimmins observed its occultation by the Moon from Australia and determined its position with an accuracy of ~ 1 arcsecond. Hazard sent this position to the Dutch astronomer Maarten Schmidt who was working at the California Institute of Technology and who had started to take spectra of the peculiar radio sources at the Mt. Wilson and Palomar Observatories. Schmidt found that the redshift of 3C273 was $z = 0.158$. It was now clear that these sources were very distant and lay outside our Galaxy. However, early attempts to find galaxies

---

[1] The name stems from one of the early catalogs of radio astronomy, which was compiled in Cambridge at the Mullard Radio Astronomy observatory and called the *Third Cambridge Catalogue*. All sources in this catalog were numbered and had the prefix 3C.

associated with quasars failed. American physicist Ed Salpeter published the idea that a massive object plowing through a galaxy and therewhile accreting gas may be the origin for the extreme energies. In 1963, Fred Hoyle and William Fowler wrote a seminal article in which they proposed that the tremendous energy for these sources could be derived from the gravitational collapse of very massive objects. These were some of the early milestones that laid the foundations of our current knowledge of active galactic nuclei.

8.1.2   *What are active galaxies?*

Active galaxies belong to the most spectacular phenomena in our universe. They are galaxies in which a significant fraction of the electromagnetic energy output is *not* contributed by stars or interstellar gas. Instead, active galaxies emit largely radiation that does not follow the typical blackbody spectra of stars and is called nonthermal (see Fig. 3.6). This implies that active galaxies emit much more energy at smaller wavelengths, that is, at higher energies, than objects whose spectrum follows that of a black body. About 3% of all galaxies are classified as active.

An active galactic nucleus (AGN) represents the central few parsec of an active galaxy. Within this relatively small volume, an AGN produces an enormous amount of energy. Besides the nonthermal emission, a generic feature of AGN is the expulsion of energy in two oppositely directed beams that are called *jets*. Further signs of nuclear activity are listed here.

Signs of nuclear activity (not all always present)

- compact ($\sim$3 pc), luminous centers,
- spectra with strong emission lines,
- strong nonthermal emission,
- strong ultraviolet emission from a compact region in the center,
- jets and double radio sources,
- variability over the whole spectrum on short timescales, and
- strongly Doppler-broadened emission lines.

Because we devote only a single chapter to AGN, we can only address some facets of these multifarious objects. The choice of topics is motivated partly by their pedagogical value and biased toward the *physics* of AGN. We have tried to concentrate on aspects that are illustrative of the AGN phenomenon and that can be explained with simple analytical estimates.

8.2   Observational facts

8.2.1   *Taxonomy of active galaxies*

AGN display a wide variety of features whose nature is in many cases poorly understood. They are observed from radio frequencies ≤ 100 MHz to extreme gamma rays at frequencies $>10^{22}$ Hz and energies >100 MeV. Because it is difficult to get a full spectral coverage in all bands, the taxonomy of AGN is a cumbersome business. Classifications that are based on features in a specific waveband, for example, in the radio band, are often hard to reconcile with features observed in another band. However, astronomers have identified a number of broad subclasses of AGN that can be characterized by certain phenomenological features. We summarize this scheme briefly, but bear in mind that this classification is far from strict and that for every subclass there are exceptions.

Quasars

Quasars are very luminous and compact centers of galaxies, which outshine the rest of their host galaxy. They are characterized by a nearly featureless spectrum that spans a wide range of wavelengths (from radio wavelengths to hard X-rays). Their luminosity ranges from about 10^{45} to 10^{49} erg s^{-1}, and their emission varies with time across the complete electromagnetic spectrum. For comparison, the typical luminosity of a field galaxy is of the order of 10^{44} erg s^{-1}. About 10^4 quasars are known, out of which about 10% show strong radio emission (so-called radio-loud quasars). Many surveys such as the SLOAN Digital Sky Survey[2] continue to discover quasars all the time.

Radio galaxies

These are galaxies that look like ordinary elliptical galaxies in the optical band but emit extreme amounts of energy in radio wavelengths. Weak radio emission, with a power at 1.4 GHz of $L_{1.4 \text{ GHz}} < 10^{30.3}$ erg s^{-1}Hz^{-1}, occurs in many galaxies. It is observed particularly in spiral galaxies, where it is probably produced by relativistic electrons accelerated by supernovae. However, the term *radio galaxy* is reserved for galaxies, typically giant elliptical galaxies, with strong radio emission with luminosities $L_{\text{radio}} \geq 10^8 L_\odot$ or $\geq 3 \cdot 10^{41}$ erg s^{-1}. The radio emission may come from only the nucleus of the galaxy, but more often it comes from a pair of more or less symmetric lobes that stretch as far as several kiloparsec on both sides

---

[2] More information is available at the Sloan Digital Sky Survey website, http://www.sdss.org/.

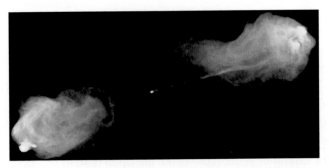

Figure 8.1   The radio galaxy Cygnus A as seen by the Very Large Array at 5 GHz, 0.5″ resolution. The galaxy is at a redshift of 0.057 (distance = 230 Mpc = 760 ly). Clearly visible are the two radio lobes that are fed by thin jets that originate in the bright spot in the center. The jets terminate in the bright spots at the outer ends of the radio lobes. These spots are called hot spots. The source extends about 150 kpc from end to end. When the galaxy is observed at optical wavelengths, its size is less than a tenth of its radio length. For details, see Carilli, C. L. and Barthel, P. D. (1996). *Astronomy and Astrophysics Reviews*, **7**, 1.

of the nucleus (Fig. 8.1). Many radio galaxies feature jets that connect from the nucleus to these lobes.

Radio galaxies can be divided into two classes that are named after the South African astronomer Bernard Lewis Fanaroff and the British astronomer Julia Margaret Riley:

Fanaroff–Riley type I (FR I): These sources are brightest close to the center of the galaxy. At the edges of the lobes, the sources become fainter, and the spectra become steeper. This type is less powerful with luminosity densities of $L_{1.4 \text{ GHz}} \leq 10^{32}$ erg s^{-1} Hz^{-1}. Jets occur in ~80% of FR I galaxies. Its jets are less collimated than those in the FR II galaxies.

Fanaroff–Riley type II (FR II): These sources have powerful, collimated jets often showing bright spots, so-called hot spots, that can be several kilopersec across (Fig. 8.1). Their luminosity per unit frequency are $L_{1.4 \text{ GHz}} \geq 10^{32}$ erg s^{-1} Hz^{-1}.

The division between these two types of radio galaxies is remarkably sharp and still not understood. It has been suggested that the difference between the two FR types may be related to the speed of the jet head. In this model FR I jets travel subsonically through the ambient medium, and FR II jets travel supersonically, thus causing shocks and strong radiation.

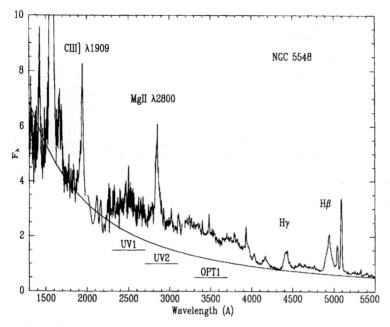

Figure 8.2 The spectrum of the Seyfert 1 galaxy NGC 5548. The prominent broad emission lines are labelled. Courtesy of Dan Maoz.

Seyfert galaxies

These are spiral galaxies with very bright, unresolved nuclei. They show line emission of highly ionized atoms that cannot be produced by stars.

In 1943, the American astronomer Carl Seyfert was the first to realize that there are several similar galaxies that form a distinct class. Seyfert selected a group of galaxies on the basis of high central surface brightness (i.e., stellar-appearing cores). Seyfert obtained spectra of these galaxies and found that the optical spectra of several of these galaxies are dominated by high-excitation nuclear emission lines (Fig. 8.2).

The luminosity of Seyfert galaxies ranges from about 10^{43} to 10^{45} erg s^{-1}, which makes the brightest Seyfert galaxies as luminous as faint quasars. As radio-quiet quasars, Seyfert galaxies tend to have only weak radio emission, and some Seyferts show signs of merging with other galaxies. Some Seyferts show broad spectral lines of permitted hydrogen lines, and these are classified as Seyfert 1, and some do not, in which case they are called Seyfert 2. Nearly 15% of Seyferts have close companions (indicative of merging or a close interaction with another galaxy) compared with only 3% in a controlled sample of non-Seyfert galaxies.

Blazars

Blazars emit polarized light with a featureless, nonthermal spectrum typical of synchrotron radiation. They are extremely luminous and highly variable sources. Blazars are characterized by high optical polarization and large X-ray and gamma-ray luminosities. On closer inspection, however, their spectra show emission and absorption lines that are swamped by the intense synchrotron radiation.

Radio astronomers sometimes group these different types of AGN into two classes: into either radio-loud or radio-quiet AGN. The distinction is made with the aid of a parameter called *radio loudness*, which is defined as

$$R_L = \log \left(\frac{F_{5\,\mathrm{GHz}}}{F_{\nu,\mathrm{B}}} \right), \tag{8.1}$$

where $F_{5\,\mathrm{GHz}}$ is the monochromatic radio flux at 5 GHz and $F_{\nu,\mathrm{B}}$ is the optical flux in the B band (see Appendix D), centered on the wavelength $\lambda = 440$ nm. If an AGN has $R_L > 1$, it is regarded as radio loud, whereas if it has $R_L < 1$, it is regarded as radio quiet. The radio-loud objects represent a small percentage (10–15%) of all AGN.

Optical astronomers, conversely, sometimes group AGN into just two categories according to their optical spectra. AGN with a bright continuum and both broad and narrow emission lines are called type 1 AGN; those with a weak continuum and only narrow emission lines are called type 2 AGN. This classification is independent of their radio properties.

In recent years, astronomers have tried to determine the sort of galaxies that play host to active nuclei and the causes for this extraordinary activity. Although this is still a much debated subject, some interesting trends have emerged. Most important, it appears that the degree of activity depends on the galaxy mass. The most active AGN are seated in the centers of the most massive galaxies. Moreover, all radio-loud AGN are found almost exclusively in elliptical galaxies. Also, the highest-power radio galaxies frequently show close companions or evidence of a merger. The conditions for triggering activity in galaxies is receiving a tremendous amount of interest because AGN activity is believed to be a critical component in the process of structure formation in the universe. We get back to this point at the end of the chapter.

The main properties of the most important types of AGN are listed in Table 8.1.

8.2.2 *Spectral features*

The optically brightest quasar 3C273 is an ideal laboratory for studying AGN. It is at a distance of about 640 Mpc ($z = 0.16$) and has been observed in great detail for more than 30 yr. It is among the few AGN detected at energies greater than 100 MeV by the Energetic Gamma-Ray Experiment Telescope (EGRET) on board

Width of spectral lines

The width of a spectral line is characterized as follows: first, the continuum emission upon which the spectral line is imprinted is interpolated across the spectral line and then subtracted from the spectrum. Then, one measures the width of the line at a level that corresponds to half of its maximum intensity. This width is called *full width at half maximum* (abbreviated to FWHM). The width is sometimes quoted in units of wavelength, for example, Ångstrom, and sometimes in units of velocity (km s^{-1} usually) if the line width is interpreted as being due to Doppler broadening. Doppler broadening of a spectral line is the result of the emitters of the spectral line moving relative to the observer. The shift of a single spectral line emitted by a source moving with a line-of-sight (LOS) velocity v_{los} is given by

$$\frac{\Delta\lambda}{\lambda} = \frac{v_{\text{los}}}{c}, \tag{8.2}$$

where λ is the wavelength, $\Delta\lambda$ the shift in wavelength, and c the speed of light. Often, the spectral line is emitted by a large number of emitters, which can be stars or blobs of gas, that move with different, line-of-sight velocities relative to the observer. Hence, there is a range in the Doppler shifts of their individual lines. The overall effect is to broaden the spectral line as a whole. The resulting width of the line is given by

$$\Delta\lambda = \langle\Delta\lambda^2\rangle^{1/2} = \lambda\frac{\langle v_{\text{los}}^2\rangle^{1/2}}{c}, \tag{8.3}$$

where $\langle\rangle$ denotes an average over all sources. In the case that the mean squares of the velocity components in every direction are approximately the same, one can write $\langle v_{\text{los}}^2\rangle \approx \langle v^2\rangle$, where v is the magnitude of the total velocity.

the Compton Gamma-Ray Observatory (CGRO). The data from about 20 000 observations in all bands of the electromagnetic spectrum covering 16 orders of magnitude in frequency from the radio to the gamma rays has been assembled in

Table 8.1   *The main properties of the most important types of AGN.*

Class	Host galaxy	Radio emission	Emission lines	Luminosity (erg s^{-1})
Blazar	E	Strong	Weak	10^{45}–10^{49}
Radio-loud quasar	E	Strong	Broad	10^{45}–10^{49}
Radio galaxy	E	Strong	Narrow	10^{43}–10^{45}
Radio-quiet quasar	S/E	Weak	Broad	10^{45}–10^{49}
Seyfert 1	S	Weak	Broad	10^{43}–10^{45}
Seyfert 2	S	Weak	Narrow	10^{43}–10^{45}

In the column *Host galaxy* the letter E stands for elliptical galaxy and S for spiral galaxy.

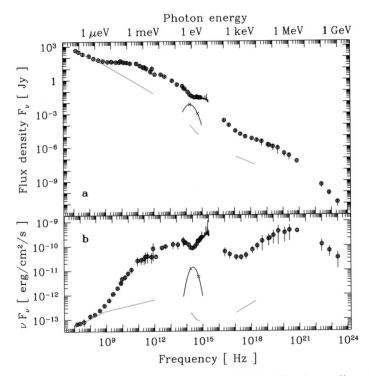

Figure 8.3   Average spectrum of the quasar 3C273. The lower line shows approximately the contribution from the outer jet according to Conway et al. (1993) for the radio domain and to Marshall et al. (2001) for the optical and X-ray parts. The contribution from the elliptical (E4) host galaxy is indicated by the bump shown by the solid line near frequencies of 10^{15} Hz. Courtesy of M. Türler, http://isdcul3.unige.ch/3c273/.

a composite spectrum shown in Fig. 8.3. The spectrum can be approximated by a power law, $F_\nu \propto \nu^{-\alpha}$, where α is the spectral index of the source and F_ν the energy flux density (or monochromatic energy flux as defined in Chapter 3) which is the energy flux received per unit frequency. In radio astronomy, the energy flux density is usually expressed in units of Jansky, named after the discoverer of cosmic radio waves, Karl Jansky. A Jansky is defined as $1 \text{ Jy} = 10^{-23} \text{ erg cm}^{-2} \text{ s}^{-1} \text{Hz}^{-1}$. The lower panel in Fig. 8.3 shows the spectral energy distribution, which is the energy flux density multiplied by the frequency, νF_ν, versus the logarithm of ν. In this representation, the power law becomes $\nu F_\nu \propto \nu^{1-\alpha}$. Plotting the spectral energy distribution this way reveals the amount of energy emitted in each equally spaced interval on the logarithmic frequency axis. The lower panel in Fig. 8.3 shows that the energy flux from 3C273 per logarithmic interval is almost constant over more than 10 orders of magnitude in frequency. This corresponds to an α of ~ 1 because then $\nu F_\nu \propto \nu^{1-\alpha}$ becomes independent of frequency. In contrast, stars emit nearly all of their energy in a frequency range spanning only three orders of magnitude. The spectrum of 3C273 has two notable features: one is the bump in the far ultraviolet near 10^{15} Hz, which is called the *big blue bump*, and the second is near 10^{20} Hz at a photon energy of about 1 MeV. The solid line in Fig. 8.3 represents the contribution of the jet and the dashed line the contribution from the host galaxy.

The continuum emission can be attributed to four main emission mechanisms.

- Radio: The radio emission is dominated by the synchrotron emission of electrons in the jet (see Section 3.8.2).
- Infrared: Thermal radiation from a warm, dusty torus heated by a central source.
- Optical/ultraviolet: Thermal emission from an accretion disk and the hot gas around it.
- X-ray/gamma-ray: The high-energy emission is caused by relativistic electrons from the hot gas surrounding the accretion disk. These electrons upscatter UV and optical photons via the inverse Compton effect to X-rays and gamma rays (see Section 3.8.3).

These different components of an AGN are described in more detail later.

On top of the continuum emission, the optical/ultraviolet spectra of quasars are distinguished by strong, broad emission lines. The strongest observed lines are the hydrogen Balmer-series lines (H_α, $\lambda = 6563$ Å, H_β, $\lambda = 4861$ Å, and H_γ, $\lambda = 4340$ Å), hydrogen Lyα ($\lambda = 1216$ Å), and prominent lines of abundant ions, such as Mg II, C III, and C IV.

The emission-line intensities and emission-line ratios contain information on the physical conditions in the line-emitting gas. The electron density and temperature, the degree of ionization and excitation, and the chemical composition can be inferred from the analysis of the line ratios. We discuss this in more detail later. AGN spectra contain lines emitted in a broad-line region and in a narrow-line region.

8.2.3 *Variability*

All types of active galaxies emit radiation that varies with time. The variability of the quasar 3C273 in four different bands is shown in Fig. 8.4. Generally, lower luminosity objects vary more rapidly than higher luminosity objects. Within an individual object, higher frequency bands vary faster than lower frequency bands. This variability can occur on timescales as short as a few minutes to years. The variablity of an AGN can be used to make inferences about the size of the emitting object because an object cannot vary in brightness faster than it takes light to cross this object. To illustrate this, imagine an object that measures 1 light-year across.

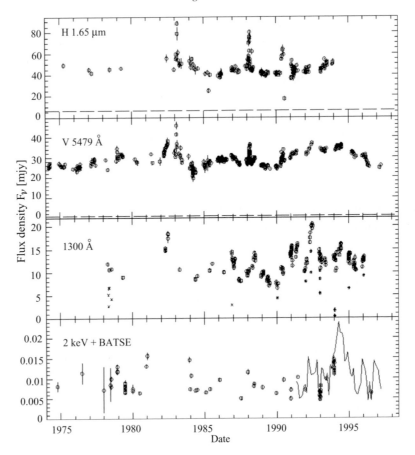

Figure 8.4   Four characteristic lightcurves of the infrared-to-X-ray behavior of 3C273 from 1974 to 1998. Three flares are clearly visible in the H- and V-band lightcurves in 1983, 1988, and 1990. The dashed line is the contribution from the host galaxy. Courtesy of M. Türler.

Suppose that at some instance the entire surface of this object emits a single flash of light. Now, the photons from that side of the object that is closest to the Earth will reach us first, whereas photons that originate at the far side of the object will arrive 1 yr later. Even though the object only emitted a single, instantaneous flash of light, we observe a signal that is stretched out over an interval of a year, that is, an interval that equals the light travel time across the object, l/c, where l is the size of the object and c the speed of light. (For a cautionary remark, see Exercise 1.2.)

Real fluctuation timescales are longer than l/c because it is rare that the entire source varies simultaneously over a scale l. More likely, an event that occurs at some point in space triggers the rise in luminosity across the source. Thus, the brightening of the source travels at a signal speed, $v_{\mathrm{signal}} \leq c$. Then, the signal is spread out over a period $\sim l/v_{\mathrm{signal}}$.

Therefore, the rapid flickering of AGN suggests that the emitting region is small. A variability over an hour as seen in some AGN implies a typical size of $l < c\Delta t \sim 3.5 \cdot 10^{-5}$ pc, which is less than the size of our Solar System (Pluto's average distance from the Sun is $1.9 \cdot 10^{-5}$ pc). This implies that a region as big as our Solar System emits as much energy as thousands of galaxies. Later, we explore possible physical processes that may be capable of producing that much energy.

Blazars exhibit the most rapid and the largest amplitude variations of all AGN. Observations suggest that any existing spectral features are swamped by a strong continuum component. The latter is most likely produced by a jet close to the line of sight, whose emission is strongly amplified by relativistic beaming (see Chapter 1).

The high density in the broad-line region implies that the recombination times are less than a few days. Consequently, if the ionizing radiation field changes, the intensities of the broad lines change, too. From the time delay between the changes in the intensities of lines and continuum, one can deduce the spatial dimensions of the broad-line region.

In the narrow-line region, the matter density is much lower, the recombination time much longer, and the mass much larger. With very few exceptions, there is no evidence for narrow-line variability in AGN. Thus, it is deduced that the gas responsible for the narrow lines is more extended and further away from the ionizing source.

8.3   What powers active galactic nuclei?

8.3.1   *Stars*

First, observations of AGN do not show stellar spectra, but even if stellar lines had been overlooked, stars are very unlikely to produce the luminosities of an AGN. Assuming that the light is provided by luminous O stars with a luminosity $L_* \sim 10^{5.5} L_\odot$, it would take $N_* = L_{AGN}/L_* \sim 8 \cdot 10^5$ stars to reproduce the luminosity of the AGN where $L_{AGN} = 10^{45}$ erg s^{-1}. The corresponding stellar density would be

$$n_* = \frac{N_*}{(4\pi/3)(l/2)^3} \geq 1.6 \cdot 10^{12} \text{ pc}^{-3}, \tag{8.4}$$

where l is the size of the emitting region, which we assumed to be $l = 10^{-2}$ pc.

This is a tremendous stellar density. The greatest known stellar densities occur in the centers of globular clusters, where stellar densities are $\sim 10^6$ pc^{-3}. From the density n_*, we can work out a mean separation between the stars of

$$l_* = \left(\frac{1}{n_*}\right)^{1/3} \approx 3800 \ R_\odot. \tag{8.5}$$

Such high stellar densities would lead to frequent collisions and an expedient collapse of the system. See also Exercise 2.3 in Chapter 2.

8.3.2   *Supernovae*

The most energetic supernova explosions reach a maximum luminosity of $10^{10}L_\odot$ (see Chapter 4). Therefore, 10^4 supernovae shining at their peak luminosity would be permanently needed to power an AGN. This would require the formation of 10^{10} stars that are permanently producing supernovae. We have shown in Section 8.3.1 that this is essentially ruled out. Moreover, no spectral lines that are indicative of supernovae have been observed in AGN spectra.

8.3.3   *Accretion*

Accretion denotes the accumulation of gas onto some object under the influence of gravity. It is a key process in the physics of AGN and other astrophysical objects (see Section 6.3) which is why it is useful to look at it in some detail. Imagine a piece of matter free-falling onto the surface of some hypothetical astrophysical object. As it slams into the surface, it comes to a halt, and the sudden deceleration produces heat. The kinetic energy of the incoming object, which started out as gravitational potential energy, is dissipated as heat on the surface of the attracting object. Thus, as matter accretes onto a massive object, it effectively converts gravitational energy into other forms of energy that can eventually produce radiation.

Consider a particle falling from infinity onto an accreting object of mass M. When it is stopped at a radius R, its kinetic energy is radiated away. If a mass ΔM is accreted within a time Δt, the rate of mass accretion is $\dot{M} = \Delta M / \Delta t$. This mass accretion generates a power

$$L = \frac{1}{2}\dot{M}v_{ff}^2 = \frac{G\dot{M}M}{R}, \tag{8.6}$$

where $v_{ff} = (2GM/R)^{1/2}$ is the free-fall velocity. It is convenient to introduce the Schwarzschild radius, which is defined as

$$R_S = \frac{2GM}{c^2} \sim 3 \cdot 10^{13}\left(\frac{M}{10^8 M_\odot}\right)\,\text{cm}, \tag{8.7}$$

see also Box on Black Holes in Chapter 6. This is the radius of the so-called event horizon of a nonrotating black hole (see Eq. [6.58]).

Inserting this into Eq. (8.6) yields

$$L = \frac{1}{2}\dot{M}c^2 \left(\frac{R_S}{R}\right) \equiv \xi\dot{M}c^2, \tag{8.8}$$

where $\xi = R_S/2R$ is the efficiency of converting gravitational energy into heat.[3] This efficiency depends on how compact a star is. For example, for a white dwarf of mass $M = 1\,M_\odot$ and $R = 5 \cdot 10^8$ cm, the efficiency is $\xi = 3 \cdot 10^{-4}$, whereas for a neutron star of the same mass but a radius of only $R = 1.5 \cdot 10^6$ cm, the efficiency is $\xi \sim 0.1$. If the central source is a black hole, the efficiency can be as high as $\xi = 0.29$. This high efficiency of turning rest-mass energy into heat is remarkable and a unique source of energy. For comparison, one can consider nuclear fusion at the center of the Sun. Fusion in the Sun proceeds mainly via the p-p chain, which denotes the chain of reactions that start with the fusion of two protons to form deuterium. The efficiency of turning rest-mass energy into heat in these fusion reactions is about $\xi = 0.007$.

8.3.4 Eddington luminosity

According to Eq. (8.8), the luminosity of an accreting source is proportional to the accretion rate. Now, does this mean that the sustained luminosity of an accreting object can be arbitrarily high, provided that it can accrete as much as it likes? This is not true, because radiation itself provides a pressure. The radiation pressure is discussed in Chapter 3.

Assume that matter is fully ionized, and for simplicity let us assume that we deal with hydrogen only. Now, consider the force on one proton–electron pair at a distance, r, from a massive object of mass M. This force is given by

$$F_{\text{grav}} = \frac{GM(m_p + m_e)}{r^2} \approx \frac{GMm_p}{r^2}, \tag{8.9}$$

because the mass of the proton is 1836 times bigger than the mass of the electron. Assume that the central source is emitting photons with a luminosity L. At radius r, this produces a number flux of photons of frequency v of

$$\mathcal{F}_v = \frac{L}{4\pi r^2 hv}, \tag{8.10}$$

assuming that the photons are emitted isotropically. Upon interaction with an electron, a photon transfers its momentum $p = hv/c$ to the electron. The effective area presented by a proton or electron to a photon is the Thomson cross section,

---

[3] The efficiency can be reduced if the gravitational energy is partly transformed into other energy forms.

σ_T, which, for a particle of charge e and mass m is

$$\sigma_T = \frac{2}{3}\left(\frac{e^2}{mc^2}\right)^2. \tag{8.11}$$

(See Section 3.7.8.) Because protons are about 1836 times more massive than electrons, the Thomson cross section of electrons is $\sim 1836^2$ greater than that of protons. Therefore, the outward radiation pressure on the infalling matter is exerted by the photons scattering off the electrons. Thus, radiation exerts an outward force of

$$F_{\text{rad}} = \sigma_T \mathcal{F}_\nu p = \frac{\sigma_T L}{4\pi r^2 c}, \tag{8.12}$$

Note that radiation pressure mainly acts on electrons, whereas gravitation acts mainly on protons. Electrostatic attraction keeps electrons and protons together so that a proton–electron pair experiences both forces. Now, equating F_{grav} and F_{rad} yields a limiting luminosity of

$$L_{\text{Edd}} = \frac{4\pi G M m_p c}{\sigma_T} \sim 3 \cdot 10^{13}\left(\frac{M}{10^9 M_\odot}\right) L_\odot, \tag{8.13}$$

which is called *Eddington* luminosity. This is the maximum luminosity that an accretion-powered source can continuously maintain. At greater luminosities, the pressure of radiation exceeds the gravitational force and gas is blown away. Note that this maximum luminosity is independent of r because both gravity and the flux of photons decrease as r^{-2}. Several assumptions have gone into this derivation:

1. We assumed the Thomson cross section for pure, ionized hydrogen. In the presence of metals or dust, the cross section with photons increases, which leads to a lower limiting luminosity.
2. We assumed spherical symmetry. If the central object radiates nonisotropically, the limiting luminosity can be higher.
3. We have assumed a stationary state.

Anisotropic emission has another important effect. The luminosity of an object is inferred from the measured flux, F, via the inverse-square law, $L = F 4\pi d^2$, where d is the distance to the object. If, however, the radiation is not isotropic but depends on the angle from which the object is observed, the true luminosity can differ substantially from the estimate based on the inverse-square law. This argument was discussed in the context of the energy of gamma-ray bursts in Chapter 7.

Let us now apply the Eddington luminosity to black holes found in the centers of galaxies.

If the Eddington luminosity is produced by gas accretion at a rate $\dot{M}$, with a radiative efficiency of $\xi = L/\dot{M}c^2 \sim 0.1$, the Eddington limit is reached at

$$\dot{M}_{Edd} \sim 22 \left(\frac{M_{BH}}{10^9 M_\odot} \right) M_\odot \, yr^{-1}, \tag{8.14}$$

which means that the black hole needs to swallow only a few tens of solar masses each year to power the most luminous quasars. Higher accretion rates are possible if the efficiency is lower.

We can also estimate the characteristic time, t_{grow}, over which the mass of the central black hole grows:

$$t_{grow} = \frac{M_{BH}}{\dot{M}} \sim 5 \cdot 10^8 \xi \left(\frac{L}{L_{Edd}} \right)^{-1} yr, \tag{8.15}$$

which suggests that the mass of black holes can grow significantly over cosmic time scales of gigayears. However, this estimate points to an important problem: the strongest constraint on the high-redshift evolution of supermassive black holes comes from the observation of luminous quasars at redshifts of $z \sim 6$ in the Sloan Digital Sky Survey. The luminosities of these quasars, well in excess of 10^{47} erg s^{-1}, imply that supermassive black holes with masses $10^9 \, M_\odot$ were already in place when the universe was only 1 Gyr ($z \sim 6$) old. The current age of the universe is 13.7 Gyr. The highest redshift quasar known at the time of writing, SDSS 1148 + 3251, at a redshift of $z = 6.4$, has an estimated black-hole mass in the range of $(2–6) \cdot 10^9 \, M_\odot$.

If a black hole has to grow from about 10 $M_\odot$ to $10^9 \, M_\odot$, using the results of Exercise 8.1, it would take $\sim \ln(10^9 M_\odot/10M_\odot) \sim 18.4$ e-folding times to assemble it or, according to Eq. (8.15) with an assumed efficiency of $\xi = 0.1, 9.2 \cdot 10^8$ yr. This is more than the age of the universe at a redshift of $z = 6.4$. So, obviously, a crucial piece is missing in our understanding of the formation of supermassive black holes.

8.3.5 *Evidence for black holes as central engines*

In recent years, the evidence for supermassive black holes at the centers of AGN has grown rapidly. *Supermassive black holes* are black holes with typical masses of 10^6 to $10^{10} \, M_\odot$. Ultimately, all evidence for them comes from the dynamical signature that a black hole leaves on the motions of surrounding matter. This mass can either be in the form of stars or in the form of gas. Because the dynamics of stars is entirely governed by gravitational forces while the dynamics of gas is also controlled by pressure and other forces, the dynamic signatures of black holes are different for stars and gas. Both the dynamics of stars and gas have been used to constrain the central mass density in AGN.

When it comes to measuring the gravitational influence of a supermassive black hole in galactic centers, a useful concept is the *sphere of influence*. It is defined as the region where the gravitational potential of the supermassive black hole is greater than the gravitational potential of the surrounding stars. Ignoring factors of order unity, we can estimate the radius of the sphere of influence by equating the kinetic energy of a star to its energy in the gravitational potential of the black hole. Its radius is given by

$$R_{\text{infl}} \sim \frac{GM_{\text{BH}}}{\sigma_*^2} \sim 11.2\,\text{pc} \left(\frac{M_{\text{BH}}}{10^8\,M_\odot} \right) \left(\frac{\sigma_*}{200\,\text{kms}^{-1}} \right)^{-2}, \qquad (8.16)$$

where σ_* is the *velocity dispersion* of the surrounding stars, $\sigma_*^2 = \langle v_*^2 \rangle$, where the average is over the stellar velocity distribution. The observed velocity dispersion is the result of the superposition of many individual stellar spectra, each of which has been Doppler shifted because of the star's motion in the gravitational potential of the black hole. Therefore, the velocity dispersion can be determined by analyzing the integrated spectrum of many stars.

In a region that lies beyond many times the Schwarzschild radius but still within the sphere of influence, we can ignore General Relativity so that masses follow essentially Keplerian orbits (see also our discussion in Section 5.5.2). Beyond the sphere of influence, the gravitational dominance of the black hole vanishes.

The Doppler shifts of stellar absorption lines in nearby AGN indicate high rotational velocities and a high velocity dispersion toward the center. From the velocity and the velocity dispersion, one can derive the enclosed mass using Kepler's laws. These measurements point to very massive objects in the centers of AGN. In the same way, one can compare the mass with the total stellar light from the same region. One finds that the ratio between mass and luminosity stays roughly contant in the outer parts of an AGN but increases drastically toward the central region. This evidence points to the existence of a very large, dark central mass concentration. In principle, this central mass concentration could either be in the form of a singularity (i.e., a black hole) or in the form of a cluster of masses that do not emit a lot of light (i.e., neutron stars).

One can compute the lifetime of a cluster of dark masses before it either collapses catastrophically or evaporates, which means that the constituents are ejected from the cluster by scattering. If we take a cluster of brown dwarfs, white dwarfs, neutron stars, and stellar-mass black holes, the maximum lifetime can be calculated. The result as a function of the cluster mass and density is shown in Fig. 8.5.

The observations of masses and densities in some galactic nuclei imply maximum lifetimes of such clusters that are smaller than the age of the galaxy, which is a strong argument in favor of the existence of black holes at the centers of galaxies.

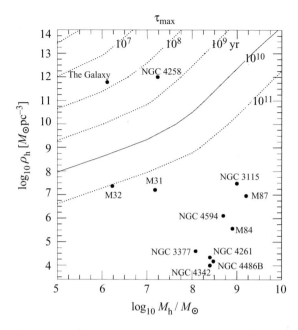

Figure 8.5   Maximum possible lifetime of a dark cluster with half-mass $M_h \equiv M/2$ and half-mass density ρ_h, which is the mean density within the cluster's half-mass radius (the radius that contains half of its entire mass), against the processes of evaporation and destruction due to physical collisions. These clusters may consist of any plausible form of nonluminous objects. The points present the current data for most of the observed black-hole candidates. We see that the lifetime of such hypothetical clusters must be much shorter than 10 Gyr only in the cases of NGC 4258 and our Galaxy, thus strongly arguing for a point mass. From Maoz, E. (1998).

Observations of stellar dynamics in centers of galaxies in order to constrain the black-hole mass can currently be done only with the space-based *Hubble Space Telescope*. As an example, we can consider our neighbor galaxy, Andromeda. Taking a typical stellar velocity dispersion of 160 km s^{-1}, and a black-hole mass of $3 \cdot 10^7 \, M_\odot$, the radius of the sphere of influence is 5.2 pc, which corresponds to an angular size of $1''$ as viewed from Earth at a distance of 770 kpc. This can just about be resolved by ground-based observatories. However, if a galaxy is only two times farther away, the only way to measure the dispersion today is with the *Hubble Space Telescope*. At a distance of the Virgo cluster, which is 15 Mpc, the sphere of influence of a supermassive black hole with a mass of $3 \cdot 10^7 \, M_\odot$ is a mere 0.07 arc seconds, which is not resolvable by any optical telescope. Thus, there are very few galaxies for which the sphere of influence can be properly resolved.

Similar kinematic studies can be done using emission lines from interstellar gas. The nearby giant elliptical galaxy M87 in the Virgo cluster, for example, contains a hot rotating disk at its center. Measurements of radial velocities with the *Hubble Space Telescope* have shown that the gas in this disk is in Keplerian motion around a mass of $2.4 \cdot 10^9 \, M_\odot$ within the inner 18 pc of the nucleus.

Strong evidence for supermassive black holes comes from interferometric observations of water masers in AGN. Masers are strong coherent sources that are powered by stimulated emission (see Section 3.6). Stimulated emission occurs whenever an electromagnetic wave of frequency $\nu = E/h$ impinges on a particle that is in an excited atomic state at energy E above some other state. The electromagnetic wave can then stimulate the emission of another photon that has the same polarization, direction, and frequency as the stimulating photon. Thus, as radiation propagates through an assembly of particles, the light is amplified. Because the emitted photons have the same characteristics as the stimulating photons, the radiation is coherent. This mechanism is called a *maser* (in the optical such a process is called *laser*, which is used, e.g., in DVD players). Water vapor molecules have energy states that can produce masers that can be received with radio telescopes. Masers produce radiation of very high intensity in a very narrow, well-defined bandwidth, which allows the precise determination of proper velocities.

Water masers from the outer, molecular part of AGN accretion disks have been resolved by very long baseline interferometry (VLBI). In this technique, radio telescopes in different parts of the world observe the same object and superimpose the detected wave trains in phase to produce an interference pattern. This interference pattern can be converted into an image using Fourier transforms. Thus, the radio telescopes act as a giant interferometer with a baseline of many thousands of kilometers. Because the resolution of a telescope goes as $\Delta\theta \approx \lambda/A$, where λ is the wavelength and A the aperture or baseline, the long baselines of VLBI allow an extremely small angular resolution (also see http://www.evlbi.org/). Unfortunately, the conditions for masers in AGN accretion disks are only rarely right, and there are not many examples. A prominent example is the maser ring in NGC 4258, shown in Fig. 8.6. The radio spectrum of the maser exhibits emission at about the velocity of the galaxy (center), red-shifted emission from the receding side of the disk (left), and blue-shifted emission from the approaching side of the disk (right). One can infer that the gas is in Keplerian motion at distances 0.14–0.28 pc from a central mass of $4 \cdot 10^7$ solar masses. This leads to inferred mass densities of more than $4 \cdot 10^9 M_\odot \, \mathrm{pc}^{-3}$, which are much larger than the density in the densest known star clusters, where the density is less than $10^6 M_\odot \, \mathrm{pc}^{-3}$. We know of no plausible alternative to hide that much mass within so little space other than in the form of a black hole.

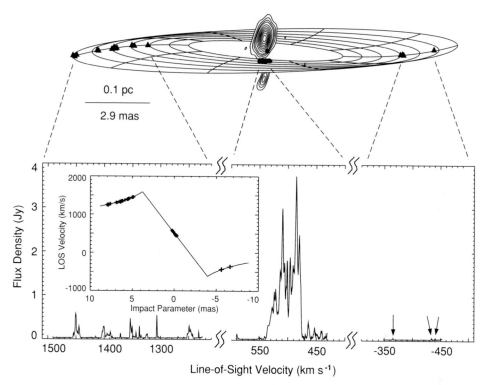

Figure 8.6   The upper panel shows the warped disk with the high-velocity maser sources marked with filled triangles. The contours in the center represent 22 GHz emission taken with the VLBA. This emission traces a subparsec-scale jet elongated along the rotation axis of the disk. The masers have frequencies that correspond to Doppler shifts of $\sim \pm 1000$ km s^{-1} with respect to the bulk velocity of the galaxy (lower panel). The inset show the line-of-sight (LOS) velocity versus distance from the center for the best-fitting Keplerian disk (shown as a line). The maser sources trace a Keplerian rotation curve to better than 1%. From Herrnstein et al. (1999).

Motions even closer to the black hole than those unveiled by masers can be probed with X-ray spectroscopy of Seyfert galaxies. Their spectra show enormously broadened emission lines from Fe ions. The profiles of these emission lines are best explained by emission from an accretion disk around a black hole. However, this method remains somewhat controversial, and we refrain from going into further detail here.

Origin of accreted matter

To sustain its luminosity, the AGN has to be supplied with a steady rate of fuel. Where does the accreted matter come from?

One of the suggestions is that a supermassive black hole grows via the accretion of stars. Let us verify whether this is a relevant mode via which a black hole can be fed. The stars that orbit the supermassive black hole have to interact in some way to get rid of their angular momentum. Only then can they fall into the black hole. In Exercise 2.3 in Chapter 2, we derived that the relaxation time of stars is

$$t_{relax} \approx \frac{v_{rms}^3}{8\pi G^2 M_*^2 n \ln \Lambda_g},$$

$$(8.17)$$

where v_{rms} is the velocity dispersion of stars, M_* the mass of the star, n the stellar number density, and $\ln \Lambda_g$ the gravitational Coulomb logarithm. The relaxation time, t_{relax}, is the time after which gravitational interactions have altered the orbits of the stars. After a typical time of $t_{collapse} \sim 100 t_{relax}$, a group of stars have interacted sufficiently so that they can collapse into a black hole. This collapse time can be written as

$$t_{collapse} \approx 9.5 \cdot 10^{12} \, \text{yr} \, \left(\frac{v_{rms}}{200 \, \text{km s}^{-1}}\right)^3 \left(\frac{n}{10^6 \, \text{pc}^{-3}}\right)^{-1} \left(\frac{M_*}{1 \, M_\odot}\right)^{-2}, \qquad (8.18)$$

for a Coulomb logarithm of 15. With the *Hubble Space Telescope*, the stellar densities and velocity dispersions near the centers of nearby galaxies have been measured. The majority of galaxies have central relaxation times that were much greater than the age of the universe. Only the smallest galaxies known to harbor massive black holes have nuclear relaxation times smaller than 10^{10} yr.

As a result, it appears more likely that AGN are powered by the infall of gas. The bulk of the gas most likely originates far from the nucleus and is driven inward by gravitational torques. This is supported by the fact that quasar activity peaks at approximately the same epoch when galaxy merging peaked. Mergers between galaxies may be efficient at driving gas into the nucleus of galaxies, which can then fuel the black hole. There is also evidence that galaxies are more likely to be active if they are interacting with a neighboring galaxy.

8.4   Basic ingredients of an AGN

By now, astronomers have identified the following main ingredients of an AGN

- black hole,
- accretion disk,
- jets,
- broad-line emission region,
- narrow-line emission region, and
- molecular/dust torus.

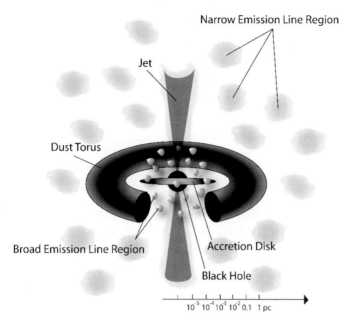

Figure 8.7 Schematic view of the central region of an AGN displaying the black hole, the accretion disk, the jets, the broad and narrow emission line regions, and the dusty or molecular torus. Note that the dimensions in this plot are not linear but logarithmic. The scale is indicated at the bottom right corner. Figure courtesy of L. Ferrarese.

These components are sketched in Fig. 8.7 and their dimensions are summarized in Table 8.2. In the following section, we summarize our present understanding of each of these components.

8.4.1 Black holes

Previously, we have reviewed the evidence for supermassive black holes. Black-hole mass, along with mass accretion rate, is a fundamental property of AGN. However, direct kinematic observations of the black-hole mass are limited by the spatial resolution (a typical AGN at redshift 2 would require nanoarcsecond resolution to probe the sphere of influence of the black hole). A further complication comes from scattered light from the bright central source that dilutes any kinematic signal from orbiting material. Therefore, in recent years, various indirect methods for estimating black-hole masses have been devised that rest on correlations between the mass of the black hole and properties of the host galaxy. These correlations have been inferred from nearby nonactive galaxies. If AGN host galaxies are similar to nonactive galaxies, this correlation should also hold for them.

Table 8.2 *Linear dimensions of the main*
components of an AGN.

Component	Typical size
Black hole (R_S for $10^9 M_\odot$)	10^{-4} pc
Accretion disk	0.01 pc
Broad-line region	1 pc
Dust torus	10 pc
Narrow-line region	1 kpc
Host galaxy	10 kpc

Without going into great detail, we mention these correlations here as they have evolved into a very widely used method to estimate black-hole masses. Studies of supermassive black holes in nearby galaxies have revealed that the masses of black holes are correlated with (a) the velocity dispersion and (b) the luminosity of the surrounding stars. However, it is still unclear why these correlations exist and whether they can readily be extended to galaxies that are far away. The existence of these correlations seems to suggest that the formation history of the host galaxy and the supermassive black hole in its center are somehow intertwined. How exactly they are intertwined is not really understood. Because of their importance for black-hole mass estimates, we quote these correlations here.

1. The M_{BH}-L_B relation[4] is

$$M_{BH} = 0.78 \cdot 10^8 \, M_\odot \left(\frac{L_{B,bulge}}{10^{10} L_{B\odot}} \right)^{1.08}, \quad (8.19)$$

where $L_{B,bulge}$ is the B-band luminosity of the bulge of the galaxy and $L_{B\odot} = 5.2 \cdot 10^{32}$ erg s^{-1} is the solar luminosity in the B-band. For a definition of the B-band, see Appendix D.

2. The $M_{BH} - \sigma$ relation[5] is

$$M_{BH} = 1.66 \cdot 10^8 \, M_\odot \left(\frac{\sigma}{200 \text{ km s}^{-1}} \right)^{4.86}. \quad (8.20)$$

Interestingly, the black-hole mass does not correlate with the luminosity of galaxy disks. Using the $M_{BH} - \sigma$ relation, the mass of supermassive black holes at the centers of galaxies can be measured with a single measurement of the bulge velocity dispersion with an accuracy of 30%. Figure 8.8 shows the $M_{BH} - L_B$ and the $M_{BH} - \sigma$ relations for all supermassive black-hole detections for which the

---

[4] Kormendy, J. and Gebhardt, K. (2001). Supermassive black holes in Galactic nuclei. In *Twentieth Texas Symposium on Relativistic Astrophysics*, ed. J. Craig Wheeler and Hugo Martel, Melville, NY: American Institute of Physics, Vol. 586, p. 363.
[5] Ferrarese, L. and Ford, F. (2005). *Space Science Reviews*, **116**, 523.

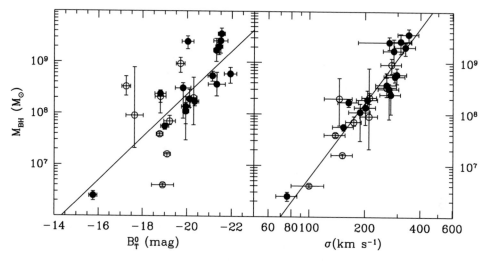

Figure 8.8   The $M_{BH} - L_B$ (L_B is expressed as magnitude B_T) (left) and $M_{BH} - \sigma$ relations for all supermassive black-hole detections for which the sphere of influence could be resolved. Filled symbols show elliptical galaxies, and open symbols show spiral galaxies. The solid lines are the best fits to the data, accounting for errors in both coordinates as well as intrinsic scatter. From Ferrarese and Ford (2005).

sphere of influence could be resolved. The graphs reveal that the correlations are pretty good and that the $M_{BH} - \sigma$ relation is tighter than the $M_{BH} - L_B$ relation. Both are widely used to estimate black-hole masses in galaxies.

On the basis of stellar velocity dispersions in host galaxies of AGN, one can try to test whether the black-hole mass is correlated with any property of the AGN, for example, its luminosity or its radio power. In recent years, astronomers[6] have collected several hundred mass estimates of AGN host galaxies and have found no such correlation. They find that there is a trend such that the Eddington luminosity given in Eq. (8.13) provides an approximate upper limit to the AGN luminosity, but apart from that, AGN of a given mass can have all luminosities (see Fig. 8.9). For a given AGN black-hole mass, the bolometric luminosity ranges over at least two, and as much as four, orders of magnitude. The ratio of the luminosity of the AGN over its Eddington luminosity, L/L_{Edd}, does not correlate with mass or luminosity either. It is sometimes stated that the most luminous AGN have high Eddington ratios, whereas low-luminosity AGN have low Eddington ratios. There may be a rough trend into this direction, but the data are not clear on this. Finally, there also does not seem to be a relation between radio power and the mass of the black hole. Observations do not require that radio-loud objects have high black-hole masses, nor do high black-hole masses imply high radio luminosities. In summary, the mass

---

[6] Woo, J. and Urry, C. M. (2002). *Astrophysical Journal*, **579**, 530.

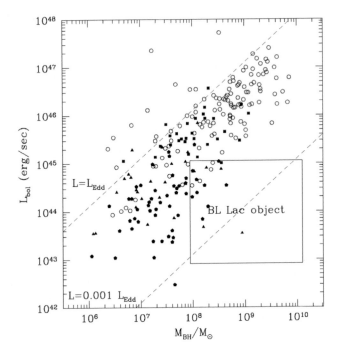

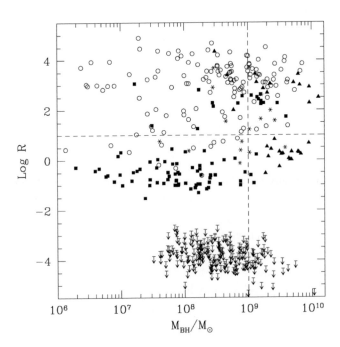

of the black hole does not appear to be a defining parameter for the appearance of an AGN. Contrary to some claims in the literature, the black-hole mass is not correlated with the bolometric luminosity or the radio power of the AGN.

8.4.2   Accretion disk

As discussed previously, AGN harbor massive black holes in their centers with masses of $M_{BH} \sim 10^6$–$10^{10} M_\odot$ that accrete matter. We expect accreting matter to form disks because matter traveling along orbits inclined to one another collide in the plane of intersection. As a result, the angular momenta of the different, inclined orbits mix, and eventually all matter orbits in a single plane, thus forming a disk. AGN black holes are surrounded by accretion disks extending to $\sim$100–1000R_S, where R_S is the Schwarzschild radius.

However, the accretion of mass onto a central black hole is not straightforward because the law of conservation of angular momentum stops matter from plunging straight into the black hole (also see Eq. [6.18]).

Assuming that the gravitational potential in the accretion disk is dominated by the black hole, M_{BH}, the gas around the black hole follows Keplerian orbits. The Kepler velocity is given by

$$v_\phi(r) = \omega_K \cdot r = \left(\frac{GM_{BH}}{r} \right)^{1/2}, \qquad (8.21)$$

with ω_K from Eq. (6.32). The specific angular momentum is given by $l = r v_\phi = \sqrt{GM_{BH}r}$. Thus, the angular velocity decreases outward, whereas the

---

Figure 8.9   Top: Bolometric luminosity versus black-hole mass for 234 AGN. There is little if any correlation. For a given black-hole mass, there is a large range of bolometric luminosities, spanning three or more orders of magnitude. The Eddington limit defines an approximate upper limit to the luminosity, but the absence of objects from the lower right of the diagram (low-luminosity, high-mass AGN) is a selection effect: this part of the diagram gets filled in simply by including lower luminosity AGN, continuously down to galaxies. Among the low-luminosity objects with large black holes are the radio galaxies and BL Lac objects (which are a certain type of blazar) for which we do not have good estimates of bolometric luminosity. The different symbols denote different data samples. From Woo, J. and Urry, C. M. (2002). *Astrophysical Journal*, **579**, 530. Bottom: Radio loudness versus black-hole mass for 452 AGN. Radio loudness is defined as the ratio between the flux density at 5 GHz and in the B-band (see Eq. [8.1]). There is no correlation between radio loudness and black-hole mass. In particular, at high mass ($M > 10^9 \, M_\odot$) there are similar distributions of black-hole masses for radio-loud ($R > 10$) and radio-quiet AGN. The different symbols denote different samples. From Woo, J. and Urry, C. M. (2002). *Astrophysical Journal*, **581**, L5.

specific angular momentum increases outward. For a ring of gas to move further in, it has to lose some of its angular momentum. As the total angular momentum is conserved, angular momentum has to be transported outward for mass to be accreted. In fluid media, transport of momentum occurs via viscosity, as we discussed in Section 2.3. If two neighboring rings of gas that are following Keplerian motion are coupled by viscous forces, the outer, slower ring brakes the inner, faster ring. Thereby, it gains angular momentum while the inner ring loses it. This leads to a spreading of the disk. The friction between the rings leads to energy dissipation. According to the virial theorem, half of the liberated potential energy is converted to kinetic energy. The other half of the potential energy is converted to internal energy. Let us attempt to quantify this a little.

If a mass M moves in from a radius $r + \Delta r$ to r, an energy

$$\Delta E = \frac{GM_{\mathrm{BH}}M}{r} - \frac{GM_{\mathrm{BH}}M}{r + \Delta r} \approx \frac{GM_{\mathrm{BH}}M\,\Delta r}{r^2} \tag{8.22}$$

is liberated, where the last term in this equation comes from expanding the previous term in a Taylor expansion for small Δr. Here, M_{BH} is the mass of the black hole, which we assume to dominate the gravitational potential in the disk. Moreover, we ignore the self-gravity of the disk. Now, half of this energy can be converted to internal energy, that is, $U = \Delta E/2$. If we assume that this energy is dissipated into heat locally, we can write for the luminosity

$$\Delta L = \frac{GM_{\mathrm{BH}}\dot{M}}{2r^2}\Delta r, \tag{8.23}$$

where $\dot{M}$ is the accretion rate onto the black hole. If the accretion disk is optically thick, every segment of the surface radiates like a black body. That is, we can write for a ring between $r + \Delta r$ and r

$$\Delta L = 2 \cdot 2\pi r\,\Delta r\sigma_{\mathrm{SB}}T(r)^4. \tag{8.24}$$

The factor 2 in this equation comes from the fact that the disk has two sides, and σ_{SB} is the Stefan–Boltzmann constant (see Eq. [3.56]). By combining Eqs. (8.23) and (8.24), we can work out the radial dependence of the temperature in the disk, which is

$$T(r) = \left(\frac{GM_{\mathrm{BH}}\dot{M}}{8\pi\,\sigma_{\mathrm{SB}}r^3}\right)^{1/4}. \tag{8.25}$$

A more careful derivation with a proper modeling of the friction, one obtains the same result, with the exception of an additional correction factor:

$$T(r) = \left(\frac{3GM_{\mathrm{BH}}\dot{M}}{8\pi\,\sigma_{\mathrm{SB}}r^3}\right)^{1/4}, \tag{8.26}$$

which is valid as long as the radius is much larger than the Schwarzschild radius, that is, $r \gg R_S$. If one expresses r in terms of R_S, one can write

$$T(r) = \left(\frac{3G M_{BH} \dot{M}}{8\pi \sigma_{SB} R_S^3} \right)^{1/4} \left(\frac{r}{R_S} \right)^{-3/4}. \tag{8.27}$$

By plugging in some typical numbers, we find

$$T(r) = 6.3 \cdot 10^5 \text{ K} \left(\frac{\dot{M}}{\dot{M}_{Edd}} \right)^{1/4} \left(\frac{M_{BH}}{10^8 M_\odot} \right)^{-1/4} \left(\frac{r}{R_S} \right)^{-3/4}, \tag{8.28}$$

where $\dot{M}_{Edd} = L_{Edd}/(0.1\,c^2)$ see Section 8.3.3. From this relationship, we can draw some interesting conclusions. First of all, the temperature profile of the disk appears to be independent of the detailed mechanism of energy dissipation. None of the equations contains the viscosity explicitly. Moreover, the model of a geometrically thin, optically thick accretion disk allows a number of quantitative predictions: The temperature of the disk decreases with radius as $r^{-3/4}$. The emission of the disk is given by a series of blackbody spectra, each of which is emitted by a ring of a certain temperature.

Consequently, the spectrum of the disk as a whole does not follow a single Planck spectrum but is much broader. For AGN, the maximum of the thermal spectrum from the disk lies in the ultraviolet band. Indeed, the continuum spectra of quasars show a clear rise in the ultraviolet–a feature that has been called *big blue bump* (Section 8.2.2) or UV bump. A generic AGN spectrum is given in Fig. 8.3. It can be observed only down to wavelengths of 912 Å because below this wavelength the photoelectric absorption of neutral hydrogen from our Galaxy blocks the radiation (until the soft X-rays, which can then be observed again).

For a given r/R_S, the temperature rises with the accretion rate $\dot{M}^{1/4}$. This is not unexpected as the radiated energy is proportional to T^4, and the dissipated energy is proportional to $\dot{M}$; thus $T \propto \dot{M}^{1/4}$. Similarly, for a given r/R_S, the temperature decreases with the mass of the black hole because $R_S \propto M_{BH}$. The larger the mass of the hole, the smaller the temperature of the surrounding disk. This seems a bit counterintuitive and comes from the fact that for a given r/R_S the tidal forces decrease with rising $\dot{M}$. This explains why the maximal temperatures of accretion disks in AGN are much lower than disks around much less massive objects such as neutron stars or stellar-mass black holes. The latter produce binaries (see Section 6.3) that radiate in the hard X-rays, whereas AGN disks peak in the ultraviolet.

The monochromatic energy flux $F_\nu(T)$ (units erg cm^{-2} Hz^{-1}) of a black body at temperature T is given by the Planck distribution, see Eqs. (3.12) and (3.56),

$$F_\nu(\nu, r) = \frac{2\pi h \nu^3}{c^2} \frac{1}{\exp[h\nu/k_B T(r)] - 1}, \tag{8.29}$$

where $T(r)$ is given by Eq. (8.26). The total monochromatic luminosity, L_ν, of the disk is now given by

$$L_\nu = 2 \cdot \int_{r_{min}}^{r_{max}} F_\nu(\nu, r) 2\pi r \, dr, \tag{8.30}$$

where the factor 2 in front of the integral comes from the disk having two sides, and r_{min} and r_{max} are the inner, and outer radii of the disk, respectively. Substituting for F_ν yields

$$L_\nu = \frac{8\pi^2 h \nu^3}{c^2} \int_{r_{min}}^{r_{max}} \frac{r}{\exp[h\nu/k_B T(r)] - 1} \, dr. \tag{8.31}$$

This integral cannot be calculated analytically but we can try to find some approximate solutions. At low frequencies (i.e., in the Rayleigh–Jeans regime of the Planck distribution), where $h\nu/k_B T \ll 1$, we can write, see Eq. (3.59),

$$\frac{1}{\exp(h\nu/k_B T) - 1} \approx \frac{k_B T}{h\nu}. \tag{8.32}$$

The lowest temperatures in the disk occur at the outer radius. There we can write for the luminosity density

$$L_\nu \approx \frac{8\pi^2 \nu^2 k_B}{c^2} \int_{r_{min}}^{r_{max}} T(r) r \, dr \approx \frac{32\pi^2}{5} \left(\frac{3GM\dot{M}k_B^4 r_{max}^5}{8\pi \sigma_{SB} c^8} \right)^{1/4} \nu^2, \tag{8.33}$$

where we have assumed $r_{min} \ll r_{max}$. Thus, at low frequencies, the spectrum from the accretion disk is a power law of the frequency that is proportional to ν^2. At high frequencies, where $h\nu/k_B T \gg 1$, we can use the Wien limit Eq. (3.61)

$$\frac{1}{\exp(h\nu/k_B T) - 1} \approx \exp\left(-\frac{h\nu}{k_B T} \right). \tag{8.34}$$

The highest temperatures occur at the inner radius of the disk, so there we can write for the luminosity density

$$L_\nu \approx \frac{8\pi^2 h \nu^3}{c^2} \int_{r_{min}}^{r_{max}} \exp\left[-\frac{h\nu}{k_B T(r)} \right] r \, dr. \tag{8.35}$$

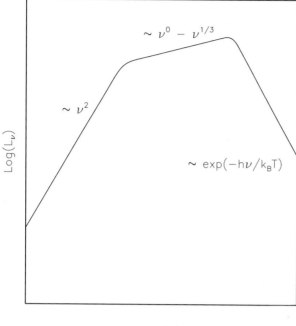

Figure 8.10   Sketch of the overall spectrum of a standard accretion disk.

This can be integrated to yield

$$L_\nu \approx \frac{4\pi^2 h\nu^3}{c^2} r_{max}^2 \exp\left[-\frac{h\nu}{k_B T(r_{min})}\right],\tag{8.36}$$

where, again, we have assumed that $r_{min} \ll r_{max}$. Hence, at high frequencies, the spectrum falls off exponentially. We have sketched the overall spectrum of an accretion disk in Fig. 8.10.

The kinematic viscosity that is caused by the thermal motions of atoms and molecules (see Section 2.3) is much too weak to explain observed accretion rates. Equation (2.45) puts the viscosity of a fully ionized plasma at

$$\eta \sim 2 \cdot 10^{-15} \left(\frac{T}{K}\right)^{5/2} (\ln \Lambda)^{-1} \, \text{g cm}^{-1}\text{s}^{-1},\tag{8.37}$$

where T is the temperature and $\ln \Lambda$ the Coulomb logarithm. We can work out the effect of this molecular viscosity by computing the Reynolds number $\text{Re} \sim R v_\phi \rho / \eta$ defined in Eq. (2.47), where R is the typical size of a turbulent eddy and v_ϕ the fluid velocity in this eddy. Simple centrifugal equilibrium, Eq. (8.21), tells us that

$$R v_\phi \sim 10^{22} \left(\frac{M}{M_\odot}\right)^{1/2} \left(\frac{R}{1\text{pc}}\right)^{1/2} \text{cm}^2\text{s}^{-1},\tag{8.38}$$

which gives

$$\text{Re} \sim 3 \cdot 10^{13} \left(\frac{M}{M_\odot} \right)^{1/2} \left(\frac{R}{1\text{pc}} \right)^{1/2} \left(\frac{n}{\text{cm}^{-3}} \right) \left(\frac{T}{K} \right)^{-5/2}. \tag{8.39}$$

For typical values in accretion disks, the Reynolds number is huge, which implies that the molecular viscosity is irrelevant. However, the high Reynolds number suggests that the fluid in the disk is turbulent. This turbulence may be able to replace the role of thermal motions of molecules in transporting momentum and may result in a *turbulent viscosity*. Now, turbulent eddies instead of molecules transport the momentum between different portions of the disk. This turbulent viscosity would be proportional to $l_{\text{turb}} v_{\text{turb}}$, where l_{turb} is the typical size of an eddy and v_{turb} its velocity. Finally, magnetic fields are also likely to have a say in the dynamics of the turbulent flow in accretion disks. In the absence of a clear physical model, the astrophysicists Nikolai Shakura and Rashid Sunyaev came up with a handy parametrization. They pointed out that the natural scale for any hydrodynamic stress is the pressure. Consequently, the simplest parametrization of the viscous stress is $-\alpha p$, where α is a proportionality constant, which can be as large as ~ 1. This parametrization of the viscosity is called *α-viscosity* and is commonly used in accretion theory.

Clearly, the radiation from the disk cannot account for the broadband, high-energy radiation from galactic nuclei. Here, more complex processes are at work, such as inverse Compton scattering and electron-positron pair production. To create an electron–positron pair, a photon needs to have a minimum energy $h\nu \sim 2m_e c^2$. The number of photons near this threshold can be estimated by dividing the energy density, $u \sim L_\gamma / 4\pi R^2 c$, of gamma-ray photons within a region of radius R, by the threshold energy $2m_e c^2$ (i.e., $n_\gamma \sim L_\gamma / 8\pi R^2 m_e c^3$). The optical depth for pair production is $\tau_{\gamma\gamma} \sim n_\gamma \sigma_T R$, see Eq. (3.46), where σ_T is the Thomson cross section. Pair production is important if its optical depth is larger than 1, that is,

$$\frac{L_\gamma \sigma_T}{8\pi R m_e c^3} \geq 1. \tag{8.40}$$

Expressing L_γ in terms of the Eddington luminosity and R in terms of the Schwarzschild radius, one can rewrite Eq. (8.40) as

$$\frac{L_\gamma}{L_{\text{Edd}}} \geq \frac{4m_e}{m_p} \left(\frac{R}{R_S} \right) \sim 2.2 \cdot 10^{-3} \left(\frac{R}{R_S} \right). \tag{8.41}$$

This condition can be met in many AGN, and pair production can be an important effect.

The geometrically thin disk that we have described has been the focus of much work and is generally referred to as the *standard disk*. Its thickness is much smaller than its radius and the radial dynamics follows Keplerian motion around the black hole. This type of disk is thought to occur when the accretion rate is less than the Eddington luminosity. When the accretion rate is so large that the accretion disk cannot cool fast enough, thermal pressure inflates the disk, making it geometrically thick. Its emission spectrum now strongly depends on the viewing angle. These types of accretion disks can be quite inefficient as thermal energy is advected with the flow into the black hole: they can have quite high accretion rates with relatively small luminosities. That is, the efficiency parameter η can be much smaller than the canonical value of 0.1. Such inefficient accretion may also occur in gamma-ray bursts.

Similarly, our simplified treatment of a thin accretion disk cannot be the whole story. The real accretion flow is bound to be much more complicated. In recent years, researchers have developed a whole zoo of intricate models for accretion flows around black holes but no clear favorite has emerged yet. The major uncertainties in the theory of these disks are the interlinked questions of viscosity and magnetic fields. For further reading, we refer the reader to Frank, King, and Raine (2002) and Balbus and Hawley (1998).[7]

8.4.3 Jets

The radio emission from active galaxies is caused by jets. Jets are collimated outflows of highly energetic plasma. They travel at extreme velocities, often close to the speed of light, with bulk Lorentz factors of up to $\sim$30. These extreme velocities give rise to a number of interesting physical phenomena, some of which we discuss in this section. The evidence for highly collimated jets in astrophysics goes back to the early radio observations of twin lobes in extended radio galaxies, of which the prototype is Cygnus A (Fig. 8.1).

We start with the apparent superluminal motion of jets and its special-relativistic explanation. Then, we look at the fundamentals of gas physics of jets starting from simple textbook models to recent computer simulations of jet flow. Finally, we discuss various mechanisms that may be responsible for the launch of the jets and end with a section on the radiation from jets and lobes.

Superluminal motion

The propagation of some prominent radio jets has been traced in real time by observing the positions of bright blobs (or knots) in the jets over the course of years (see Fig. 8.11). Surprisingly, some jets appear to be moving at speeds greater than

---

[7] Balbus, S. A. and Hawley, J. F. (1998). Instability, turbulence, and enhanced transport in accretion disks. *Review of Modern Physics*, **70**, 1.

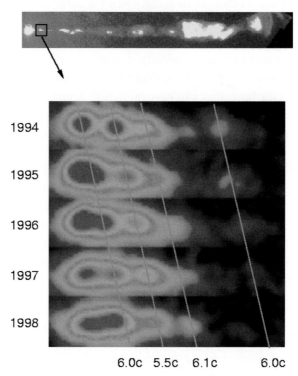

Figure 8.11   Sequence of *Hubble Space Telescope* images showing the jet of
the galaxy M87, which is moving at an apparent speed of six times the speed of
light. The slanting lines track the moving features. Courtesy of J. Biretta, Hubble
Heritage.

the speed of light. This apparent which superluminal motion is in no contradiction
to the theory of relativity and can be explained by the geometry of the jet. Take a
blob of jet material that is ejected at speed v at an angle θ to the line of sight as
shown in Fig. 8.12. If the transverse distance covered within the interval Δt is Δy,
then

$$\frac{\Delta y}{\Delta t} = v \sin \theta. \tag{8.42}$$

However, this is not the tranverse velocity of the blob that we observe. As the
blob moves, it is coming toward us, and the interval between the reception of two
photons is less than the interval between their emission, that is,

$$\Delta t_{\text{obs}} = \Delta t - \frac{\Delta x}{c} = \Delta t(1 - \beta \cos \theta), \tag{8.43}$$

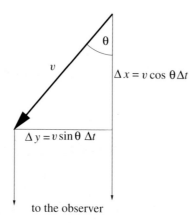

Figure 8.12   Graph showing the geometry of superluminal motion.

where $\beta = v/c$. Thus, the apparent velocity across the plane of the sky is given by

$$v_{\text{app}} = \frac{\Delta y}{\Delta t_{\text{obs}}} = \frac{v \sin \theta}{1 - \beta \cos \theta}. \tag{8.44}$$

This equation states that the apparent speed is larger than the speed of light for all angles if $\gamma \geq 3$. Hence, apparent superluminal motion occurs even for modest values of γ if the angle to the line of sight is small (note: the apparent velocity decreases again for very small angles). The maximum of v_{app} is found by solving

$$\frac{\partial v_{\text{app}}}{\partial \theta} = 0, \tag{8.45}$$

which yields $\cos \theta_{\text{max}} = \beta$. At this angle, $\sin \theta = \sqrt{1 - v^2/c^2} = 1/\gamma$ and

$$v_{\text{app,max}} = \frac{v/\gamma}{1 - v^2/c^2} = v\gamma. \tag{8.46}$$

Doppler boosting

The radiation emitted by a blob of jet matter is relativistically enhanced as it approaches the observer. This effect is called *Doppler boosting*, and we briefly explain how it works here.

As shown in Eq. (3.22), specific intensity divided by the frequency cubed, I_ν/ν^3, is Lorentz invariant. Moreover, the relativistic Doppler formula, Eq. (1.22), relates the observed to the emitted frequency of a moving source. So, if the emitted intensity is I_ν^{em} and the observed one I_ν^{em}, we have from Eq. (3.22)

$$\frac{I_\nu^{\text{obs}}}{(\nu^{\text{obs}})^3} = \frac{I_\nu^{\text{em}}}{(\nu^{\text{em}})^3}. \tag{8.47}$$

From the relativistic Doppler formula, Eq. (1.22), we find

$$\nu^{obs} = \mathcal{D} \, \nu^{em}, \tag{8.48}$$

where $\mathcal{D}$ is the Doppler factor

$$\mathcal{D} = \frac{\sqrt{1 - \beta^2}}{1 - \beta \cos\theta}, \tag{8.49}$$

where $\beta = v/c$ and θ is the angle with the line of sight. Thus, we find

$$I_\nu^{obs}(\mathcal{D}\nu) = \mathcal{D}^3 \, I_\nu^{em}(\nu). \tag{8.50}$$

Even for mildly relativistic jets with $\gamma \sim 4$, which corresponds to a speed of 97% of the speed of light, $\mathcal{D}^3$ can be as high as 1000. This means that the flux from the forward part of the jet is boosted by a factor of 1000 while the flux from the receding side of the jet is diminished by 1000. This is confirmed by observations that often show only one side of the jet. As we will elaborate later, the radiation from jets is primarily synchrotron radiation, which usually follows a power law in the form $I_\nu^{em} \propto \nu^{-\alpha}$. Thus the intensity at the same frequency transforms

$$I_\nu^{obs} = \mathcal{D}^{3+\alpha} \, I_\nu^{em}. \tag{8.51}$$

If the spectrum in the rest frame of the jet is a power law, then it remains a power law in the observer's frame. Every frequency gets shifted by the factor $\mathcal{D}$, and thus the slope of the power law remains unchanged.

Gas dynamics of the jets

A fundamental uncertainty concerns the material composition of the jets. The synchrotron radiation emitted by the jets suggests the presence of electrons. As the synchrotron power is proportional to B^2/m^2, see Eq. (3.162), the magnetic field required to produce the same frequency with protons would be excessively high. However, it is not clear whether overall charge neutrality is provided by protons or positrons, the antiparticles of electrons. For this discussion it does not matter whether the jets consists of electrons and baryons or of electrons and positrons.

As the jet forces its way through the ambient medium, it develops a shock front. A primary shock develops ahead of the jet and moves through the intergalactic medium as the jet pushes its way out of the galaxy. Jet material at the tip of the jet is decelerated by the external medium and, consequently, cannot move as fast as material near the source of the jet. As a result, a *reverse shock* develops that moves backward against the motion of the jet material. Further shocks may be produced within the jet because of variations in the jet speed or changes in the medium through which the jets propagate. For example, if the ambient pressure changes more rapidly than the internal pressure of the jet can adjust, then the jet may develop internal shocks. Similar shocks are observed in rocket exhausts and in gamma-ray bursts, see Section 7.4.

A puzzling thing about jets is their very narrow width. The earliest models for the collimation of jets were derived from simple gas physics. It is simplest to approximate the jet as a one-dimensional flow of variable cross-sectional area A. The rate at which mass is injected into the jet is $\dot{M}$. The flow inside the jet has a velocity v. Moreover, let us assume that the jet expands adiabatically and that it is stationary, that is, the mass injection rate, given by $\dot{M} = \rho A v$, is constant. For stationary flow, and neglecting gravity, Bernoulli's equation (Eq. [2.78]) states that

$$h + \frac{1}{2}v^2 = h_0, \tag{8.52}$$

where h is the specific enthalpy. The subscript 0 refers to values at the point where $v = 0$. The enthalpy of an ideal gas is given by

$$h = c_p T = \frac{p}{\rho} \frac{\Gamma}{\Gamma - 1} = \frac{c_s^2}{\Gamma - 1}. \tag{8.53}$$

Here $\Gamma = c_p/c_v$, where c_p and c_v are the specific-heat capacities at constant pressure and volume, respectively, and c_s is the adiabatic sound speed. If we insert this into Bernoulli's equation (8.52), we obtain

$$\frac{p}{\rho} \frac{\Gamma}{\Gamma - 1} + \frac{1}{2}v^2 = \frac{c_{s,0}^2}{\Gamma - 1}. \tag{8.54}$$

Now, in a polytropic gas, pressure and density are related via

$$p = p_0 \left(\frac{\rho}{\rho_0}\right)^\Gamma. \tag{8.55}$$

We can thus eliminate the pressure from Eq. (8.54) and find

$$\rho = \rho_0 \left(1 - \frac{\Gamma - 1}{2} \frac{v^2}{c_{s,0}^2}\right)^{1/(\Gamma - 1)}, \tag{8.56}$$

where we have used $c_{s,0}^2 = \Gamma p_0/\rho_0$. Similarly, we can start from Eq. (8.54) and solve it for v as a function of p. We need only to substitute the density via the polytropic relationship

$$\rho = \rho_0 (p/p_0)^{1/\Gamma}. \tag{8.57}$$

So, we can write Eq. (8.54)

$$v^2 = \frac{2\Gamma}{\Gamma - 1} \frac{p_0}{\rho_0} \left[1 - \left(\frac{p}{p_0}\right)^{(\Gamma - 1)/\Gamma}\right]. \tag{8.58}$$

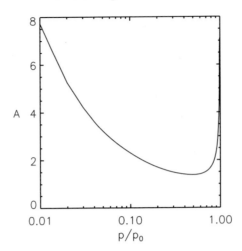

Figure 8.13   Cross-sectional area A as a function of pressure p/p_0 for a one-dimensional, adiabatic jet as given in Eq. (8.60). Here we chose $\dot{M} = 1$ and $\rho_0 = 1$. The units of A are irrelevant here.

By combining Eqs. (8.58) and (8.56), we find

$$\rho v = \left(\frac{p}{p_0}\right)^{1/\Gamma} \left\{\frac{2\Gamma}{\Gamma-1}\rho_0 p_0 \left[1 - \left(\frac{p}{p_0}\right)^{(\Gamma-1)/\Gamma}\right]\right\}^{1/2}. \tag{8.59}$$

Using $A = \dot{M}/(\rho v)$, we can express the cross-sectional area in terms of the pressure

$$A = \dot{M} \left(\frac{p}{p_0}\right)^{1/\Gamma} \left\{\frac{2\Gamma}{\Gamma-1}\rho_0 p_0 \left[1 - \left(\frac{p}{p_0}\right)^{(\Gamma-1)/\Gamma}\right]\right\}^{-1/2}. \tag{8.60}$$

In Fig. 8.13, we plot A as a function of p/p_0 assuming $\Gamma = 5/3$. We see that the cross-sectional area passes through a minimum, which occurs where p is roughly half of the so-called *stagnation pressure*, p_0.

This minimum occurs when the jet becomes supersonic, that is, at the point where the jet speed equals the local sound speed. If a jet accelerates to supersonic speeds, the flow converges as if it was flowing through a nozzle. This phenomenon is called a *De Laval nozzle*. It can be shown that in the subsonic portion of the flow, the pressure and density are approximately constant. So, according to $\rho A v =$const., the area decreases inversely proportional to the velocity, that is, $A \propto 1/v$. In the supersonic regime, however, the velocity is almost constant. This implies that the area increases as $A \propto \rho^{-1} \propto p^{-3/5}$ for $\Gamma = 5/3$. In many galaxies, the pressure profile scales roughly with the inverse of the square of the radial coordinate, r, measured from the center of the galaxy, that is, $p \propto r^{-2}$. As a result, the opening angle decreases as $A^{1/2}r^{-1} \propto r^{-2/5}$. Thus, a supersonic jet can be collimated as the pressure diminishes, regardless of the expansion of its cross-sectional area.

Earlier, we established that many jets are relativistic. So, we can attempt to compare our simple nonrelativistic model with a fully relativistic model. We assume that the fluid inside the jet is an ultrarelativistic plasma with an equation of state $p = u/3 \propto n^{4/3}$, where u is the proper internal energy per unit proper volume and n the number of particles per unit proper volume. According to relativistic fluid dynamics, the relativistic power L can be written

$$L = 4p\gamma^2 vA = \text{const.,} \tag{8.61}$$

where γ is the bulk Lorentz factor of the fluid. It follows from the continuity equation that

$$n\gamma vA = \text{const.} \tag{8.62}$$

Combining these two relations implies that

$$A \propto \frac{\gamma^2}{v}. \tag{8.63}$$

As in the nonrelativistic case, the area A is again minimized at the point where the flow is transonic. This model was first applied to extragalactic jets by Roger Blandford and Martin Rees in 1974 and was termed the *twin-exhaust model*.

Real astrophysical jets do not satisfy most of the assumptions made previously. This is because in a real jet, there is a velocity shear across the jet, which probably gives rise to turbulence. Furthermore, the jet material suffers from radiative losses. Internal shocks cause dissipation and the acceleration of relativistic particles. Not even the mass density flux is constant along the jet because the jets entrain gas from the surrounding environment. Most important, the twin-exhaust model runs into a severe problem: very well-resolved radio images of the core regions of jets reveal that the collimation occurs on very small scales. Consequently, the pressure of the collimating gas must be so high that the gas suffers catastrophic radiative cooling. Even if it was reheated somehow, the radiation associated with this cooling would be visible, and its absence creates a problem.

Instead, the collimation of the jets is thought to be produced by magnetic confinement. The essence of the so-called *pinch effect* is the following: if the jets represent a strong electrical current in the direction of their flow, by Ampere's law, they produce a toroidal magnetic field, B_ϕ, around the jets. This magnetic field gives rise to a magnetic pressure, $p_{\text{mag}} \sim B_\phi^2/8\pi$, which decreases with distance from the jet axis. In other words, it produces a pressure gradient that collimates the flow into thin jets.

In addition to the problem of collimation, another puzzle is how the jets, once launched, survive the large distances that are observed. Stability analyses of hydrodynamical jets revealed that Kelvin–Helmholtz instabilities (see Section 2.9.3) and other instabilities such as so-called kink instabilities quickly disrupt the jet. The major shortcoming of all of these analytical models is that they can only compute the linear growth rates of various instabilities under initially regular conditions; if taken at face value, probably no analytically computed jet could propagate stably for 100 kpc or more, yet of course many such extended extragalactic sources are observed. Therefore, numerical simulations are required to explore the nonlinear effects that can provide saturation of the linear instabilities.

In the recent past, advances in the theory of relativistic jets have been made using sophisticated hydrodynamical computer simulations. These computations have been able to reproduce many morphological features of observed jets. For example, in highly supersonic jets, the shocked jet material is deflected backward to form a sort of *cocoon*, which envelops the jet (see, e.g., Fig. 7.16). At kiloparsec scales, the implications of relativistic flow speeds and/or relativistic internal energies for the morphology and dynamics of jets have been the subject of a number of articles in recent years. Beams with large internal energies show little internal structure and relatively smooth cocoons, allowing the terminal shock (the hot spot in the radio maps) to remain well defined during the evolution. Their morphologies resemble those observed in naked quasar jets such as 3C273. Figure 8.14 shows a snapshot of the time evolution of a light, relativistic jet with large internal energy.

Highly supersonic models, in which kinematic relativistic effects dominate, have extended, overpressured cocoons. These overpressured cocoons can help to confine the jets during the early stages of their evolution and even cause their deflection when propagating through nonhomogeneous environments. The cocoon overpressure causes the formation of a series of oblique shocks within the beam in which the synchrotron emission is enhanced. In long-term simulations, the evolution is dominated by strong deceleration phases, during which large lobes of jet material start to inflate around the jet's head. These simulations reproduce some properties observed in powerful extragalactic radio jets (lobe inflation, hot-spot advance speeds and pressures, deceleration of the beam flow along the jet) and can help to constrain the values of basic parameters (such as the particle density and the flow speed) in the jets of real sources.

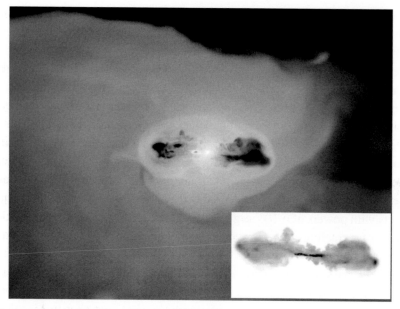

Figure 8.14   Snapshots of the gas density in a simulation of a jet in a cluster of galaxies. One can see how the jet drives shocks into the intracluster medium. The inset shows the simulated radio emission from this jet. The morphology is very similar to a FR II source such as Cygnus A, shown in Fig. 8.1. Courtesy of Sebastian Heinz and Marcus Brüggen.

What produces the jets?

Any disk structure near a black hole provides a pair of preferred directions along the rotation axis. Presently, the most promising idea for extracting energy from an accretion flow is via magnetic fields because fairly strong magnetic fields are likely to develop in all accretion flows. The advantage of magnetic acceleration mechanisms is that they can simultaneously produce relativistic velocities, narrow jets, and large momentum fluxes.

An enormous number of variants on magnetically accelerated jet models have been put forward over the past quarter century, and we cannot even begin to summarize them here. The majority of them fundamentally rely on either extracting energy and angular momentum through magnetic fields anchored in the disk[8] or extracting the spin energy of the black hole itself, through magnetic fields threading its horizon.[9]

How exactly does the magnetic field tap the energy of a rotating black hole? Even though recent magnetohydrodynamic simulations have begun to shed some

[8] Blandford, R. D. and Payne, D. G. (1982). *Monthly Notices of the Royal Astronomical Society*, **199**, 883.
[9] Blandford, R. D. and Znajek, R. L. (1977). *Monthly Notices of the Royal Astronomical Society*, **179**, 433.

light on this, the exact mechanism is far from clear. Moreover, a detailed study of the physics involved would go beyond the scope of this text. Nonetheless, we try to convey the basic ideas and make some rough estimates here.

The mechanism to extract energy from the spin energy of the black hole relies on a remarkable analogy between black holes and electrically charged conductors. The solution of Maxwell's equations in a Schwarzschild metric shows that the electric field lines cross the event horizon of the black hole perpendicularly. In this respect, the event horizon of a black hole behaves like an electrically conducting surface. Suppose we embed a black hole rotating at an angular velocity, ω, in a uniform magnetic field B. Because the event horizon acts like a conducting surface, unipolar induction produces an electric potential difference. This potential difference can drive jets with powers of $\sim 10^{45}$ erg s^{-1} for fast-rotating, supermassive black holes. This way of extracting spin energy from the black hole is called the Blandford and Znajek mechanism.

The other class of jet-formation mechanisms accelerates the jets via magnetic fields that are anchored in the accretion disk. These mechanisms work, crudely speaking, as follows: if the gas is sufficiently ionized such that the assumptions of ideal magnetohydrodynamics are valid, gas is effectively tied to the magnetic field lines. If magnetic field lines stick out of the disk, then the gas above and below the disk is forced to co-rotate with the field. Because the Lorentz force acts only perpendicular to the field lines, the gas can only move freely along the field lines, like beads on a wire. At the foot point of the field line, the inward force of gravity just balances the centrifugal force because we assume a Keplerian disk. Along the field line, the centrifugal force increases with distance from the axis. When the component of the centrifugal force along the field line exceeds the gravitational pull, the gas tied to the field line is accelerated outward. In this way the gas is accelerated. At a certain height above the disk, the field lines no longer co-rotate with the disk because the inertia of the gas forces the magnetic field lines to twist. The field lines are subsequently wrapped around the rotation axis, forming a spiral structure that looks like a twisted rope. The loops in the spiral form a predominantly toroidal field whose tension forces point to the axis and thus collimate the gas. A good introduction to the physics of magnetohydrodynamical jet production mechanisms can be found in the review by Spruit.[10]

Radiation from jets

The radio emission from the jets is presumably synchrotron radiation, which comes from relativistic electrons that gyrate in a magnetic field. The fact that synchrotron

---

[10] Spruit, H. C. (1996). *Evolutionary Processes in Binary Stars*, NATO ASIC Proc. 477. Kluwer: Dordrecht, p. 249.

emission does not provide spectral lines means that even the simplest facts about the properties of jets, such as their velocities and compositions, have remained controversial. The emission in the jets is often linearly polarized, as one would expect for synchrotron radiation in a relatively ordered magnetic field. The polarization direction can be used to infer the direction of the magnetic field projected onto the plane of the sky. Such measurements have revealed that the magnetic field is predominantly perpendicular to the jet axis at larger distances. This can be explained by the conservation of magnetic flux. Even if, initially, the magnetic field in the jet is tangled, conservation of magnetic flux causes a different scaling of the field components parallel and perpendicular to the jet axis. The field component parallel to the jet axis is inversely proportional to the cross section of the jet, which is given by $A \propto (r\theta)^2$, where r is the distance along the jet and θ the opening angle of the jet. Hence, if the magnetic flux is conserved, we have

$$B_{\parallel} \propto (r\theta)^{-2}. \tag{8.64}$$

Similarly, the field component perpendicular to the jet axis scales like the inverse of the circumference of the jet, that is, like

$$B_{\perp} \propto r^{-1}. \tag{8.65}$$

Eventually, at large r, one would expect the field to be predominantly perpendicular to the jet axis. However, there are some powerful jets that show magnetic fields parallel to their axes, even at large distances. They require a more sophisticated model for the evolution of the magnetic field along their axis.

Although there are exceptions to this rule, it is well established that, in general, the magnetic field in a FR II jet remains aligned with the jet along most of its length, whereas in a FR I jet the magnetic field is predominantly transverse on multikiloparsec scales. The detailed magnetic field patterns are best explained if the jets in these sources consist of a narrow "spine" of relativistic flow with a predominantly transverse magnetic field, surrounded by a slower moving "sheath," probably contaminated by entrained material where the magnetic field is stretched into a predominantly longitudinal configuration.

The prime example of an extended radio source is Cygnus A, which is the second-brightest radio source on the northern sky with a luminosity of $L \sim 10^{11} L_{\odot}$ (see Fig. 8.1). The radio emission comes from two *radio lobes* that are outside the optical limits of the galaxy and extend to 0.1–0.5 Mpc. The density in these lobes is very low ($n_e \sim 10^{-4}$–10^{-3} cm^{-3}). Because of the low density, the electrons collide so infrequently that they never reach a thermal Maxwell–Boltzmann distribution. As a result, one can have a large number of high-energy electrons that lose their energy by synchrotron radiation rather than by collisions with low-energy electrons. The magnetic fields in the lobes appear to range between 10 and 100 μG.

The spectrum of the radio emission follows a power law, $F_\nu \propto \nu^{-\alpha}$, where the power-law index, α, is typically around 0.6. Clearly, the form of the electromagnetic spectrum depends on the energy spectrum of the particles that produce the radiation. Assuming that this radiation is indeed synchrotron radiation, we can work out the underlying energy spectrum of the electrons. Supposing that the electron energies follow a power law of index ζ, $n(E) \propto E^{-\zeta}$, we showed in Eq. (3.73) that the spectral index of the source, α, is given by

$$\alpha = \frac{(\zeta - 1)}{2}. \tag{8.66}$$

This is a useful equation that permits us to infer the relativistic electron spectrum from synchrotron emission of a distant source.

Back to the synchrotron emission in our radio lobes. Observations show that the power-law steepens (i.e., α increases) in the direction from the hot spots to the center of the galaxy. This implies that the relativistic electrons that are freshly accelerated at the shock in the hot spot lose their energy by emitting synchrotron radiation in the magnetic field of the lobe. The electrons are said to *age* in the magnetic field (see Fig. 3.12). The polarization direction reveals that, in the radio lobes, the magnetic field is predominantly parallel to the axis and somewhat chaotic in the hot spots. The energy produced by an electron gyrating in a magnetic field is, in the relativistic case, $\beta \approx 1$, given by (see Eq. [3.162])

$$\left(\frac{dW}{dt}\right)_{\text{sync}} = \frac{4}{3}\sigma_T c \left(\frac{E}{m_e c^2}\right)^2 \left(\frac{B^2}{8\pi}\right) = 1.1 \cdot 10^{-15} \text{ erg s}^{-1}\gamma^2 \left(\frac{B}{G}\right)^2. \tag{8.67}$$

The associated loss time (see Eq. [3.163]) is given by

$$\tau_{\text{sync}} = \frac{E}{(dW/dt)_{\text{sync}}} = \frac{6\pi m_e^2 c^3}{\sigma_T} B^{-2} E^{-1} = 635 \text{ s} \left(\frac{B}{G}\right)^{-2} \left(\frac{E}{\text{erg}}\right)^{-1}. \tag{8.68}$$

So, the larger the magnetic fields and the larger the energies, the quicker the electrons lose their energy through synchrotron radiation.

Synchrotron radiation has a characteristic frequency

$$\nu_c = \frac{\omega_c}{2\pi} = \frac{eB}{2\pi m_e c}\left(\frac{E}{m_e c^2}\right)^2, \tag{8.69}$$

where we used Eq. (3.169) for the characteristic synchrotron frequency. This allows us to rewrite the synchrotron loss time in Eq. (8.68) as

$$\tau_{\text{sync}} = \frac{6(\pi m_e c e)^{1/2}}{2^{1/2}\sigma_T} B^{-3/2}\nu_c^{-1/2} \tag{8.70}$$

Let us now work out the total energy in the lobes. In a source of radio luminosity L, the energy in the electrons must be at least Lt_{sync}:

$$U_e = Lt_{\text{sync}} = L\frac{6(\pi m_e c e)^{1/2}}{2^{1/2}\sigma_T}B^{-3/2}v_c^{-1/2}. \tag{8.71}$$

For the total energy, we should add the energy of the magnetic field, which is $U_{\text{mag}} \sim VB^2/8\pi$, where V is the volume of the source. To get a feeling for the magnitude of this energy, we can express the field energy, plugging in typical values for radio lobes, as $U_{\text{mag}} = 1.2 \cdot 10^{63}(V/\text{kpc}^3)(B/\text{G})^2$ erg.

Finally, we need to add the energy of the protons (or positrons). It is not clear whether the protons are accelerated to the same energies as the electrons. So, we parametrize our ignorance and put $U_p = kU_e$, where k is an unknown factor. It is difficult to determine k directly. A clue may come from high-energy particles that impact Earth and are known as cosmic rays. In cosmic rays, protons carry about 100 times as much energy as the electrons. So a typical value for k may be around 100. For the total energy of the source, we can thus write

$$U = (1 + k)U_e + U_{\text{mag}}. \tag{8.72}$$

Now, it is very difficult to measure the magnetic field directly. By looking at Eq. (8.72), we can note that the first term is a decreasing function of B, whereas the second term is an increasing function of B. Consequently, there must be a minimum in the total energy to produce the observed luminosity L. The magnetic field that minimizes this energy is found by setting its derivative with respect to B to zero (i.e., $dU/dB \equiv 0$), which gives

$$U_{\text{mag}} = \frac{3}{4}(1 + k)U_e. \tag{8.73}$$

This can be re-arranged to estimate a minimum-energy magnetic field,

$$B_{\text{min}} \sim 1300\ G\ L^{2/7}\ v_{\text{max}}^{2/7}\ V^{-2/7}\ (1 + k)^{2/7}, \tag{8.74}$$

where in the last equation L is in erg s^{-1}, v_{max} in Hz, and V in cm^3. At this minimum field, the energy in the particles is nearly equal to the energy in the magnetic field. It is a simple exercise to show that at B_{min}

$$\left(\frac{U_{\text{particles}}}{U_{\text{mag}}}\right)_{B_{\text{min}}} = \frac{4}{3}. \tag{8.75}$$

So, alternatively, one could estimate the magnetic field in the lobes by postulating that the energy was equally divided between particles and magnetic fields. This is called *equipartition of energy* and obviously leads to nearly the same result.

Table 8.3   *Features of the radio-emitting parts*
of radio galaxies.

Parameter	Core	Jet	Hot spot	Lobe
Size (kpc)	10^{-3}	$2 \cdot 10^3$	5	$50\text{--}10^3$
B (Gauss)	N/A	$\leq 10^{-3}$	10^{-3}	10^{-5}
n_e cm^{-3}	N/A	$10^2\text{--}10^{-5}$	$\leq 10^{-2}$	$\leq 10^{-4}$
Polarization (%)	≤ 2	0–60	15	0–60
Spectral index	0.0	0.6	0.6	0.9
v/c	~ 1	10^{-1}	10^{-3}	10^{-3}

For a typical FR II radio source, such as Cygnus A, the minimum energy magnetic field is $5 \cdot 10^{-5}$ G and the required energy 10^{59} erg. At the observed luminosity of $5 \cdot 10^{44}$ erg s^{-1}, the synchrotron lifetime is $t_{\text{sync}} \sim 10^4$ yr. This time is much less than the light travel time across a lobe, $t \sim 50$ kpc$/c \sim 1.6 \cdot 10^5$ yr. Consequently, the electrons in the lobes cannot have acquired all their energy from the jets that feed them. Some re-acceleration of electrons must have taken place inside the lobes. As the spectrum steepens away from the hot spots in FR II sources, this suggests that the re-acceleration takes place in the hot spots, from where the fresh electrons diffuse into the lobe. The re-acceleration probably occurs in the termination shock by a mechanism called Fermi acceleration (see Section 2.8).

Surprisingly, when the latest X-ray observatories such as CHANDRA made spatially resolved observations of the jets, it was found that, on top of the radio emission, the jets emit a significant amount of X-rays. Moreover, the X-ray and radio emissions were well correlated, which implies that both types of radiation are produced in the same location. The explanation lies in another emission mechanism for relativistic electrons, which is Compton scattering (see Section 3.8.3). The same electrons that are responsible for the radio emission produce X-rays via Compton scattering. In Compton scattering, low-energy photons scatter off the relativistic electrons and are thus boosted to higher energies. If a photon has a frequency ν, it acquires a frequency $\nu' \sim \gamma^2 \nu$, see Eq. (3.184), after scattering off an electron with a Lorentz factor of γ. The characteristic electron Lorentz factors (not to be confused with bulk Lorentz factors) in jets can be as high as 10^4. Hence, low-frequency radio photons can be converted to high-frequency X-ray photons. This effect is called *synchrotron self-Comptonization*. The main features of the radio-emitting parts of radio galaxies are summarized in Table 8.3.

8.4.4   *Broad-line region*

The broad-line region is the origin of broad, permitted emission lines in the optical spectra of quasars and Seyfert1s. Photoionization is the most likely source of excitation for the emission lines in AGN. This is mainly because there is nothing to prevent the radiation from the strong, central continuum source from interacting with the surrounding gas in the broad-line region. Further evidence comes from the observed correlations between the variations in the line and continuum emission in many Seyfert1s. The line widths of the emission lines are several thousand km s^{-1}(see the box about width of spectral lines in this chapter). This line width cannot be due to the thermal motions of the ions because this would require temperatures of $k_B T \sim m_p (\Delta v)^2/2$ or $T \sim 10^{12}$ K. At this temperature, one would not expect any emission lines as all atoms would be completely ionized. If the line widths are not caused by thermal motions, then bulk motions are necessary to explain the large widths of the emission lines, that is, the lines are broadened by the Doppler effect of the moving gas. The large line width indicates that the gas has velocities of up to several 10 000 km s^{-1}. If these velocities are caused by motions in a gravitational field of a central object, this object must be very massive, and the broad-line region must be close to the nucleus. The strength of certain spectral lines can be used to deduce the density of gas in the broad-line region, and we now briefly sketch how this is done.

Electron densities are determined from the intensity ratio of spectral lines. For a density determination, one needs spectral lines that arise from the decay of two closely spaced excited states to a common lower state. An often-used example for such a line pair occurs in the sulphur ion S II with wavelengths of 6717 nm and 6713 nm. The basic principle of the density determination can be demonstrated with a simple two-level atom that produces a line by decaying from level 2 to level 1. The two levels are separated by an energy gap ΔE. The emissivity of the line is given by (see Eq. [3.78])

$$j_{21} = n_2 A_{21} \frac{h\nu_{21}}{4\pi}, \tag{8.76}$$

where n_2 is the number density of atoms in level 2, A_{21} is the Einstein coefficient for a spontaneous radiative transition from 2 to 1, and $h\nu_{21}$ is the energy of a photon resulting from this transition. The rate at which level 2 is populated by collisional excitation is $r_{12} = \langle \sigma_{12} v \rangle$, where σ_{12} is the velocity-dependent cross section for collisional excitation of level 2, and the average, denoted by $\langle \rangle$, is over the electron velocity distribution.

The mean free path for an electron to collide with an ion is (see Eq. [2.2])

$$l = \frac{1}{n_i \sigma}, \tag{8.77}$$

where n_i is the number density of ions and σ their cross section. The mean time between collisions between an electron and an ion is thus $t = l/v = 1/n_i \sigma v$. Hence, the rate at which collisions occur per unit volume is $n_e/t = n_e n_i \sigma v$. Because σ is velocity dependent, the mean rate is found by averaging over the electron velocity distribution.

The collisional excitation is balanced by the rate at which the level is depopulated by subsequent collisions and by spontaneous radiative transitions to level 1, that is,

$$\langle \sigma_{12} v \rangle n_e n_1 = n_2 A_{21} + \langle \sigma_{21} v \rangle n_e n_2, \tag{8.78}$$

where σ_{21} is the cross section for collisional de-excitation from level 2. This equation can be solved for n_2 and inserted into Eq. (8.76) to yield

$$j_{21} = n_e n_i \langle \sigma_{12} v \rangle \frac{A_{21}}{A_{21} + n_e \langle \sigma_{21} v \rangle} \frac{h \nu_{21}}{4\pi}. \tag{8.79}$$

The principle of detailed balance states that

$$\langle \sigma_{12} v \rangle = \langle \sigma_{21} v \rangle \frac{g_2}{g_1} \exp\left(-\frac{\Delta E}{k_B T_e}\right), \tag{8.80}$$

where g_1 and g_2 are the statistical weights of levels 1 and 2, respectively, and T_e is the electron temperature. If the electron density is low, the radiative de-excitation rate is much higher than the collisional de-excitation rate. This implies that all collisional excitations immediately lead to radiative de-excitations. In this case, $n_e \langle \sigma_{21} v \rangle \ll A_{21}$, and we can simplify Eq. (8.79) to

$$j_{21} \approx n_e n_1 \langle \sigma_{12} v \rangle \frac{h \nu_{21}}{4\pi} = n_e n_1 r_{12} \frac{h \nu_{21}}{4\pi}. \tag{8.81}$$

This equation states that $j \propto n^2$. In the opposite limit of high electron densities, $n_e \langle \sigma 21 v \rangle \gg A_{21}$, and Eq. (8.79) becomes

$$j_{21} \approx n_e n_1 A_{21} \frac{\langle \sigma_{12} v \rangle}{n_e \langle \sigma_{21} v \rangle} \frac{h \nu_{21}}{4\pi} = n_1 A_{21} \frac{h \nu_{21}}{4\pi} \frac{g_2}{g_1} \exp\left(-\frac{\Delta E}{k_B T_e}\right), \tag{8.82}$$

which shows that $j \propto n$ in this regime.

One can now define a critical density, which defines the transition between the low- and high-density cases:

$$n_{\text{crit}} = \frac{A_{21}}{\langle \sigma_{21} v \rangle} = \frac{A_{21}}{r_{21}}. \tag{8.83}$$

At the critical density, the line emissivity goes from $j \propto n^2$ to $j \propto n$. If the critical density for a pair of lines is such that one line is still in the low-density limit while the other one lies in the high-density limit, the ratio of their emissivities is

$$\frac{j_{\lambda_1}}{j_{\lambda_2}} \propto \frac{n_e}{n_e^2} \propto n_e^{-1}. \tag{8.84}$$

Thus, the flux ratio of these lines depends on the electron number density, and a measurement of the flux ratio provides a measurement of n_e.

This method reveals that in broad-line regions the electron number density is around $n_e \sim 10^9$–10^{10} cm^{-3}. The volume-filling factor of the broad-line region is very low (10^{-6} to 10^{-5}) and the total corresponding mass only a few solar masses.

A very powerful technique has been invented to probe the dynamics and the geometry of the broad-line region. It works roughly as follows: it is believed that the gas in the broad-line region is illuminated and photoionized by the high-energy photons from the central engine. When the ionizing flux varies, so does the amount of photoionization of the broad-line region gas. By measuring the time delay between the continuum fluctuations and the emission line response, one can obtain a measure of the light travel time from the black hole to the broad-line region. Effectively, the gas in the broad-line region recombines instantly, so it is only the light travel time that causes the delay. Typical time delays range from 1 day to several hundred days for low-luminosity AGN. The idea of echo mapping (also known as *reverberation mapping*) was invented in 1982, but it took until the 1990s to get data good enough to use it in practice. In addition, the response of different parts of the line profile provides a measure of the velocity flow of the echoing gas. Thus, structure at microarcsecond scales can be measured. For example, for the Seyfert 1 galaxy NGC 5548, an observed time delay of 20 days for the H$_\beta$ line corresponds to an angular resolution of 50 microarcsec.

From these observations, the following picture for the broad-line region has emerged: clouds of thick gas with electron number densities around $n_e \sim 10^9$–10^{10} cm^{-3} are moving with $v \sim 10^4$ km s^{-1} around the black hole and extend to ~ 0.1–1 pc (see Fig. 8.7). These clouds are presumably heated and ionized by the continuum radiation of the central source. By comparing the continuum emission to the line emission from the broad-line region, one obtains the fraction of ionizing photons that are absorbed by the broad-line region clouds. This fraction turns out to be of the order of 10%. Assuming that the clouds are optically thick, this fraction is also the fraction of the solid angle (as subtended from the source) that these clouds occupy. From the filling factor, the typical distance from the source, and the occupied solid angle, one can estimate the characteristic sizes of these clouds. Typical sizes turn out to be $\sim 10^{11}$ cm or slightly more than a solar radius.

However, the nature of the broad-line region clouds remains unclear. Because of their high temperatures, one would expect that they evaporate on relatively short timescales. Thus, the clouds either need to be resupplied at a steady rate or they need to be confined. This confinement could be achieved by the pressure of an ambient medium or magnetic fields, or maybe even gravity. It is also not clear what the overall geometry of the broad-line region is, in particular whether it is spherical or flattened similar to an accretion disk.

Alternatives to scenarios of discrete emitters, several broad-line region models, have been suggested that involve accretion disk winds, hydrodynamical flows, and the shock interaction between a disk wind and the wind of the central continuum source. See Fig. 8.7 for an illustration of the broad-line region.

8.4.5   *Narrow-line region*

In addition to the broad emission lines, most AGN show much narrower emission lines. These lines have widths of 200–2000 km s^{-1}, which, by the way, is still larger than a typical line width in ordinary galaxies. In analogy with the broad-line region, the region where these lines are produced is the called narrow-line region (see Fig. 8.7). The narrow emission lines are forbidden lines, which means that quantum-mechanical selection rules disallow simple radiative transitions from the excited state to a lower state (see Section 3.6). The corresponding transition probabilities are low and the life times of the excited states long. The timescale in forbidden transitions can be of the order of seconds, whereas for allowed transition it is of the order of 10^{-8} s. The occurrence of forbidden lines implies that the atoms do not suffer frequent collision that would bump the electrons into other states from which they could make an allowed transition. Forbidden lines are thus indicative of low number densities (10^2–10^4 cm^{-3}). The narrow-line region is spatially resolved in many Seyfert galaxies. Its size ranges from 50 pc to 200 kpc. With a volume-filling factor of around 10^{-6}, the narrow-line region contains a total mass of several million solar masses. The temperature of the gas ranges from 10 000 to 25 000 K. The morphology is axisymmetric rather than spherically symmetric. It appears as if the ionization of the narrow-line region by the radiation from the nucleus is not isotropic but is directed along two cones that are perpendicular to the plane of the disk. Thus, the narrow-line region consists of clouds of thin gas with electron number densities of $n_e \leq 10^5$ cm^{-3} that are moving with $v \sim 10^2$–10^3 km s^{-1} around the black hole and extend to several pc.

8.4.6   *Molecular torus*

The spectra of quasars show a bump in the infrared that most likely originates from the thermal emission of a torus of dust or molecular gas (see Fig. 8.7). The dust in the torus absorbs the high-energy radiation from the black hole and re-emits the energy with a blackbody spectrum at the temperature of the heated dust (20–80 K), which corresponds to a peak wavelength of 60–150 μm. The IR bump has a minimum wavelength of 1 μm or a frequency of $3 \cdot 10^{14}$ Hz, which corresponds to a temperature of $T = h\nu_{max}/2.82k_B = 5100$ K (Eq. [3.66]). This minimum can be explained by the properties of dust. Dust sublimates at temperatures greater than 2000 K, and hence the bump does not go much lower than $\lambda \leq 1\,\mu$m.

We can roughly estimate the size of the dusty torus, assuming that it comes from a thin disk with a surface of $2\pi R_{dust}^2$, so that the Stefan–Boltzmann law yields

$$L_{IR} = 2\pi R_{dust}^2 \sigma_{SB} T_{dust}^4, \tag{8.85}$$

where L_{IR} is the total luminosity in the infrared. For the parameters of quasars one finds

$$R_{dust} = \left(\frac{L_{IR}}{2\pi \sigma_{SB} T_{dust}^4}\right)^{1/2} = 0.4\,\mathrm{pc}\left(\frac{L_{IR}}{10^{16}\mathrm{erg\ s^{-1}}}\right)^{1/2}\left(\frac{T_{dust}}{2000\,\mathrm{K}}\right)^{-2}. \tag{8.86}$$

Typically, the torus has an inner radius of $\sim$1 pc and outer radius of $\sim$50–100 pc.

The torus appears to be larger than the broad-line region and smaller than the narrow-line region, which for certain aspect angles absorbs all radiation from the nucleus. As a result, ionizing radiation escapes anisotropically from the AGN. An example for a dusty torus that has been discovered with the *Hubble Space Telescope* is shown in Fig. 8.15.

8.4.7   *Unified Models*

There is a whole range of ideas to unify the zoo of active galaxies that go by the term *unified models*. Unification schemes of AGN are based on the assumption that AGN belonging to different classes may be intrinsically identical but appear different because we see them from different angles. The differences between the separate classes of AGN are explained mainly by two factors: (i) relativistic beaming and (ii) obscuration. These effects can make AGN look different from different directions.

There are really two main unification schemes: the first one is based on radio observations and explains the difference between quasars and radio galaxies in terms of the orientation of their jet axis with respect to the line of sight. The jets in radio galaxies are weaker than in quasars by a factor of 10–100, which may be explained by relativistic beaming. In this sense, quasars are radio galaxies whose jets have a small angle to our line of sight, whereas what we classify as radio galaxies have jets that lie more in the plane of the sky. This idea has some observational support. Quasars never appear to lie in the plane of the sky. Moreover, some radio galaxies show an extended polarization structure that comes

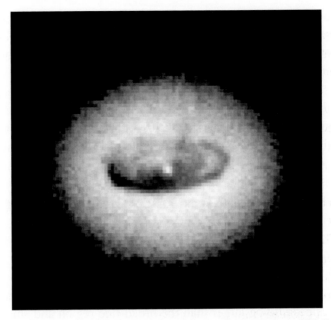

Figure 8.15   This is a *Hubble Space Telescope* image of an 800-light-year-wide disk of dust fueling a massive black hole in the center of galaxy, NGC 4261, located 100 million light-years away in the direction of the constellation Virgo. Courtesy of L. Ferrarese (National Research Council of Canada), H. C. Ford (John Hopkins University), and NASA.

from a reflection nebula. This nebula scatters the light not seen directly and points to a very strong nuclear source that is not observed directly.

The second unification scheme is based on optical and infrared data. Obscuration forms the essence of this orientation unification scheme. As we have seen previously, much evidence, especially in Seyfert galaxies, indicates that most nuclei are surrounded by dusty, massive tori. These tori of gas and dust are optically thick to the ionizing radiation from the nucleus and are thought to girdle the continuum source and the broad-line region. Looking down the axis of the torus, we can see right into the broad-line region and accretion disk. However, when viewed edge-on, the torus blocks our view of the broad-line region and absorbs hard X-rays, and so the AGN looks like a type 2. The tori are much larger than the broad-line region and, therefore, it is conceivable that Seyfert1s and Seyfert2s are actually the same objects.

In this simple unification scenario, Seyfert1s are viewed face-on, whereas in narrow-lined Seyfert2s, the broad emission line region, the soft X-rays, and much of the optical/ultraviolet emission from the accretion disk are hidden by the dust. The spectra of Seyfert2s have less prominent big blue bumps and strong soft X-ray absorption than their type 1 counterparts. Comparisons between type 1 and

2s at low redshift, where most known Seyfert2s reside, are in accordance with these expectations. Their relatively low luminosities result in their optical continua including a significant contribution from the host galaxy within which the central black hole and AGN reside.

Support for this scheme comes from spectropolarimetry of many Seyfert 2 nuclei. They show broad-line emission in polarized light. This polarized emission is interpreted as light that is scattered into the line-of-sight. Also, the narrow-line region configuration is often quite elongated, typically with the highest ionization material in a biconical configuration centered at the nucleus. This is interpreted as the illumination pattern of radiation escaping the torus. Further evidence for obscuration by a torus comes from the observation of the X-ray background. AGN are thought to constitute the bulk of the sources that produce this background emission. This X-ray background is very hard: it rises sharply in the 2- to 10-keV band and peaks near 40 keV. Seyfert 1s, however, have much softer X-ray spectra, and a simple summation of all the unobscured AGN spectra would not produce the observed emission. Instead, it is believed that photo-electric absorption by an obscuring torus hardens the X-ray spectrum.

If the difference between Seyfert 1 and Seyfert 2 objects is the angle of observation and if the difference between Seyfert 1 and quasars is the luminosity, the question arises whether there are luminous counterparts of Seyfert 2s. Recent observations by X-ray satellites such as CHANDRA and XMM-NEWTON have found a number of type 2 quasars, in which the soft X-rays are absorbed by the photoelectric effect of dust.

So how do blazars fit into this picture? Let us remind ourselves of the main features of blazars. Blazars frequently show superluminal motion and no or very weak emission lines. Also, the emission is strongest in the center, highly variable and polarized. An attractive explanation for these features is that in blazars we are actually looking down along the axis of the jet. The jet material that is coming toward us at almost the speed of light is so strongly Doppler boosted that it swamps all features such as emission lines. The synchrotron emission from the jets dominates all other sources of radiation, which explains the strong polarization of blazars (because synchrotron emission is polarized). The strong beaming factor also explains the rapid variability: small changes in velocity or orientation of the jet lead to significant changes in the Doppler factor and thus to the total luminosity.

More ambitious unification schemes aim to explain why some AGN are radio loud while others are radio quiet. It has been suggested that the crucial physical difference may be the spin of the black hole. In this picture, radio-loud AGN have a black hole of high spin at their center, and jets extract the spin energy from the black hole (as discussed in Section 8.4.3). Radio-quiet objects, on the other hand, are conjectured to have low-spin black holes at their centers. They have no jets, and their spectra are produced primarily by the accretion disk.

8.5   Cosmological significance of AGN

8.5.1   *Impact of AGN on structure formation*

AGN are believed to play a crucial role in the formation of structure in the universe. Galaxies and clusters of galaxies form as gas cools inside the potential wells of clouds of dark matter. This cooling proceeds through the emission of radiation that can be observed. In the largest gravitationally bound structures in the universe, clusters of galaxies, the gas is very hot (more than 10 million Kelvin, and often hotter than the center of the Sun). The dilute gas in clusters, the so-called intracluster medium (ICM), radiates X-rays, mainly through thermal *bremsstrahlung* (see Section 3.8). This cooling should lead to a slow condensation of gas in the centers of these clusters where the gas can form stars and galaxies. However, observations have shown that the gas appears to produce fewer stars and less cold gas in the centers than expected. This problem has been termed the *cooling-flow problem*.

Hence, the gas must be heated in some fashion as to keep the ICM at a fairly steady temperature. The prime candidate for heating the gas is through the enormous energy that is deposited in the intracluster gas through AGN. How exactly this heating proceeds and how the activity level of AGN is related to the thermal state of the intracluster gas in which it resides is currently unclear.

The lifetimes of sources in AGN can be estimated in a number of ways. The spatial extent of some radio jets is equal to or more than 1 Mpc. From this length scale, one can infer a minimum age for the activity of these objects. Even if the jets propagate with the speed of light, their age must be more than or equal to 10^7 yr.

Luminous quasars have luminosities of equal to or more than 10^{47} erg s^{-1}. Assuming that this luminosity has been constant over the lifetime of the quasar, one can derive a total energy of the source via

$$E \geq 10^{47} \text{ erg s}^{-1} \cdot 10^7 \text{ yr} \sim 3 \cdot 10^{61} \text{ erg.} \tag{8.87}$$

This is an enormous amount of energy that is likely to have a significant effect on the thermal state of the ambient gas and will thus affect the accretion of gas onto the host galaxy.

8.5.2   *AGN as cosmological probes*

The fact that AGN are such strong sources of light that can be seen from large distances allows astronomers to use the light from AGN to probe the properties of galaxies along the line of sight to them. One way to do this is to measure the absorption in the spectra of the AGN caused by intervening matter. For example, clouds of hydrogen cause absorption, most prominently in the rest wavelength of the Lyman-α line. The absorption in the AGN spectrum takes place at the wavelength of the Lyman-α line shifted by the redshift of the absorbing cloud. Thus, every cloud along the line of sight leads to its own dip in the AGN spectrum, leading to a sequence of absorption lines that are called the Lyman-α forest. If the AGN is at a great distance, the light we receive from it samples a large fraction of the observable universe and carries information about the larger part of the history of the universe. In recent years, the Lyman-α forest has become a valuable source of information to study the evolution of galaxies.

In addition to hydrogen absorption lines, absorption lines by metal ions, such as those from magnesium, iron, and carbon in intervening galaxies, can also be used to trace the chemical history of the gas in galaxies. The physics of absorption-line systems is a whole field by itself, and we refer the reader to Krolik (1999).

Finally, we mention another important use of AGN for cosmology. The light from AGN at high redshifts is frequently gravitationally lensed by massive clusters of galaxies. The lensed image depends on the distribution of matter in the lens plane, and thus studies of lensing can be used to measure the distribution of dark matter in clusters. Finally, lensing can be used to infer cosmological parameters. For more details, the reader is referred to the bibliography at the end of this chapter.

8.6   **Exercises**

8.1 **Black-hole growth**

The growth of a black hole can be described by

$$\frac{dM}{dt} = \frac{L}{c^2 \eta}. \tag{8.88}$$

Show that the growth time can be written as

$$t_{\text{growth}} = \frac{\eta c \sigma_{\text{T}} L_{\text{Edd}}}{4\pi G \mu_e L}. \tag{8.89}$$

8.2 **Eddington luminosity**

(i) In this chapter, we derived the Eddington luminosity for pure hydrogen. Show that for a general composition the Eddington luminosity can be expressed as

$$L_{\text{Edd}} = \frac{4\pi GMc}{\kappa},\tag{8.90}$$

where κ is the absorption coefficient per unit mass (with units cm^2 g^{-1}). What is the Eddington limit for a plasma completely composed of ionized helium?

(ii) The cross section of positrons with photons is also the Thomson cross section. What is the Eddington luminosity for an electron–positron plasma?

(iii) If the luminosity is greater than the Eddington luminosity, gas placed in the system will be ejected. If the material starts from rest at a distance R from the central object, show that the terminal velocity of the material is given by

$$v_t = \left[\frac{2GM}{R} \left(\frac{L\kappa}{4\pi GMc} - 1 \right) \right]^{1/2}.\tag{8.91}$$

8.3 Sphere of influence

Studies of stellar dynamics in the active galaxy NGC 3115[11] have revealed a central black-hole mass of $M_{\text{BH}} \sim 9.2 \cdot 10^8\, M_\odot$. Given that the velocity dispersion is 278 km s^{-1}, compute the radius of the sphere of influence of the central black hole. NGC3115 has a distance of 9.7 Mpc. What angle does the sphere of influence subtend here on Earth?

8.4 Jet angles

Show that the probability, $P(\theta)$, that a jet makes an angle smaller than θ to the line of sight with an observer is given by

$$P(\theta) = 1 - \cos\theta.\tag{8.92}$$

For large Lorentz factors, γ, show that this can be expressed as $P(\theta) \sim 1/2\gamma^2$.

8.5 Superluminal motion

A blob of emission is observed near the nucleus of a quasar. The quasar is known to be at a distance of 18 Mpc. One year later, the same blob has now moved by an angular distance of 0.003 arcsec from the nucleus. How fast does it appear to be moving? At what angle is it moving relative to our line of sight if $\gamma \sim 1000$?

8.6 Emission from jets

The quasar 3C273 features a thin jet, which is about 64 kpc long. This jet emits a continuous spectrum of radiation from the radio to the optical bands. This radiation can be produced by synchrotron radiation or, if the magnetic field is too weak, by inverse Compton scattering with the photons from the Cosmic Microwave Background with an energy density of $u_{\text{CMB}} = aT^4 \sim 4.2 \cdot 10^{-13}$ erg cm$^{-3}\sim 0.26$ eV cm^{-3}. What is the greatest possible age of electrons emitting radiation with a frequency of 10^{15} Hz? What can you deduce about the acceleration of the electrons?

---

[11] Emsellem, E. et al. (1999). *Monthly Notices of the Royal Astronomical Society*, **303**, 495.

8.7 **Emission from radio lobes I**

If the magnetic field in a radio lobe is $B = 10 \ \mu G$ and an electron is emitting synchrotron radiation at 5 GHz, what is the Lorentz factor of the electrons?

8.8 **Emission from radio lobes II**

The energy spectrum of electrons in a radio lobe can be expressed in terms of their Lorentz factors (not bulk Lorentz factors) as $n_e(\gamma) = A\gamma^{-p}$, where γ lies in the range between $0 < \gamma < \gamma_c$, A is a constant, and $p \leq 2$. Here, $n_e(\gamma)$ is the number of electrons per unit volume whose Lorentz factors lie within an interval $d\gamma$ around γ.

(i) If the magnetic field in the lobe is $B = 10 \ \mu G$ and the temperature of the Cosmic Microwave Background is 2.7 K, show that synchrotron radiation losses dominate over inverse Compton losses.

(ii) Show that the luminosity of synchrotron radiation per unit volume is given by

$$\frac{dL}{dV} = \left(\frac{4}{3}\right)\left(\frac{c\sigma_T u_{mag} A\gamma_c^{3-p}}{3-p}\right), \tag{8.93}$$

where u_{mag} is the energy density of the magnetic field and the other constants have their usual meaning.

8.9 **Accretion disks**

Show that for an accretion disk accreting at the Eddington limit, the temperature distribution as a function of radius, r, is given by

$$T(r) = \left(\frac{G^2 M_{BH}^2 m_p}{2\xi \sigma_{SB} \sigma_T c r^3}\right)^{1/4}, \tag{8.94}$$

where ξ is the radiative efficiency, M_{BH} is the mass of the black hole, and all other symbols have their usual meaning. Assuming that the disk extends down to the last stable circular orbit of a nonrotating black hole whose radius is given by

$$R_{ISCO} = \frac{6GM_{BH}}{c^2}, \tag{8.95}$$

show that the maximum temperature of the disk is given by

$$T_{max} = \left(\frac{m_p c^5}{432\xi GM_{BH}\sigma_{SB}\sigma_T}\right)^{1/4}. \tag{8.96}$$

What is the maximum temperature for a black hole with $M_{BH} = 10^8 \ M_\odot$? At what wavelength does the spectrum from this part of the disk peak?

8.10 **Line widths**

Interpret the width of emission lines from the broad-line region as due to the superposition of many Doppler-shifted lines from sources rotating around a central black hole. Assume that the sources are rotating in stable orbits at a distance r around a point-like black hole. Use the virial theorem to infer the mass of the black hole as

$$M_{BH} \approx \frac{3rc^2}{G}\left(\frac{\Delta\lambda}{\lambda}\right)^2, \tag{8.97}$$

where λ is the wavelength of the center of the line, $\Delta\lambda$ the line width, c the speed of light, and G the gravitational constant.

8.7   Further reading

Balbus, S. A. and Hawley, J. F. (1998). Instability, turbulence, and enhanced transport in accretion disks. *Review of Modern Physics*, **70**, 1.

Blandford, R. D., Netzer, H., and Woltjer, L. (1990). *Active Galactic Nuclei*. Saas Fee Advanced Course, 20. Berlin: Swiss Society for Astrophysics and Astronomy/Springer.

Ferrarese, L. and Ford, H. (2005). Supermassive black holes in galactic nuclei: Past, present and future research. *Space Science Reviews*, **116**(3–4), 523–624.

Frank, J., King, A., and Raine, D. (2002). *Accretion Power in Astrophysics*. Cambridge: Cambridge University Press.

Kembhavi, A. K. and Narlikar, J. V. (1999). *Quasars and Active Galactic Nuclei*. Cambridge: Cambridge University Press.

Krolik, J. H. (1999). *Active Galactic Nuclei*. Princeton: Princeton University Press.

Longair, M. (1992). *High Energy Astrophysics*. Cambridge: Cambridge University Press.

Padmanabhan, T. (2002). *Galaxies and Cosmology*, vol. III of *Theoretical Astrophysics*. Cambridge: Cambridge University Press, Chapters 1 and 8.

Peterson, B. M. (1997). *An Introduction to Active Galactic Nuclei*. Cambridge: Cambridge University Press.

Robson, I. (1996). *Active Galactic Nuclei*. Chichester, UK: Wiley.

Spruit, H. C. (1996). Magnetohydrodynamic jets and winds from accretion disks. In *Evolutionary Processes in Binary Stars*. NATO ASI Series C., vol. 477. Dordrecht, The Netherlands: Kluwer, pp. 249–286.

Véron-Cetty, M. P. and Véron, P. (2000). The emission line spectrum of active galactic nuclei and the unifying scheme. *Astronomy and Astrophysics Review*, **10**, 81.

Web resources

Marti, J. M. and Müller, E. (2003). Numerical hydrodynamics in special relativity. Living Reviews: http://www.livingreviews.org/lrr-2003-7.

Appendix A

Some recent high-energy astrophysics instruments

In this appendix, we give a very brief overview of some current observatories and instruments that are particularly relevant for high-energy astrophysics. This list is by no means complete. A synopsis of all missions that have contributed to this broad and diverse subject would go far beyond the scope of this text. Hence, we decided to concentrate on current or recent projects. Even if we restrict ourselves to recent projects, it is virtually impossible to decide whether a given observatory constitutes a high-energy astrophysics observatory or not. For example, the *Hubble Space Telescope* has made a tremendous contribution to the physics of black holes, compact stars, supernovae, AGN, and so forth. Yet, we decided not to include it in this list and instead focus on observatories that operate in wavebands other than the optical, ultraviolet, or infrared wavelengths. This is a somewhat arbitrary decision, and we are aware that our list is similarly arbitrary. Still, we hope that this list and the hyperlinks to further information provide a useful source of information.

A.1   AMANDA and IceCube

The AMANDA-II (Antarctic Muon and Neutrino Detector Array) telescope is a high-energy neutrino detector that consists of 19 strings of photomultiplier modules buried between 1 and 1.5 miles beneath the ice under the geographic south pole. It has been in operation since 2000 and was designed to detect very high-energy neutrinos. It works by detecting the muons that are produced when the neutrino collides with the nucleus of an oxygen atom in the ice. Because the muon is electrically charged, it generates Cherenkov radiation, which is emitted when a charged particle is traveling in excess of the speed of light in matter. The direction of the muon is nearly the same as the neutrino that created it. Therefore, by measuring the direction of the muon, which is relatively easy to do, we know the direction of the neutrino.

IceCube is the extension of the AMANDA detector with a detector volume of 1 km^3. Its assembly was started in January 2005, and it will detect of the order of 10^6 neutrinos over 10 years with energies in the range between 100 GeV and 10 PeV.

More information is available at http://www.amanda.uci.edu/public_info. html.

A.2   ASCA

The Advanced Satellite for Cosmology and Astrophysics, ASCA, was successfully launched on February 20, 1993, and operated successfully until it re-entered the Earth's atmosphere on March 2, 2001, after seven and half years of scientific observations. ASCA was the first satellite to use CCD detectors for X-ray astronomy and the first X-ray mission to combine imaging capability with broad pass band, good spectral resolution, and a large effective area. Some of the scientific highlight of the mission were the following:

1. measurement of broad iron lines from AGN, probing the strong gravity near the central engine;
2. spectroscopy of interacting binaries;
3. detection of nonthermal X-rays from the supernova remnant SN 1006, a site of cosmic ray acceleration; and
4. measurement of the abundances of heavy elements in clusters of galaxies, consistent with Type II supernova origin.

More information is available at http://heasarc.gsfc.nasa.gov/docs/asca/ascagof.html.

A.3   BeppoSAX

BeppoSAX was a major program of the Italian Space Agency with participation from the Netherlands Agency for Aerospace Programs. It was launched in 1996 and operated for 6 years. It was the first X-ray mission with a scientific payload covering more than three decades of energy – from 0.1 to 300 keV – with a relatively large effective area, medium energy resolution, and imaging capabilities in the range of 0.1–10 keV. Its instrumentation consists of proportional counters (0.1–10 keV), scintillation counters (3–120 keV), and phoswich detectors (15–300 keV). Some scientific highlights were the following:

1. first position of GRBs with an accuracy of arc minutes and position determination on a rapid timescale,
2. first X-ray follow-up observations and monitoring of the GRB, and
3. broadband spectroscopy of different classes of X-ray sources.

More information is at available http://www.asdc.asi.it/bepposax.

A.4   CGRO

The Compton Gamma-Ray Observatory was launched on April 5, 1991. CGRO has four instruments that cover an unprecedented six orders of magnitude in energy, from 30 keV to 30 GeV. Over this energy range, CGRO has an improved sensitivity over previous missions of a full order of magnitude. It operated for almost 9 years, and the mission ended on June 4, 2000. It contained four experiments:

1. the Burst and Transient Source Experiment (BATSE), an all-sky monitor, 20–1000 keV;
2. the Oriented Scintillation Spectrometer Experiment (OSSE), 0.05–10 MeV energy range;
3. the Compton Telescope (Comptel), 0.8–30 MeV, capable of imaging 1 steradian; and
4. Energetic Gamma Ray Experiment Telescope (EGRET), 30 MeV–10 GeV.

   Its major highlights were the following:

1. the discovery of an isotropic distribution of the gamma-ray burst events,
2. discovery of blazar AGN as primary source of the highest energy cosmic gamma rays, and
3. discovery of the bursting pulsar.

More information is available at http://cossc.gsfc.nasa.gov/docs/cgro/index.html.

A.5   CHANDRA

NASA's CHANDRA X-ray Observatory, named after the Indian astrophysicist Subrahmanyan Chandrasekhar, was launched and deployed by the space shuttle *Columbia* on July 23, 1999. The combination of high resolution, large collecting area, and sensitivity to higher energy X-rays makes it possible for CHANDRA to study extremely faint sources in crowded fields. CHANDRA's energy range lies between 0.1 and 10 keV. Its spatial resolution of less than 1 arcsecond makes it the most powerful X-ray observatory to date. CHANDRA was boosted into an elliptical high Earth orbit that allows long-duration uninterrupted exposures of celestial objects. CHANDRA observations have revealed the presence of many previously undetected stellar black holes in nearby galaxies. Very long CHANDRA exposures have shown that there may be twice as many active supermassive black holes in the universe as previously thought. Other highlights include these:

1. evidence for shock acceleration in SNRs,
2. detection of X-rays from AGN jets, for example, in Cygnus A, and
3. detection of ghost bubbles in clusters of galaxies.

More information is available at http://chandra.harvard.edu/.

A.6   H.E.S.S.

H.E.S.S. is a system of imaging atmospheric Cherenkov telescopes for the investigation of cosmic gamma rays in the 100-GeV energy range. The name H.E.S.S. stands for High-Energy Stereoscopic System and should also remind us of Victor Hess, who received the Nobel Prize in physics for his discovery of cosmic radiation in 1936. H.E.S.S. is located in Namibia and consists of four telescopes that went into operation in September 2004.

A Cherenkov telescope works as follows: an incident high-energy gamma ray interacts high up in the atmosphere and generates an air shower of secondary particles. The number of shower particles reaches a maximum at about 10 km in height, and the shower dies out deeper in the atmosphere. Because the shower particles move at essentially the speed of light, they emit Cherenkov light, a faint blue light. The Cherenkov light is beamed around the direction of the incident primary particle and illuminates an area on the ground of about 250 m in diameter, often referred to as the Cherenkov light pool. For a primary photon at TeV energy (10^{12} eV), only about 100 photons per m^2 are seen on the ground. They arrive within a very short time interval of a few nanoseconds. A telescope located somewhere within the light pool observes the air shower, provided that its mirror area is large enough to collect enough photons. The image obtained with the telescope shows the track of the air shower, which points back to the celestial object where the original gamma ray originated. The intensity of the image is related to the energy of the gamma ray. With a single telescope providing a single view of a shower, it is difficult to reconstruct the exact geometry of the air shower in space. To achieve this, multiple telescopes are used that view the shower from different points and allow a stereoscopic reconstruction of the shower geometry.

More information is available at http://www.mpi-hd.mpg.de/hfm/HESS/.

A.7   HETE-2

HETE-2 stands for High-Energy Transient Explorer and is a collaboration between the United States, Japan, France, and Italy. Launched in 2000, HETE-2 is designed to detect and localize gamma-ray bursts (GRBs). The suite of instruments onboard allows simultaneous observations of GRBs in soft and medium X-ray and gamma-ray energies. HETE-2 computes the location of the GRB and transmits the coordinates as soon as they are calculated. These coordinates are quickly distributed to ground-based observers to allow detailed studies of the initial phases of GRBs. HETE-2 also performs a survey of the X-ray sky. One of the highlights of HETE-2 may be the discovery of a possible new type of X-ray burst that may be related to gamma-ray bursts.

More information is available at http://heasarc.gsfc.nasa.gov/docs/hete2/.

A.8 INTEGRAL

The International Gamma-Ray Astrophysics Laboratory (INTEGRAL) of the European Space Agency was launched on October 17, 2002. INTEGRAL is producing a complete map of the sky in the high-energy gamma-ray waveband and is performing high spectral and spatial observations in gamma rays. The observatory is also equipped with X-ray and optical detectors to provide simultaneous observations in these wavebands. INTEGRAL is surveying the sky continuously using four instruments: two main gamma-ray instruments dedicated to imaging (IBIS) and spectroscopy (SPI), an X-ray instrument (JEM-X), and an optical monitor (OMC). The mission is also a gamma-ray-burst detector and has been spotting one burst per month in its field of view. INTEGRAL recorded the closest and faintest gamma-ray burst to date, suggesting the existence of an entire population of subenergetic gamma-ray bursts that have been unnoticed so far in the Universe. Other highlights include the following:

1. evidence for a molecular torus around AGN,
2. detection of X-ray emission from the Galactic center, and
3. discovery of new class of X-ray binaries.

More information is available at http://integral.esac.esa.int/integral.html.

A.9 Pierre Auger Observatory

The Pierre Auger Project is an international collaboration with the objective to study the highest energy cosmic rays. This project involves the construction of two detector arrays, each one covering 3000 km^2. One is located in the northern and the other in the southern hemisphere.

The southern site is located in Mendoza Province in western Argentina. A matching site will also be built in southeastern Colorado, providing nearly uniform coverage of the skies in the northern and southern hemispheres.

The Auger Observatory is a hybrid detector that employs two independent methods to detect and study high-energy cosmic rays. One technique is ground based and detects high-energy particles through their interaction with water. The other technique tracks the development of air showers by observing ultraviolet light emitted high in the Earth's atmosphere.

The first detection method uses the observatory's main visible feature – the 1600 water tanks that cover an enormous section of the Pampa and serve as particle detectors. Each 3000-gallon tank, separated from each of its neighbors by 1.5 km, is completely dark inside – except when particles from a cosmic-ray air shower pass through it. These energetic particles are traveling faster than the speed of light in water when they reach the detectors; therefore, their electromagnetic shock waves produce Cherenkov light. Slight differences in the detection times at different tank positions help scientists determine the trajectory of the incoming cosmic ray.

The charged particles in an air shower also interact with atmospheric nitrogen, causing it to emit ultraviolet light via a process called fluorescence, which is invisible to the human eye – but not to the Auger Observatory's optical detectors. The observatory's second detection method uses these detectors to observe the trail of nitrogen fluorescence and track the development of air showers by measuring the brightness of the emitted light.

More information is available at http://www.auger.org/observatory/observatory.html.

A.10   RXTE

The Rossi X-ray Timing Explorer, RXTE, was launched on December 30, 1995. RXTE is designed to facilitate the study of time variability in the emission of X-ray sources with moderate spectral resolution. Timescales from microseconds to months are covered in a broad spectral range from 2 to 250 keV. It is designed for a required lifetime of two years, with a goal of 5 years. Some of its scientific discoveries include the following:

1. discovery of spin periods in low-mass X-ray binaries,
2. detection of X-ray afterglows from gamma-ray bursts,
3. extensive observations of the soft state transition of Cyg X-1, and
4. observations of the bursting pulsar over a broad range of luminosities, providing a stringent test of accretion theories.

More information is available at http://heasarc.gsfc.nasa.gov/docs/xte/.

A.11   Suzaku

Suzaku, a U.S.–Japanese collaboration, is a satellite-based X-ray observatory that was launched on July 10, 2005. It covers the energy range 0.2–600 keV with two instruments, X-ray CCDs (X-ray imaging spectrometer, XIS) and the hard X-ray detector (HXD). Suzaku also carries a third instrument, an X-ray microcalorimeter (X-ray spectrometer, XRS), but the XRS lost all its cryogen before routine scientific observations could begin.

Its three types of instruments are designed to do the following:

- have high X-ray spectral resolution throughout the 0.2- to 10-keV energy band where the bulk of K-shell lines of astrophysically abundant elements (O–Ni) exist,
- image spectroscopy of extended sources using nondispersive spectrometers,
- have a large collecting area for high sensitivity, and

• have very large simultaneous bandwidth to enable disentangling complex, multicomponent spectra.

More information is available at http://suzaku.gsfc.nasa.gov/docs/suzaku/about/overview.html.

A.12 SWIFT

SWIFT is a NASA mission, developed in collaboration with the U.K. and Italy and launched in 2004. The primary scientific objectives are to determine the origin of gamma-ray bursts and to pioneer their use as probes of the early universe. SWIFT is a multiwavelength observatory carrying three instruments: the burst alert telescope (BAT, gamma ray), the X-ray telescope (XRT), and the ultraviolet/optical telescope (UVOT). The SWIFT key characteristics are the rapid response to newly detected gamma-ray bursts and rapid data dissemination. As soon as the BAT discovers a new gamma-ray burst, SWIFT rapidly relays its 1- to 4-arcminute position estimate to the ground and triggers an autonomous spacecraft slew to bring the burst within the field of view of XRT and UVOT to follow up the afterglow. SWIFT is expected to provide redshifts for the bursts and multiwavelength lightcurves for the duration of the afterglow. The BAT will also perform a high-sensitivity hard X-ray sky survey.

More information is available at http://swift.gsfc.nasa.gov/docs/swift/swiftsc.html.

A.13 VLA

The Very Large Array (VLA) is one of the world's most powerful astronomical radio observatories and consists of 27 radio antennas situated in New Mexico. Each antenna is 25 m in diameter. The data from the antennas are combined electronically to form interference patterns. The interference patterns can be converted into maps of the radio sky using Fourier transforms. Thus, the VLA has the resolution of an antenna 36 km across, with the sensitivity of a dish 130 m in diameter. The receivers at the VLA can receive frequencies between 73 MHz to 50 GHz. The VLA has made major contributions in the physics of AGN, in particular jets, and has made many discoveries in all fields of high-energy astrophysics, such as pulsars and gamma-ray bursts.

More information is available at http://www.vla.nrao.edu/.

A.14 Very Long Baseline Interferometry

Very long baseline interferometry (VLBI) is a type of astronomical interferometry used in radio astronomy, in which the data received at each antenna in the array is paired with timing information, usually from a local atomic clock, and then stored for later analysis on magnetic tape or hard disk. At that later time, the data are correlated with data from other antennas similarly recorded to produce the resulting image. The resolution achievable using interferometry is proportional to the distance between the antennas furthest apart in the array. VLBI uses radio dishes spread over different continents. It consists of 10 dishes, each 25 m in diameter, spread from Hawaii to the Virgin Islands. The baseline is thus thousands of kilometers long.

A.15   XMM-NEWTON

The European Space Agency's X-ray Multi-Mirror Mission (XMM-NEWTON) was launched on December 10, 1999. It carries high-throughput X-ray telescopes with an unprecedented effective area and an optical monitor, the first flown on a X-ray observatory. The large collecting area and ability to make long uninterrupted exposures provide highly sensitive observations. XMM-NEWTON carries two cameras and one reflection grating spectrometer. Some scientific highlights are these:

1. evidence for a molecular torus around AGN,
2. detection of heavy elements in the spectrum of the afterglow of gamma-ray bursts, and
3. mapping of elements in SNRs.

More information is available at http://sci.esa.int/xmm.

Appendix B

Physical constants

Physical quantity	Symbol	Value	Units
Boltzmann constant	k_B	$1.3807 \cdot 10^{-16}$	erg/deg (K)
Elementary charge	e	$4.8032 \cdot 10^{-10}$	statcoulomb (statcoul)
Electron mass	m_e	$9.1094 \cdot 10^{-28}$	g
Proton mass	m_p	$1.6726 \cdot 10^{-24}$	g
Gravitational constant	G	$6.6726 \cdot 10^{-8}$	dyne cm^2/g^2
Planck constant	h	$6.6261 \cdot 10^{-27}$	erg sec
	$\hbar = h/2\pi$	$1.0546 \cdot 10^{-27}$	erg sec
Speed of light in vacuum	c	$2.9979 \cdot 10^{10}$	cm/sec
Atomic mass unit	m_u	$1.6605 \cdot 10^{-24}$	g
Standard temperature	T_0	273.15	deg (K)
Proton/electron mass ratio	m_p/m_e	$1.8362 \cdot 10^3$	
Electron charge/mass ratio	e/m_e	$5.2728 \cdot 10^{17}$	statcoul/g
Rydberg constant	$R_\infty = 2\pi^2 m_e e^4/ch^3$	$1.0974 \cdot 10^5$	cm^{-1}
Bohr radius	$a_0 = \hbar^2/m_e e^2$	$5.2918 \cdot 10^{-9}$	cm
Atomic cross section	πa_0^2	$8.7974 \cdot 10^{-17}$	cm^2
Classical electron radius	$r_e = e^2/m_e c^2$	$2.8179 \cdot 10^{-13}$	cm
Thomson cross section	$(8\pi/3)r_e^2$	$6.6525 \cdot 10^{-25}$	cm^2
Compton wavelength of	$h/m_e c$	$2.4263 \cdot 10^{-10}$	cm
electron	$\hbar/m_e c$	$3.8616 \cdot 10^{-11}$	cm
Fine-structure constant	$\alpha = e^2/\hbar c$	$7.2974 \cdot 10^{-3}$	
	α^{-1}	137.04	
Stefan–Boltzmann constant	σ_B	$5.6705 \cdot 10^{-5}$	erg/cm^2 sec deg^4
Wavelength associated with 1 eV	λ_0	$1.2398 \cdot 10^{-4}$	cm
Frequency associated with 1 eV	ν_0	$2.4180 \cdot 10^{14}$	Hz
Wave number associated with 1 eV	k_0	$8.0655 \cdot 10^3$	cm^{-1}

Energy associated with 1 eV	$1.6022 \cdot 10^{-12}$	erg
Energy associated with 1 cm^{-1}	$1.9864 \cdot 10^{-16}$	erg
Energy associated with 1 Rydberg	13.606	eV
Energy associated with 1 deg Kelvin	$8.6174 \cdot 10^{-5}$	eV
Temperature associated with 1 eV	$1.1604 \cdot 10^{4}$	deg (K)

Astronomical quantity	Symbol	Value	Units
Solar mass	$M_\odot$	$1.989 \cdot 10^{33}$	g
Solar radius	$R_\odot$	$6.955 \cdot 10^{10}$	cm
Solar absolute luminosity	$L_\odot$	$3.85 \cdot 10^{33}$	ergs^{-1}
Parsec	pc	$3.085 \cdot 10^{18}$	cm
Astronomical unit	AU	$1.496 \cdot 10^{13}$	cm

Multiple	Prefix	Symbol
10^{-1}	deci	d
10^{-2}	centi	c
10^{-3}	milli	m
10^{-6}	micro	μ
10^{-9}	nano	n
10^{-12}	pico	p
10^{-15}	femto	f
10^{-18}	atto	a
10	deca	da
10^{2}	hecto	h
10^{3}	kilo	k
10^{6}	mega	M
10^{9}	giga	G
10^{12}	tera	T
10^{15}	peta	P
10^{18}	exa	E

Source: http://www.spp.astro.umd.edu/Formulary/toc.html.

Appendix C

Distances

Astronomers deal with distances that are so large that they call for special units. For relatively small distances, from the point of view of an astronomer, a frequently used unit is the *astronomical unit*, which is defined as the mean radius of the Earth's orbit around the Sun. It is

$$1 \text{ AU} = 1.49578 \cdot 10^{13} \text{ cm.} \tag{C.1}$$

For distances beyond our Solar System, this unit quickly becomes cumbersome, too. The most commonly used unit of distance in Galactic and extragalactic astronomy is the parallax-second, or parsec (pc). It is defined to be the distance at which the mean distance between the Earth and Sun subtends an angle of one second of an arc. Thus, we can write

$$\frac{1 \text{ AU}}{1 \text{ pc}} \approx 1'', \tag{C.2}$$

which yields

$$1 \text{ pc} = 3.01 \cdot 10^{18} \text{ cm.} \tag{C.3}$$

Often it is convenient to express distances in kiloparsecs (1 kpc = 1000 pc), megaparsecs (1 Mpc = 10^6 pc), or even gigaparsecs (1 Gpc = 10^9 pc).

A distance unit that is sometimes used, especially in the popular literature, is the light-year. One light-year is the distance light travels in vacuum in a year. It is

$$1 \text{ lyr} = 9.4605 \cdot 10^{17} \text{ cm.} \tag{C.4}$$

and thus

$$1 \text{ pc} = 3.26 \text{ lyr.} \tag{C.5}$$

%

Appendix D

Luminosity, brightness, magnitude, color

Luminosity is the amount of energy that is radiated per unit time by an object in all directions. It is thus an intrinsic property of an object and independent of its distance to the observer. It has units of power: in cgs units, it is expressed in erg s^{-1}.

Like all intrinsic properties of astronomical objects, the luminosity is not a directly observable quantity. On Earth we observe the amount of energy per unit time and unit area that is received by a detector (e.g., a CCD camera at the focus of a telescope). This quantity is called *brightness* and is a flux with units of erg s^{-1}cm^{-2}. The brightness of an object depends on its luminosity *and* its distance from the observer. For an isotropically radiating source, the brightness or flux, F, is related to the luminosity, L, and distance, d, via

$$F = \frac{L}{4\pi d^2},$$

(D.1)

which follows from energy conservation and is sometimes called the inverse-square law; see Section 3.3.

In optical astronomy, brightnesses are measured in somewhat archaic units called magnitudes. This system is not particularly intuitive, but it is unlikely to be changed any time soon. It is motivated by the properties of the human eye, which senses light on a logarithmic rather than a linear scale. Our eyes can clearly distinguish the brightness of two objects only if their brightnesses vary by a factor of roughly 2.5.

The magnitude system goes back to the Greek astronomer Hipparchos, who called the brightest stars first-magnitude stars, stars half as bright second-magnitude stars, and so on. Nineteenth-century astronomers then realized that for the already-classified stars, a difference of five magnitudes corresponds to a brightness ratio of 100. Hence, the brightness and the apparent magnitude, m, are related by $F \propto 10^{m/2.5}$. This way,

$$F_1/F_2 = 10^{(m_1-m_2)/2.5},$$

(D.2)

so that when $m_1 - m_2 = 5$, the brightness ratio is 100, as required. Thus, the brightness ratio between objects that differ by one magnitude is $10^{1/2.5} \approx 2.512$. Now, the proportionality factor has to be determined by measuring the flux of a standard star of zero magnitude. The result is

$$\left(\frac{F}{\text{erg cm}^{-2} \text{ s}^{-1}} \right) \approx 2.75 \cdot 10^{-5} \, 10^{-m/2.512}. \tag{D.3}$$

It follows that objects that are brighter have smaller magnitudes, with very bright objects having negative magnitudes. The apparent magnitude of the Sun is -26.73; of the full Moon, -12.6; of Venus at maximum brightness, -4.7. The faintest objects that the *Hubble Space Telescope* can detect have apparent magnitudes of 30.

Absolute magnitudes are a measure for the luminosity of an object. It is equal to the apparent magnitude if the source was placed at 10 parsec from the observer. Using the inverse-square law, the apparent magnitude, m, and the absolute magnitude, M, are related via

$$m - M = 5 \log r - 5, \tag{D.4}$$

where r is measured in parsecs. The quantity $m - M$ is called *distance modulus*. Magnitudes can refer to the energy radiated within a certain band. The most common system of filters in which quantities such as brightness or luminosity are measured follow a standardized system. The most common system is the UBV system, in which magnitudes are measured using a set in the U(ltraviolet), B(lue), and V(isual) bands. The central wavelengths, λ_0, and bandwidths, $\delta\lambda$, are given in this table.

Band	U	B	V
λ_0 (nm)	365	440	550
$\delta\lambda$ (nm)	69	96	90

Magnitudes that refer to the energy radiated in all wavelengths are called *bolometric magnitudes*.

$$\log \left(\frac{L}{L_\odot} \right) = \frac{1}{2.512} (M_{\text{bol},\odot} - M_{\text{bol}}). \tag{D.5}$$

The luminosity of the Sun is $3.8 \cdot 10^{33}$ ergs^{-1}, and its bolometric magnitude is 4.6.

The spectrum of a source is a rich source of information about the emission mechanisms at work. Instead of obtaining a very finely resolved spectrum of the source, it is often useful to measure the brightness of the source in various well-defined bands. The spectral properties of the source are defined in terms of color indices.

A *color index* is the difference in brightness as measured in certain wavebands and gives a rough indication of the spectral shape of the source. For example, the color index CI_{UV} denotes the difference in apparent brightnesses measured in the U and V band, that is, $m_U - m_V$, and CI stands for *color index*. Because of the negative sign in the definition of the magnitude, the smaller algebraically its value of $(U - B)$ or $(B - V)$, the more ultraviolet or blue, respectively, an object is. Because magnitudes are logarithms of fluxes, the difference of two magnitudes is the ratio of the respective fluxes and thus is independent of the distance to the object.

Index